Studies in Systems, Decision and Control

Volume 79

Series editor

Janusz Kacprzyk, Polish Academy of Sciences, Warsaw, Poland
e-mail: kacprzyk@ibspan.waw.pl

About this Series

The series "Studies in Systems, Decision and Control" (SSDC) covers both new developments and advances, as well as the state of the art, in the various areas of broadly perceived systems, decision making and control- quickly, up to date and with a high quality. The intent is to cover the theory, applications, and perspectives on the state of the art and future developments relevant to systems, decision making, control, complex processes and related areas, as embedded in the fields of engineering, computer science, physics, economics, social and life sciences, as well as the paradigms and methodologies behind them. The series contains monographs, textbooks, lecture notes and edited volumes in systems, decision making and control spanning the areas of Cyber-Physical Systems, Autonomous Systems, Sensor Networks, Control Systems, Energy Systems, Automotive Systems, Biological Systems, Vehicular Networking and Connected Vehicles, Aerospace Systems, Automation, Manufacturing, Smart Grids, Nonlinear Systems, Power Systems, Robotics, Social Systems, Economic Systems and other. Of particular value to both the contributors and the readership are the short publication timeframe and the world-wide distribution and exposure which enable both a wide and rapid dissemination of research output.

More information about this series at http://www.springer.com/series/13304

Nabil Derbel · Jawhar Ghommam
Quanmin Zhu
Editors

Applications of Sliding Mode Control

Springer

Editors
Nabil Derbel
University of Sfax
Sfax
Tunisia

Quanmin Zhu
University of the West of England
Bristol
UK

Jawhar Ghommam
University of Tunis Carthage
Urbain Nord
Tunisia

ISSN 2198-4182 ISSN 2198-4190 (electronic)
Studies in Systems, Decision and Control
ISBN 978-981-10-2373-6 ISBN 978-981-10-2374-3 (eBook)
DOI 10.1007/978-981-10-2374-3

Library of Congress Control Number: 2016948266

Printed on acid-free paper

This Springer imprint is published by Springer Nature
The registered company is Springer Nature Singapore Pte Ltd.
The registered company address is: 152 Beach Road, #22-06/08 Gateway East, Singapore 189721, Singapore

Preface

This book as titled is prepared with a clear aim to bridge exiting representative theoretical results to the development of design prototype, algorithms, and applications. The readers/users are targeted for a wide range of groups from university, research, and development institute to industry, but for the reference of top academic peers to write their theoretical papers. It is not our intention to cover all disciplines of electrical engineering systems of recent control-related contributions, but priority has been placed on the application of recently reported techniques on sliding mode techniques, their significant impact on stability robustness, and their challenges over the last decade. The integrated approaches on sliding model control contribute to some sort of a state of the art in modeling and controlling of complex dynamic systems with feasible and concise solutions. This book is made of 16 invited chapters have been written by leading researchers covering recent theoretical developments and applications of the sliding mode techniques to class of systems from the electrical engineering fields.

The book presents theoretical explorations on several fundamental problems for several kinds of systems. By integrating fresh concepts and state-of-the-art results to form a systematic approach for controlling complex systems, fundamental theoretical approaches and practical framework have been established.

This book is intended for readers from a wide spectrum of specialties in electrical engineering fields and shall educate them about the fundamental advances in SMC techniques. It is expected that the readers will require limited background knowledge to understand various concepts and results outlined in the book. Furthermore, a notable feature of this book is to provide not only theoretical results and techniques, but also experimental case study on a test bed of robotic system, which is anticipated to highly motivate young students and researchers.

The book has been organized into the following chapters which are regrouped into three parts.

- The first part contains four chapters detailing fundamentals of sliding mode control. Then, the book is opened with a review chapter presenting a state of the art of the sliding mode control (SMC) techniques by providing some generalities, notions, and classifications of different SMC control strategies accompanied with explanatory examples and detailed discussion. The rest of the chapters covers a broad scope of topics in sliding modes from theoretical investigations to significant applications. Chapter 2 presents disturbance rejection for discrete first, second order and repetitive sliding mode controllers. Chapter 3 gives robust exponential higher-order sliding mode controllers. The interest of chapter concerns four sliding functions using LMIs approach for time-delay systems.

- The second part is dedicated to sliding mode controller applied to robotic systems. It contains six chapters. Chapter 5 applies the sliding mode control to underactuated quadrotor. Chapter 6 applies the sliding mode control of an inverted pendulum. Chapter 7 considers the autonomous surface vessel. Chapter 8 consists to apply the first and the second order sliding mode controls with time-delay control for robot manipulators. Chapter 9 is an application of sliding mode control for a 5DOF serial robot for tele-echography. Chapter 10 applies the sliding mode control for an active exoskeleton robot. Chapter 11 is an application to multiagent systems
- The third part consists to apply the sliding mode control to power systems. Chapter 12 considers the application of this technique to photovoltaic power systems. Chapter 13 considers the sliding bifurcations for a two-cell DC/DC buck converter. Chapter 14 applies the control by sliding mode of induction motor drives. Chapter 15 applies the sliding mode control for a fault induction machine. Finally, Chap. 16 studies an electrohydraulic system controlled by sliding mode control.

We wish to express our sincere gratitude to many colleagues who have contributed to this book. First of all, we are particularly indebted to our colleagues who have contributed their excellent research in order to bring the valuable materials for graduate students, researchers, and practitioners. We greatly appreciate the contributors to this book for their patience and time taken to collaborate with us to finally complete this book. Next, particular thanks go to students and colleagues who helped engaging in the preparation and assisted us to improve this book. Finally, we would like to sincerely express our deepest gratitude to the Springer editorial staff for their continuous support, assistance, and significant improvement in the manuscript. Without their help, the book would not be published as scheduled.

Sfax, Tunisia — Nabil Derbel
Urbain Nord, Tunisia — Jawhar Ghommam
Bristol, UK — Quanmin Zhu
October 2016

Contents

Chapter 1
On the Sliding Control

Ahmed Said Nouri, Fatma Abdelhedi Bouazi and Nabil Derbel

Abstract In this chapter, we give an overview of the sliding mode control. Firstly, We are interested to the sliding mode control and different solutions for reducing the chattering phenomenon have been given. Then, we have introduced the generalized sliding mode control where the discontinuity is applied to the highest derivative of the control if the system presents zero dynamics. Finally, we present the high order sliding mode control as a solution to reduce the chattering phenomenon while keeping the robustness against external disturbances and parametric variations compared to that provided by the first order sliding mode control approach.

Keywords Sliding mode · Generalized sliding mode · High order sliding mode · Chattering · Robustness

1.1 Introduction

Variable Structure Systems (VSS) have been introduced after 1950 in the Soviet Union after the works of Filippov (1960) and Popovski (1950) on differential equations with discontinuous second member. Since the 60th years, the VSS have been exploited by Emelyanov et al. (1970), Utkin (Emelyanov et al. 1970; Utkin 1972) and Neymark (1957) for controlling discontinuous dynamic systems or accepting discontinuous inputs.

Controller parameters of variable structure systems can vary allowing under certain conditions, to maintain in the phase space, the evolution of the system on a

A.S. Nouri (✉)
Research Unit CONPRI, ENIG, University of Gabes, Gabes, Tunisia
e-mail: ahmedsaid.nouri@enig.rnu.tn

F.A. Bouazi · N. Derbel
Control & Energy Management Laboratory (CEMLab), Sfax Engineering School, University of Sfax, Sfax, Tunisia
e-mail: fatma.abdelhedi@live.fr

N. Derbel
e-mail: n.derbel@enis.rnu.tn

N. Derbel et al. (eds.), *Applications of Sliding Mode Control*,
Studies in Systems, Decision and Control 79, DOI 10.1007/978-981-10-2374-3_1

hypersurface a priori chosen. The closed loop system is evolving in sliding mode on the considered hypersurface. In this situation, the system becomes insensitive to external disturbances and to variations of system parameters (Lopez and Nouri 2006).

This type of control has proven itself in both stabilization and trajectory tracking problems (Emelyanov 1967; Lopez and Nouri 2006; Utkin 1978). But it has been often criticized because of the chattering problem caused by the discontinuous-type control. Improvements have been proposed: the equivalent control (Utkin 1978, 1992), the continuous function in a band around the surface (Slotine and Li 1991), the generalized control (Fliess 1990; Lopez and Nouri 2006) and the control whose structure and sliding mode are of higher order (Fridman and Levant 1999; Levant 1993; Lopez and Nouri 2006).

The insensitivity of the sliding mode control with respect to parameter variations and external disturbances is valid only if the invariance condition called matching condition is verified (Utkin 1978) when the phase trajectory is on the sliding surface, that is to say that the system is in sliding mode, in contrary to the reaching phase surface, the control is sensitive.

Several research studies have been developed to reduce the time of the reaching phase by imposing of the reachability time, the sliding function is a solution of a differential equation of the first order (Sira-Ramirez et al. 1992); by a change in the surface slope (case of a second order system) (Medhafar et al. 2005) or by choosing a nonlinear sliding surface starting from the initial point (the sliding surface is composed of several segments).

This chapter consists of 3 parts. In the first part, we introduce different notations and definitions of sliding modes. We present the sliding conditions and the invariance conditions. Different forms of sliding surfaces and of control laws with sliding mode existing in the literature are also presented.

In the second part, we explain the generalized sliding modes considered as a solution to reduce the reluctance phenomenon by applying the discontinuity on the highest control derivative. The effective control is obtained after several integrations making the signal virtually continuous.

In the third part, the Higher Order Sliding modes are introduced. In addition to the sliding function, the higher order sliding modes (of order r) allow to keep equal to zero successive derivatives of the sliding function (until the derivative $(r-1)$), the control is obtained after $(r-1)$ integrations.

1.2 Classical Sliding Mode

For differential equations with discontinuous second members (Lopez and Nouri 2006), the theory of ordinary differential equations ceases to be valid since it no longer checks the classical conditions of existence and uniqueness of solutions of the Cauchy–Lipschitz theorem. This has been caused by the existence of the discontinuity.

1.2.1 Notion of Sliding Mode

Consider a single input–single output dynamic system, linear in the input (affine nonlinear system), described by the following equation:

$$\dot{x}(t) = f(x, t) + g(x, t)\, u(x, t) \tag{1.1}$$

where:

- $t \in \mathbb{R}$ is the time,
- $x \in \mathbb{R}^n$ is the state vector,
- $u \in \mathbb{R}$ presents the control vector,
- f and g are vector fields.

Assume that the control variable $u(x, t)$ undergoes a discontinuity on the sliding surface: $S(x) = 0$.

$$u(x, t) = \begin{vmatrix} u^{+}(x, t) & \text{si } S(x) > 0 \\ u^{-}(x, t) & \text{si } S(x) < 0 \end{vmatrix} \tag{1.2}$$

Definition 1 The system defined by (1.1) and (1.2) is called a variable structure system.

If the control variable $u(x, t)$ further satisfies the attractiveness condition $S(x)\dot{S}(x) < 0$, the control is said in *sliding mode*.

Definition 2 A sliding mode exists on $S(x) = 0$ if and only if the phase trajectory is on the sliding surface and the attractiveness condition is verified, i.e.:

$$x(t) \in \{x/S(x) = 0\} \quad \text{and} \quad S(x)\dot{S}(x) < 0$$

The sliding mode is said *ideal* if $S(x) = 0$ and $\dot{S}(x) = 0$. In the case where only the attractiveness condition is verified ($S(x)\dot{S}(x) < 0$), the sliding is said *real*.

1.2.1.1 Ideal Sliding Mode and Equivalent Control

The equivalent control is defined as the control allowing to have an ideal sliding mode i.e., it satisfies the condition $S(x) = 0$ et $\dot{S}(x) = 0$.

In the following, we assume the general case of a sliding 'hypersurface' depending on the state and on the time, which is denoted by $S(x, t)$.

By considering the ideal sliding conditions, we have:

$$\dot{S}(x,t) = \left(\frac{\partial S}{\partial x}\right)^T \frac{dx}{dt} + \frac{\partial S}{\partial t}$$

Using the system state Eq. (1.1) and replacing $u(x,t)$ by $u_{eq}(x,t)$, we get:

$$\dot{S}(x,t) = \left(\frac{\partial S}{\partial x}\right)^T \left[f(x,t) + g(x,t)\, u_{eq}(x,t)\right] + \frac{\partial S}{\partial t} = 0$$

if $\left(\frac{\partial S}{\partial x}\right)^T g(x,t)$ is non singular, the equivalent control is:

$$u_{eq}(x,t) = -\left[\left(\frac{\partial S}{\partial x}\right)^T g(x,t)\right]^{-1} \left\{\left(\frac{\partial S}{\partial x}\right)^T f(x,t) + \frac{\partial S}{\partial t}\right\} \tag{1.3}$$

Physically, the equivalent control represents the average value of the control variable $u(t)$ that maintains the system state on the sliding surface $S(x,t) = 0$.

Remark

If the hypersurface is independent of time and if we assume that $f = (f_1, \ldots, f_n)^T$ and $g = (g_1, \ldots, g_n)^T$, the equivalent control will be expressed as:

$$\begin{aligned} u_{eq}(x) &= -\left[\left(\frac{\partial S}{\partial x}\right)^T g(x,t)\right]^{-1} \left\{\left(\frac{\partial S}{\partial x}\right)^T f(x,t)\right\} \\ &= -\frac{\sum_{i=1}^{n} \frac{\partial S}{\partial x_i} f_i(x,t)}{\sum_{i=1}^{n} \frac{\partial S}{\partial x_i} g_i(x,t)} \quad \text{if } \sum_{i=1}^{n} \frac{\partial S}{\partial x_i} g_i(x,t) \neq 0 \end{aligned} \tag{1.4}$$

The equation of the ideal mode may be deduced by replacing $u_{eq}(t)$ in Eq. (1.1) in the case of a time dependent surface:

$$\begin{aligned} \dot{x}(t) = &\left[I - g(x,t)\left[\left(\frac{\partial S}{\partial x}\right)^T g(x,t)\right]^{-1} \left(\frac{\partial S}{\partial x}\right)^T\right] f(x,t) \\ &- g(x,t)\left[\left(\frac{\partial S}{\partial x}\right)^T g(x,t)\right]^{-1} \left(\frac{\partial S}{\partial t}\right)^T \end{aligned} \tag{1.5}$$

If the sliding surface is independent of time, the sliding mode dynamics is described by:

$$\dot{x}(t) = f(x,t) + g(x)\frac{\sum_{i=1}^{n} \frac{\partial S(x)}{\partial x_i} f_i(x,t)}{\sum_{i=1}^{n} \frac{\partial S(x)}{\partial x_i} g_i(x,t)} \tag{1.6}$$

1.2.1.2 Application to Linear Systems

Consider a linear system defined in the state space by:

$$\dot{x}(t) = Ax(t) + Bu(t) \tag{1.7}$$

with $A \in \mathbb{R}^{nxn}$ and $B \in \mathbb{R}^{n}$.

The surface is selected linear and defined by:

$$S(x,t) = Cx(t) = 0 \quad \forall t \geq t_0 \tag{1.8}$$

where t_0 is the time for which the sliding mode is reached.

The equivalent control is deduced from Eq. (1.4) as:

$$\begin{aligned} \dot{S}\,(x,t) &= \left(\frac{\partial S}{\partial x}\right)^T Ax(t) + \left(\frac{\partial S}{\partial x}\right)^T B\,u_{eq}(t) = 0 \\ \dot{S}\,(x,t) &= CA\,x(t) + CB\,u_{eq}(t) = 0 \end{aligned}$$

If $CB \neq 0$, then the equivalent control becomes:

$$u_{eq}(t) = -\,(CB)^{-1}\,CA\,x(t) \tag{1.9}$$

The dynamical behavior of the closed loop system in sliding mode is described by the following differential equation:

$$\dot{x}(t) = \left[I - B\,(CB)^{-1}\,C\right] A\,x(t) \tag{1.10}$$

1.2.2 *Invariance Terms of the Sliding Mode*

The essential property of the sliding mode operating is its insensitivity and its robustness against a certain class of modeling errors or disturbances. The system is called invariant if the dynamic of the closed loop system in the sliding mode is independent of modeling errors and external disturbances.

Invariance conditions are given for an external disturbance in the case of systems which are linear with respect the control input, and other invariance conditions in the case of parametric variations for linear systems (Utkin and Li 1986). We give below invariance conditions in the case of a disturbed linear system affected by a parametric variation. Therefore, consider the following linear system:

$$\dot{x}(t) = (A + \Delta A)\ x(t) + B\,u(t) + h(t) \tag{1.11}$$

where $x(t) \in \mathbb{R}^n$ represents the system state, $u(t) \in \mathbb{R}$ is the control variable, $A \in \mathbb{R}^{n\times n}$ is the state matrix, $\Delta A \in \mathbb{R}^{n\times n}$ is the uncertainty of the state matrix, $B \in \mathbb{R}^n$ is a column vector and $h(t)$ is an external perturbation.

The invariance property of the sliding mode is ensured if certain assumptions called 'matching conditions' are verified with respect to external disturbances and parametric variations.

Definition 3 ΔA and $h(t)$ verify the matching conditions if it exists $\Delta\bar{A} \in \mathbb{R}^{n\times n}$ and $\Delta\bar{h} \in \mathbb{R}^n$ such as:

$$\Delta A = B\,\Delta\bar{A} \qquad h(t) = B\,\Delta\bar{h} \tag{1.12}$$

In other words, the sliding mode is invariant if external disturbances and parametric variations belong to the space generated by B, i.e.:

$$\Delta A \in \text{Span}(B), \qquad h(t) \in \text{Span}\{B\} \tag{1.13}$$

1.2.2.1 Application to an Additive Disturbance

Consider the following nonlinear system:

$$\begin{cases} \dot{x}_i = x_{i+1} \quad i = 1 \ldots n-1 \\ \dot{x}_n = f\,(x,t) + g\,(x,t)\,u\,(x,t) + h\,(x,t) \end{cases} \tag{1.14}$$

where $x = (x_1, \ldots, x_n)^T \in \mathbb{R}^n$ represents the system state, $u \in \mathbb{R}$ is the control variable, f and g are nonlinear functions and h is a bounded disturbance such as:

$$|h\,(x,t)| \le H$$

Let us choose a linear sliding surface described by:

$$S\,(x) = \sum_{i=1}^{n} c_i x_i\ ; \quad c_n = 1$$

An adequate control satisfying the attractiveness condition of the surface and providing a sliding mode for the non perturbed system can be taken as:

$$u(x,t) = u_{eq}(x,t) - K \text{ sign}(S(x))$$

The equivalent control ensures an ideal sliding mode on $S(x) = 0$ of the non perturbed system. Suppose that:

$$\dot{S}(x) = \sum_{i=1}^{n-1} c_i x_{i+1} + f(x,t) + g(x,t)\, u_{eq}(t) = 0$$

Then, if $g(x,t) \neq 0$:

$$u_{eq}(x,t) = \frac{1}{g(x,t)}\left[-\sum_{i=1}^{n-1} c_i x_{i+1} - f(x,t)\right]$$

Let us now seek adequate conditions for which this control variable is robust with respect to the additive disturbance $h(x,t)$. Then, it is sufficient to determine the attractiveness conditions for the perturbed system, either:

$$\begin{aligned}
S(x)\,\dot{S}(x) &= S(x)\left[\sum_{i=1}^{n-1} c_i x_{i+1} + f(x,t) + g(x,t)\, u(x,t) + h(x,t)\right] \\
&= S(x)\left[-K \text{ sign}(S(x)) + h(x,t)\right] \\
&= -K|S(x)| + S(x)h(x,t) \\
&\le -K|S(x)| + |S(x)| \times |h(x,t)| \le -K|S(x)| + |S(x)|H \\
&\le -|S(x)|(K - H)
\end{aligned}$$

So if $K > H$, the attractiveness condition of the area is fulfilled.

In the case of an additive disturbance, the sliding mode is invariant (robust with respect to the additive disturbance) if the discontinuous part gain K exceeds the upper bound of the disturbance.

1.2.2.2 Example

Consider the following example:

$$\begin{cases} \dot{x}_1(t) = x_2(t) \\ \dot{x}_2(t) = -3x_1(t) - 4x_2(t) + u(t) + \varepsilon(t) \end{cases}$$

where $\varepsilon(t)$ is a sinusoidal disturbance type defined by $\varepsilon(t) = \sin 10t$.

For this example, the system is a second order one, and the sliding surface is chosen linear whose equation is $S(x) = 6\ \ x_1 + x_2$. The control variable is $u(t) = -M \text{ sign}(S(x))$ with $M = 10$.

Based on the previous paragraph results, the control is robust if the gain M exceeds the additive perturbation magnitude, either in our example, the control is robust if $M > 1$. Simulation results are given in Figs. 1.1 and 1.2.

Figure 1.1 represents the trajectory evolution in the phase plane and Fig. 1.2 shows the evolution of the sliding function. Note that for an amplitude perturbation ($\varepsilon_{max} \leq 10$), the answer is unchanged and the controller is robust against the additive disturbance, but in the case where the disturbance amplitude becomes larger than the gain $M = 10$ ($\varepsilon_{max} = 11 > 10$), the controller becomes sensitive to this disturbance.

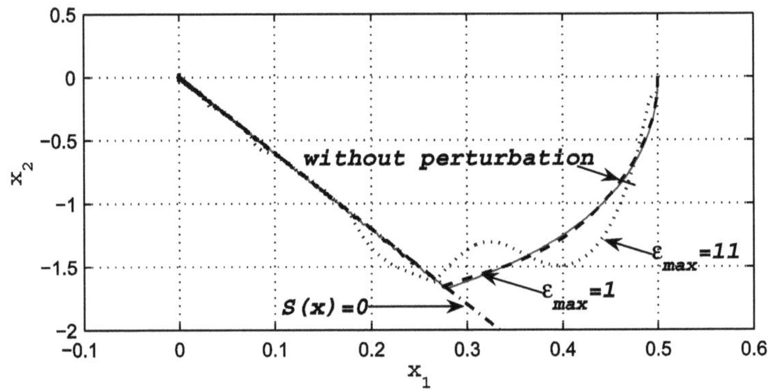

Fig. 1.1 Trajectory evolution in the phase plane

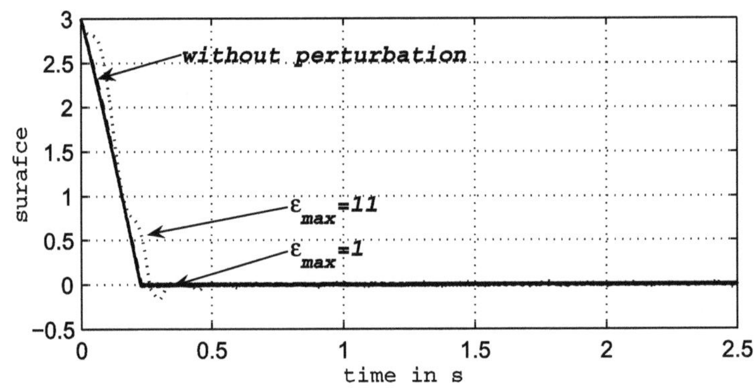

Fig. 1.2 Evolution of the sliding function with respect to time

1.2.3 Reaching Phase and Sliding Phase

The evolution of the state trajectory is divided into two parts:

- from the initial state to the intersection with the sliding surface, this part is called "the reaching phase",
- from the intersection with the sliding surface to the origin, this part is called "the sliding phase".

Both phases are represented by Fig. 1.3 in the case of a second order system.

During the reaching phase, the attractiveness condition of the sliding surface $\dot{S}(x)S(x) < 0$ is verified. This condition is global but it does not guarantee a finite sliding time.

To ensure a finite sliding time, the condition becomes:

$$\dot{S}(x)S(x) < -\varepsilon \quad \text{if} \;\; S(x) \neq 0 \;\; \text{and} \;\; \varepsilon > 0 \tag{1.15}$$

This sliding time can be imposed by choosing a sliding surface, that is a solution of a differential equation (Sira-Ramirez et al. 1992) of the form:

$$\dot{S}(x) = \;\; -\mu S(x) - \mu \Omega \; \text{sign}\,(S(x)) \tag{1.16}$$

The convergence of $S(x)$ to zero is done in a finite time t_g (cf. Fig. 1.3):

$$t_g = \frac{1}{\mu} \ln \left(1 + \frac{|S_0|}{\Omega} \right) \tag{1.17}$$

where S_0 is the value of $S(x)$ at $t = 0$.

During the sliding phase ($S(x) = 0$ and $\dot{S}(x)S(x) < 0$), the closed loop system has the same behavior than $S(x) = 0$.

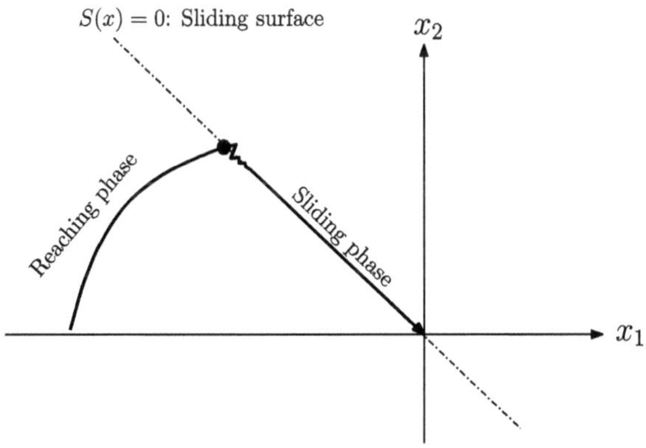

Fig. 1.3 Reaching and sliding phases

If S is a linear state function, and during the sliding phase, the closed loop system will have the same behavior as a linear system of a free and reduced order.

For a system of order n which described in the state space by:

$$\begin{cases} \dot{x}_i = x_{i+1} \quad i = 1, \ldots, n-1 \\ \dot{x}_n = f(x,t) + g(x,t)u(t) \end{cases} \tag{1.18}$$

the linear sliding surface according to the state has the form:

$$S(x) = Cx = \sum_{i=1}^{n} c_i x_i, \quad (c_n = 1) \tag{1.19}$$

In the sliding phase, the closed loop system will have the same behavior as the system described by:

$$\begin{cases} \dot{x}_i = x_{i+1} \quad i = 1, \ldots, n-2 \\ \dot{x}_{n-1} = -\sum_{i=1}^{n-1} c_i x_i \end{cases} \tag{1.20}$$

This is a linear system of order $(n-1)$. The equation $\dot{x}_{n-1} = -\sum_{i=1}^{n-1} c_i x_i$ constitutes a *linear feedback with sliding mode* (Lopez and Nouri 2006).

Remark

In the literature, there are other forms of the sliding surface:

- A nonlinear function of the state (Hamerlain 1993):

$$S(x) = x_n + H(x_1, x_2, \ldots, x_{n-1}) \tag{1.21}$$

  H is a nonlinear fonction.
- A sliding function in 3 or 4 parts in the case of a second order system (Lopez and Nouri 2006):

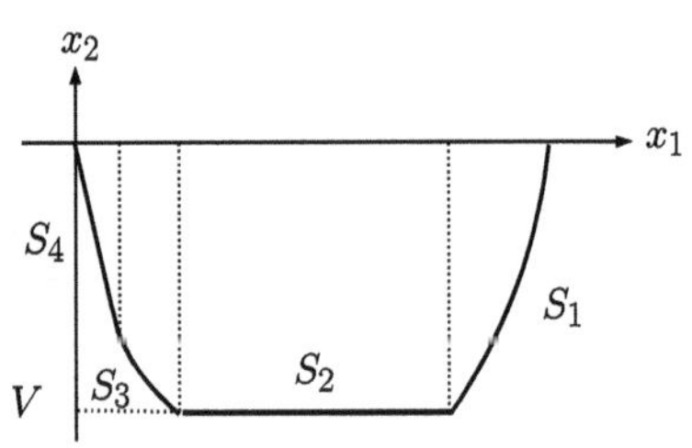

Fig. 1.4 Sliding surface in 4 parts

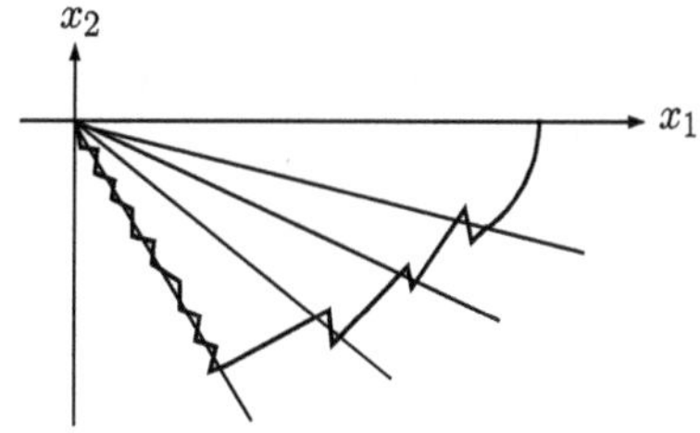

Fig. 1.5 Variable slope sliding surface

- A part S_1 with a constant acceleration defined by: $S_1(x) = \alpha x_2^2 + x_1 - x_0$
- A part S_2 with a constant speed defined by: $S_2(x) = x_2 - V$
- A part S_3 with a constant deceleration defined by $S_3(x) = -\alpha x_2^2 + x_1$
- A part S_4 with a conventional sliding mode for a second order system defined by $S_4(x) = c_1 x_1 + x_2$

- A variable slope sliding function (in the case of a second order system) (Lopez and Nouri 2006), this variation can be performed by a fuzzy supervisor (Medhafar et al. 2005) (Figs. 1.4 and 1.5).

1.2.4 Synthesis of the Sliding Mode Control

The synthesis of the sliding mode control is accomplished by satisfying the attractiveness condition. The used condition will be selected depending on the desired sliding time. In several cases, it is interesting to impose a control structure and to determine adequate parameters while respecting the adequate attractiveness condition.

The most used structures are the following:

- Relay control: the control form is (Lopez and Nouri 2006):

$$u(x,t) = \begin{cases} u^+(x,t) & \text{if } S(x) > 0 \\ u^-(x,t) & \text{if } S(x) < 0 \end{cases}$$

  The most commonly used form is $u = -M \text{ sign }(S(x))$, where M is determined to verify the selected attractiveness condition.
- Linear feedback with switched gains: the control form is given by Vernhes (1971):

$$u(x,t) = \psi(x)\,x(t) \quad \text{with} \quad \psi_i(x) = \left| \begin{array}{ll} \alpha_i & \text{if } S(x)x_i > 0 \\ \beta_i & \text{if } S(x)x_i < 0 \end{array} \right.$$

  Parameters α_i and β_i are selected so that an attractiveness condition is verified.
- Control variable $u_{eq} + \Delta u$: the control form is given by Utkin (1992):

$$u(x,t) = u_{eq} + \Delta u$$

where u_{eq} the equivalent control giving an ideal sliding mode. The term Δu is a discontinuous term of high frequencies which may take the following different classical forms:

1. $\Delta u = -k\,\mathrm{sign}\,(S)$, (Utkin 1978)
2. $\Delta u = -k\,\mathrm{sat}\left(\frac{S}{\phi}\right)$, (Slotine and Coestsee 1986) where:

$$\mathrm{sat}\left(\frac{S}{\phi}\right) = \left|\begin{array}{ll} \frac{S}{\phi} & \text{if } |S| \leq \phi \\ \mathrm{sign}\left(\frac{S}{\phi}\right) & \text{if } |S| > \phi \end{array}\right.$$

3. $\Delta u = -k\,\mathrm{cont}\left(\frac{S}{\phi}\right)$, with:

$$\mathrm{cont}\left(\frac{S}{\phi}\right) = \left|\begin{array}{ll} \frac{S}{|S| + \delta} & \text{if } |S| \leq \phi \\ \mathrm{sign}\left(\frac{S}{\phi}\right) & \text{if } |S| > \phi \end{array}\right. \tag{1.22}$$

and:

$$\delta = \left|\begin{array}{lll} 0 & \text{if} & |S| > \phi \\ \delta_0 + |\eta| & \text{if} & |S| \leq \phi \end{array}\right. \qquad \eta = \left|\begin{array}{lll} 0 & \text{if} & |S| > \phi \\ \eta_0 + \int_0^t S\,d\tau & \text{if} & |S| \leq \phi \end{array}\right.$$

4. Another expression has been proposed for u in the phase space, in order to reduce the steady state error by introducing an integral form: (Asada and Slotine 1986)

$$\Delta u = -k\,\mathrm{sign}\,(S) \tag{1.23}$$

with:

$$S = \left(\frac{d}{dt} + \lambda\right)^n \left(\int_0^t x(\tau)\,d\tau\right) \tag{1.24}$$

or:

$$S = \left(\frac{d}{dt} + \lambda\right)^{n-1} x \tag{1.25}$$

1.3 Generalized Sliding Mode

In this paragraph, we are interested in the problem of the discontinuous control with sliding mode of nonlinear systems, whose dynamic is described using the control concept and the generalized observability canonical forms introduced by Fliess (1990) and Sira-Ramirez (1988).

1.3.1 Feedback Linearization

Consider a single input–single output dynamic system (where u represents the input and y is the output) of dimension m, which is represented by the following open-loop system equations:

$$\begin{cases} \dfrac{d\eta}{dt} = f\,(\eta,\ u) \\ y = h\,(\eta) \end{cases} \tag{1.26}$$

where η refers to the state with m state variables $(\eta_1,\ \eta_2,\ldots,\eta_m)$. $f = (f_1, f_2, \ldots, f_m)^T$ is a vector field; $y \in \mathbb{R}$ and $u \in \mathbb{R}$; η and $f \in \mathbb{R}^m$.

The nonlinear state representation (1.26) can be transformed, using the formalism of implicit equations, around a non-singular point of the state space, into a system of three equations: the first one describes the dynamics of the non observable part; the second one represents observable states as a function of inputs, outputs and their derivatives; and the third equation describes the external dynamic system behavior. Assuming that the observable part of the system (1.26) is of dimension n, its external dynamics can also be represented by the implicit differential equation (1.27) involving the input as well as the output and their derivatives:

$$\mathrm{c}\left(y, \dot{y}, \ldots, y^{(n)}, u, \dot{u}, \ldots, u^{(\alpha)}\right) = 0 \tag{1.27}$$

(If the system is without zero dynamics $\alpha = 0$).

Equations (1.26) and (1.27) can be studied by several approaches which are specific for nonlinear systems (Lyapunov 1966; Mira 1990). Another approach consists in combining these equations of explicit generalized canonical forms, that generalize the Kalman form because they keep, a priori, a nonlinear character (Fliess 1990). These canonical forms are well adapted to the local description of the operation of the system in the state space and easily lead to the development of a feedback linearization via correctors that allow to enslave the system output.

1.3.2 Observable Single Input–Single Output Systems Case with Zero Dynamics

Using the differential algebra formalism, we have defined a generalized dynamic, associated with (1.26) which is assumed to be as single input–single output observable, and characterized by the two canonical forms which are valid at least locally. Suppose that:

$$x_1 = y \qquad x_2 = \dot{y} \qquad \ldots \qquad x_n = y^{(n-1)}$$

and by using the theorem of implicit functions, we can obtain:

$$(\text{GOCF}) \quad \left\{ \begin{array}{l} \left. \begin{array}{l} \dot{x}_i = x_{i+1} \quad i = 1, \ldots, (n-1) \\ \dot{x}_n = C\left(x_1, \ldots, x_n, u, \dot{u}, \ldots, u^{(\alpha)}\right) \end{array} \right\} (\text{GCCF}) \\ y = x_1 \end{array} \right. \tag{1.28}$$

The first and the second equations of this system give the generalized controllability canonical form (GCCF). With the third equation, the generalized observability canonical form (GOCF) is obtained. Equation (1.28) defines a generalized dynamic that can be associated with system (1.26) assuming its locally open loop observability (Messager 1992). This dynamic is called non-degenerate when the input derivatives involved.

1.3.2.1 Continuous Feedback Linearization

From the explicit generalized canonical forms (1.28), Fliess introduced feedback loops that linearize the closed loop system dynamic (Fliess 1990).

For example, consider a feedback linearization type:

$$C\left(x_1, \ldots, x_n, u, \dot{u}, \ldots, u^{(\alpha)}\right) = \sum_{i=1}^{n} a_i x_i + \sum_{j=1}^{\alpha} b_j v^{(j)} \tag{1.29}$$

where a_i, b_j are coefficients and v is a new control variable. If there is a command $u(t)$ solution of Eq. (1.29), then the closed loop system dynamic is linearized and is written as:

$$\left\{ \begin{array}{l} \dot{x}_i = x_{i+1} \qquad i = 1, \ldots, (n-1) \\ \dot{x}_n = \sum_{i=1}^{n} a_i x_i + \sum_{j=1}^{\alpha} b_j v^{(j)} \\ y = x_1 \end{array} \right. \tag{1.30}$$

The closed loop system is described firstly by Eq. (1.29) which expresses the control variable $u(t)$ (this equation appears as the corrector equation) and secondly by Eq. (1.30) that reflects the new dynamic of the closed loop system: so there are changes in the dynamic behavior by the feedback linearization since the system dynamic is replaced due to the feedback loop by the dynamic (1.30). We note that the closed loop system presents (with such a feedback linearization), the same order as the original dynamic open loop system.

Consider the following non-degenerate single input–single output system of dimension n:

$$\dot{x}_n = C\left(x_1, \ldots, x_n, u, \dot{u}, \ldots, u^{(\alpha)}\right) = f(x) + g(x)\,h(u, \dot{u}, \ldots, u^{(\alpha)}) \tag{1.31}$$

where functions $f(x)$, $g(x)$ and $h(.)$ can be nonlinear. Consider that:

$$h(u, \dot{u}, \ldots, u^{(\alpha)}) = g(x)^{-1}\left(\sum_{j=1}^{\alpha} b_j\, v^{(j)}(x) - f(x)\right)$$

The generalized canonical form is written as:

$$\begin{cases} \dot{x}_i = x_{i+1} \quad i = 1, \ldots, (n-1) \\ \dot{x}_n = \sum_{j=1}^{\alpha} b_j\, v^{(j)}(x) \\ y = x_1 \end{cases} \tag{1.32}$$

The closed loop system shows now the linearized and canonical dynamic of order n that is described by (1.32), the control variable u given as a solution of the feedback loop equation yields:

$$\sum_{j=1}^{\alpha} b_j\, v^{(j)}(x) = f(x) + g(x)\,h(u, \dot{u}, \ldots, u^{(\alpha)}) \tag{1.33}$$

If we set:

$$\sum_{j=1}^{\alpha} b_j\, v^{(j)}(x) = -k_0\, x_1 - k_1\, x_2 - \cdots - k_{n-1}\, x_n$$

the closed loop system is linear, free and of order n. The stability is ensured if the roots of the characteristic polynomial $P(p)$:

$$P(p) = p^n + k_{n-1}p^{n-1} + \cdots + k_1 p + k_0$$

have negative real parts. This requires a suitable choice of gains k_i. The transmittance of the closed loop system $F(p)$ is:

$$F(p) = \frac{1}{p^n + k_{n-1}p^{n-1} + \cdots + k_1 p + k_0}$$

There is new dynamical behavior without modifying the order by the feedback loop. Remember that the open loop dynamic is described (cf. Eq. (1.28)) by the following equations:

$$\begin{cases} \dot{x}_i = x_{i+1} \quad i = 1, \ldots, (n-1) \\ \dot{x}_n = f(x) + g(x)\, h(u, \dot{u}, \ldots, u^{(\alpha)}) \\ y = x_1 \end{cases} \tag{1.34}$$

Several types of feedback linearization are proposed in the literature (Fliess and Messager 1991; Slotine and Li 1991). The new dynamic resulting from these feedback loops presents the order of the system to be controlled.

In the next paragraph, a linearizing loop allowing to have a linear closed loop system of an order lower than that of the open loop system is proposed.

1.3.2.2 Feedback Linearization with Sliding Mode and Discontinuous Control (Variable Structure)

The feedback linearization with the sliding mode mode decreases the order of the closed loop system, due to the introduction of a sliding surface in the state space.

Introduction of a Sliding Surface

The stability of the closed loop system should be ensured. Then, consider the classical presentation (Fliess and Messager 1990; Utkin 1978) in the state space(x_1,..., x_n), and in particular, consider the case of the phase space with the state: (y, $\dot{y}$,..., $y^{(n-1)}$).

Consider a surface defined in the phase space by the sliding function:

$$S(x) = \mathbf{C}^{\mathrm{T}}\mathbf{x} = \sum_{i=1}^{n} c_i x_i \tag{1.35}$$

where $C^T = (c_1, c_2, \ldots, c_n)$ and $x = (x_1, x_2, \ldots, x_n)^T$. Suppose that $c_n = 1$.

The surface $S(x) = 0$ may be linear or nonlinear (Mira 1990; Lopez and Nouri 2006; Utkin 1992). In this case, linear surfaces defined in the phase space have been considered.

The introduction of the sliding function and the discontinuity of the control can be considered in several ways:

- $S(x)$ is a solution of a differential equation in S (Sira-Ramirez 1990),
- $\dot{S}(x) = -k\text{sign}\,(S(x))$ (Lopez and Nouri 2006),
- by introducing the concept of u_{eq} (Lopez and Nouri 2006).

In the dynamic approach used by Sira-Ramirez (1990), the sliding function $S(x)$ is a solution of the following differential equation:

$$\dot{S}(x) + \mu S(x) = -\mu\,\Omega\ \text{sign}\,(S(x)) \tag{1.36}$$

where:

$$\text{sign}\,(S) = \begin{vmatrix} +1 \text{ if } S(s) > 0 \\ 0 \text{ if } S(x) = 0 \\ -1 \text{ if } S(x) < 0 \end{vmatrix}$$

The first convergence of $S(x)$ towards zero (the trajectory $x(t)$ starting from the initial state $x_0(t = 0)$) is done in a finite time t_g whose expression is:

$$t_g = \mu^{-1} \ln \left(1 + \frac{|S(0)|}{\Omega}\right) \tag{1.37}$$

This reaching time t_g, (Slotine and Coestsee 1986), the sliding time (cf. Fig. 1.3), or the required time to reach the surface $S(t, x) = 0$, corresponds to the duration of the transient mode from initial conditions until the beginning of the actual sliding.

The $S(x)$ derivative can be written as:

$$\dot{S}(x) = \dot{x}_n + \sum_{i=1}^{n-1} c_i\, x_{i+1}$$

replace the derivative of $S(x)$ by its expression in Eq. (1.36):

$$\dot{x}_n + \sum_{i=1}^{n-1} c_i\, x_{i+1} = -\mu S(x) - \mu\Omega\,.\text{sign}\,(S(x)) \tag{1.38}$$

By substituting to $S(x)$ its expression (1.35), it results for $\dot{x}_n$ an expression which is function of the (sign($S(x)$)):

$$\dot{x}_n = -\sum_{i=1}^{n-1} c_i\, x_{i+1} - \mu \sum_{i=1}^{n} c_i\, x_i - \mu\Omega\,.\,\text{sign}\,(S(x)) \tag{1.39}$$

Control Equation Ensuring the Sliding Mode

Equations (1.28) and (1.39), expressing $\dot{x}_n$, from the system representation or based on the sliding function, a feedback linearization with variable structure is deduced:

$$C\left(x_1, \ldots, x_n, u, \dot{u}, \ldots, u^{(\alpha)}\right) = -\sum_{i=1}^{n-1} c_i x_{i+1} - \mu \left[\sum_{i=1}^{n} c_i\, x_i + \Omega \text{ sign}\,(S)\right] \tag{1.40}$$

This differential equation gives a variable structure control. Equation (1.40) corresponds to a discontinuous feedback loop with a dynamic state feedback. The knowledge of function $C\left(x_1, \ldots x_n, u, \ldots u^{(\alpha)}\right)$ resulting from the modeling of the system, allows the resolution of this differential equation with respect to the control u.

The convergence of the state to the sliding surface verifies $S\,\dot{S} < 0$ (Utkin 1972). The open-loop dynamic system described by the system (GCCF, GOCF) described by Eq. (1.28), with the feedback linearization defined by Eq. (1.40) becomes the reduced dynamic system (of order $n-1$). It is defined in the phase space by:

$$\begin{cases} \dot{x}_i = x_{i+1}, & i = 1, \ldots, (n-2) \\ \dot{x}_{n-1} = x_n = -\sum\limits_{i=1}^{n-1} c_i x_i \\ y = x_1 \end{cases} \tag{1.41}$$

The control variable $u(t)$ is the solution of the differential equation (1.40).

Thus, when the sliding mode is reached, the operating point stays on the surface $S(x) = 0$. Therefore, the closed loop system becomes insensitive to parameter variations, and its behavior is defined by Eq. (1.41), and depend only on coefficients c_i. Equations (1.41) are equivalent to the equation of the surface $S(x) = 0$. The discontinuity is applied on the highest derivative of the controller. The control is obtained after (α) integrations giving practically a continuous signal.

1.4 Higher Order Sliding Mode

The major drawback of the sliding mode control is the chattering phenomenon. Among several proposed solutions to overcome this problem, we present in this section the higher order sliding control, which consists in moving the discontinuity on the highest control derivatives.

This concept has been introduced by Levantovsky (1986) and Emelyanov et al. (1986). Other works have been proposed and several applications have proved the robustness of the control with respect to the reduction of the chattering phenomenon.

1.4.1 Position of the Problem

Consider a dynamic system described by:

$$\dot{x}(t) = f(t, x, u) \tag{1.42}$$

where:

- $x = [x_1, \ldots, x_n]^T \in X \subset \mathbb{R}^n$ represents the state vector,
- $u \in \mathbb{R}$ is the control variable, $t \in \mathbb{R}$ presents the time,
- f is a sufficiently differentiable function.

Consider a sliding function $S(t, x)$, which is $r-1$ times differentiable with respect to time (r is an integer number).

The objective is to synthesize a higher order sliding mode control law enabling the system trajectory to evolve on the sliding surface and to maintain $S(x)$ and its successive $(r-1)$ derivatives equal to zero. For a sliding order r, it is necessary that:

$$S(t,x) = \dot{S}(t,x) = \ddot{S}(t,x) = \cdots = S^{(r-1)}(t,x) = 0$$

In the case where $r = 1$, the sliding mode corresponds to the classical ideal sliding mode.

1.4.2 *Principle*

The standard sliding mode is based on annulling the first derivative with respect to time of the sliding function. A sliding mode of order r acts on the $(r-1)$ first successive derivatives of the sliding function. In this case, the convergence of the system is provided with a precision of order r. The calculation of the successive derivatives of the sliding variable $S(t,x)$ results in a discontinuity for the order r derivation. Then, it is possible to classify sliding modes of higher order by r successive derivatives of the surface. This determines the smoothness degree of the system. The rth order of the system is determined by the equation:

$$S(t,x) = \dot{S}(t,x) = \cdots = S^{(r-1)}(t,x) = 0 \tag{1.43}$$

One of the major problems of the implementation of the r-sliding algorithms is that required information increase proportionally with the sliding mode order. For example, when using a sliding algorithm of order 3 with respect to $S(t,x)$, it is required to have the knowledge of $S(t,x)$, $\dot{S}(t,x)$ and $\ddot{S}(t,x)$. The Twisting and the Super Twisting algorithms, which are second order sliding modes (Emelyanov et al. 1986; Levantovsky 1985), and which require only the knowledge of the function $S(t,x)$, have been proposed as solutions to this issue.

In Emelyanov et al. (1986) and Levantovsky (1985), several examples ensuring second order sliding concepts for the system (1.42) can be found. In the following, we present the algorithm proposed by Levantovsky (1985) and Emelyanov et al. (1986):

1.4.3 *The "Twisting" Algorithm*

The goal is to force the sliding function to satisfy the condition $S(x) = 0$. Then, consider the following conditions:

- Regarding the sliding function and system (1.42), we assume that $|u| \leq K$, function f is a continuously differentiable function, the sliding function is twice continuously differentiable and the state belongs a variety X. Each solution of Eq. (1.42) is defined for a continuous control variable $u(t)$ satisfying $|u(t)| \leq K$ for all t.

- Suppose that there exists u_1 in [0,1] such that for every continuous function $u(t)$ with $|u(t)| > u_1$ for all t, then we have $S(x)u(t) > 0$ for a finite set of time t.
- There exists positive constants α_0, K_m, K_M, u_0 with $u_0 < 1$ such that:

$$\text{If } \; |S(x)| < \sigma_0 \quad \text{Then } \forall u: \quad 0 < K_m \leq \frac{\partial S}{\partial u} \leq K_M$$
$$|u| > u_0 \Rightarrow \dot{S}\, u > 0$$

Levantovsky (1985) has shown that the following algorithm (twisting type) ensures the convergence of the sliding function and its derivative to zero in a finite time.

$$\dot{u} = \left| \begin{array}{lll} -u & \text{if} & |u| > 1 \\ -\alpha_m \text{sign}\,[S(x)] & \text{if} & S(x)\dot{S}(x) \leq 0 \;\text{ and }\; |u| \leq 1 \\ -\alpha_M \text{sign}\,[S(x)] & \text{if} & S(x)\dot{S}(x) > 0 \;\text{ and }\; |u| \leq 1 \end{array} \right. \tag{1.44}$$

where $\alpha_M > 0$ and $\alpha_m > 0$ satisfying:

$$\alpha_m > 4K_m/\sigma_0, \quad \alpha_m > C_0/K_m, \;\text{ et }\; K_m\alpha_M - C_0 > K_M\alpha_m + C_0$$

This algorithm uses the derivative of the sliding function. In the following, we consider another algorithm ensuring a second order sliding without considering the model of the system (Levantovsky 1985):

$$u = u_1 + u_2 \tag{1.45}$$
$$\dot{u}_1 = \left| \begin{array}{ll} -u & \text{if} \;\; |u| > 1 \\ -\alpha \;\text{sign}\,(S(x)) & \text{if} \;\; |u| \leq 1 \end{array} \right.$$
$$\dot{u}_2 = \left| \begin{array}{ll} -\lambda\, |S_0|^{\rho}\, \text{sign}\,(S(x)) & \text{if} \;\; |S(x)| > S_0 \\ -\lambda\, |S|^{\rho}\, \text{sign}\,(S(x)) & \text{if} \;\; |S(x)|S(x)| \leq S_0 \end{array} \right.$$

where α and λ are strictly positive and with $\eta \in [0\ \ 1]$. The initial condition $u_1(t_0)$ is chosen such that

$$|u| = |u_1(t_0) + u_2(t_0, x_0)| \leq K.$$

Remark:

- For the generalized higher order sliding modes, the discontinuity is applied at the highest derivative of the control variable.
- For the higher order control (of order r) in a finite time, we have:

$$S = \dot{S} = \cdots = S^{(r-1)} = 0$$

The operating point remains on the surface and the convergence is done in a finite time.

- For the generalized sliding mode, the required condition is $\dot{S}(x)S(x) < 0$ in order to obtain an asymptotic convergence.

1.4.4 Second Order Sliding Mode Control

The second order sliding control ($r = 2$) (called 2-sliding) allows to eliminate or to reduce the chattering phenomenon. Its main purpose is to generate a second order sliding mode on a selected sliding surface $S(t, x)$. This can be done by imposing:

$$S(t,x) = \dot{S}(t,x) = 0$$

This is achieved by considering a second order system whose states are the sliding function $S(t, x)$ and its derivative $\dot{S}(t, x)$. To have a second order sliding mode, i.e. $S(t, x) = 0$ and $\dot{S}(t, x) = 0$, a sliding mode control is applied to this new second order system. Therefore, we obtain the finite time convergence of $S(t, x)$ and of its derivative $\dot{S}(t, x)$ to zero.

Consider a dynamic system described by Eq. (1.42) expressed by:

$$\dot{x}(t) = f(t,x,u)$$

The derivative of $S(t, x)$ is written as:

$$\frac{d}{dt}S(t,x) = \frac{\partial}{\partial t}S(t,x) + \frac{\partial}{\partial x}S(t,x)\frac{\partial x}{\partial t}$$

Then:

$$\dot{S}(t,x) = \frac{\partial}{\partial t}S(t,x) + \frac{\partial}{\partial x}S(t,x)f(t,x,u) \tag{1.46}$$

The second derivative of $S(t, x)$ is written as:

$$\frac{d}{dt}\dot{S}(t,x) = \frac{\partial}{\partial t}\dot{S}(t,x,u) + \frac{\partial}{\partial x}\dot{S}(t,x)\frac{\partial x}{\partial t} + \frac{\partial}{\partial u}\dot{S}(t,x)\frac{\partial u}{\partial t}$$

This gives:

$$\frac{d}{dt}\dot{S}(t,x) = \frac{\partial}{\partial t}\dot{S}(t,x,u) + \frac{\partial}{\partial x}\dot{S}(t,x)f(t,x,u) + \frac{\partial}{\partial u}\dot{S}(t,x)\dot{u}(t) \tag{1.47}$$

Suppose that:

$$\begin{cases} \theta(t,x) = \dfrac{\partial}{\partial t}\dot{S}(t,x,u) + \dfrac{\partial}{\partial x}\dot{S}(t,x)f(t,x,u) \\ \varsigma(t,x) = \dfrac{\partial}{\partial u}\dot{S}(t,x) \end{cases} \tag{1.48}$$

Consider now the new system in which state variables are the sliding function $S(t,x)$ and its derivative $\dot{S}(t,x)$ which are noted respectively by $y_1(t,x)$ and $y_2(t,x)$:

$$\begin{cases} y_1(t,x) = S(t,x) \\ y_2(t,x) = \dot{S}(t,x) \end{cases} \tag{1.49}$$

Using Eqs. (1.48) and (1.49), we obtain:

$$\begin{cases} \dot{y}_1(t,x) = y_2(t,x) \\ \dot{y}_2(t,x) = \theta(t,x) + \varsigma(t,x)\dot{u}(t) \end{cases} \tag{1.50}$$

The system described by (1.50) is a second order one. We propose for this new system a new sliding function:

$$\sigma(t,x) = y_2(t,x) + \alpha y_1(t,x) = \dot{S}(t,x) + \alpha S(t,x) \tag{1.51}$$

The system whose input is $\dot{u}(t)$ and whose output is $\sigma(t,x)$ is of relative degree 'one', so a sliding mode can exist on $\sigma(t,x)$ (Sira-Ramirez 1990). So the input $\dot{u}(t)$ can be taken as:

- $\dot{u}(t) = -M \text{ sign } (\sigma(t,x))$.
- $\dot{u}(t) = \dot{u}_{eq(t)} - k \text{ sign } (\sigma(t,x))$.

The term $\dot{u}_{eq}(t)$ is obtained by:

$$\dot{\sigma}(t,x) = \dot{y}_2(t,x) + \alpha\dot{y}_1(t,x) = \ddot{S}(t,x) + \alpha\dot{S}(t,x) \tag{1.52}$$

with: $\ddot{S}(t,x) = C^T\ddot{x}(t)$.

$\ddot{x}(t)$ is deduced from the derivative of the controlled system:

$$\ddot{x}(t) = \frac{\partial}{\partial t}f(t,x,u) + \frac{\partial}{\partial x}f(t,x,u)\dot{x}(t) + \frac{\partial}{\partial u}f(t,x,u)\dot{u}(t) \tag{1.53}$$

Then, the equivalent control variable for the new system is:

$$\dot{u}_{eq}(t) = \frac{1}{C^T\dfrac{\partial}{\partial u}f(t,x,u)}\left(C^T\frac{\partial}{\partial t}f(t,x,u) + C^T\frac{\partial}{\partial x}f(t,x,u)\dot{x} + \alpha\dot{S}(t,x)\right) \tag{1.54}$$

The control expression of the new system is:

$$\dot{u}(t) = \dot{u}_{eq}(t) + \dot{u}_{dis}(t) \tag{1.55}$$

where:

$$\dot{u}_{eq}(t) = \frac{1}{C^T \frac{\partial}{\partial u} f(t, x, u)} \left(C^T \frac{\partial}{\partial t} f(t, x, u) + C^T \frac{\partial}{\partial x} f(t, x, u)\dot{x} + \alpha \dot{S}(t, x) \right)$$
$$\dot{u}_{dis}(t) = -k \ \text{sign}\ (\sigma(t, x))$$

It is the integral of this control variable that must be applied to the system to be controlled:

$$u(t) = \int [\dot{u}_{eq}(t) + \ \dot{u}_{dis}(t)]dt \tag{1.56}$$

If we consider a system with a relative order equal to 1 with respect to $S(t, x)$, the control algorithm is convergent if there exists Γ_m, Γ_M, Φ and S_0, positive constants, such that in a neighborhood $|S(t, x)| \leq S_0$, the following conditions are satisfied:

$$\begin{array}{l} 0 < \Gamma_m \leq \varsigma(t, x) \leq \Gamma_M \\ |\theta(t, x)| \leq \Phi \end{array} \tag{1.57}$$

1.4.5 Example

Consider the following linear system:

$$\begin{cases} \dot{x}_1 = x_2 \\ \dot{x}_2 = -a_1 x_1 - a_2 x_2 + u \\ y = x_1 \end{cases}$$

The desired output y_d is constant.
Letting $e_1 = y - y_d$, we can obtain the following error system:

$$\begin{cases} \dot{e}_1 = e_2 \\ \dot{e}_2 = -a_1 e_1 - a_2 e_2 + u + a_1 y_d \\ e_1 = y - y_d \end{cases}$$

where $a_1 = 8.86$ and $a_2 = 2.58$.

The aim is to synthesize a second order sliding mode control law in order to have the convergence of $S(e)$ and $\dot{S}(e)$ to zero in a finite time. The sliding function $S(e)$ is chosen as:

$$S(e) = c_1 e_1 + e_2 \quad c_1 = 2.$$

Twisting Algorithme

This algorithm has been described by Eq. (1.44):

$$\dot{u} = \left| \begin{array}{lll} -u & \text{if} & |u| > 1 \\ -\alpha_m \text{ sign}(S(e)) & \text{if} & S(e)\dot{S}(e) \le 0 \text{ and } |u| \le 1 \\ -\alpha_M \text{ sign}(S(e)) & \text{if} & S(e)\dot{S}(e) > 0 \text{ and } |u| \le 1 \end{array} \right.$$

where: $\alpha_m = 5$ and $\alpha_M = 20$.

Algorithm 2-sliding

This algorithm has been described in the previous paragraph, which has been applied to a linear second order system.

In this case, we define the new system whose state variables are the sliding function $S(e)$ and its derivative $\dot{S}(e)$:

$$\begin{cases} y_1(t, e) = S(e) \\ y_2(t, e) = \dot{S}(t, e) \end{cases}$$

This gives:

$$\begin{cases} \dot{y}_1(t, e) = y_2(t, e) \\ \dot{y}_2(t, e) = \ddot{S}((t, e) = c_1\ddot{e}_1 + (-a_1\dot{e}_1 - a_2\dot{x}_2 + \dot{u}) \end{cases}$$

which can be written as:

$$\begin{cases} \dot{y}_1(t, e) = y_2(t, e) \\ \dot{y}_2(t, e) = \theta(t, e) + \varsigma(t, e)\dot{u} \end{cases}$$

where:

$$\theta(t, e) = c_1\ddot{e}_1 - a_1\dot{e}_1 - a_2\dot{x}_2 \quad \varsigma(t, e) = 1$$

For this new system, we associate a new sliding function $\sigma(t)$ defined by:

$$\sigma(t) = \eta y_1 + y_2$$

The chosen control variable for this new system is of the form $-M$ sign $(\sigma(t))$. So we have:

$$\dot{u} = -M \text{ sign}(\sigma(t)) \Rightarrow u(t) = \int \dot{u} dt$$

In this case, we have $\eta = 10$ and $M = 1$.

Simulation results of the two control types (Twisting and 2-sliding) are given in Figs. 1.6 and 1.7.

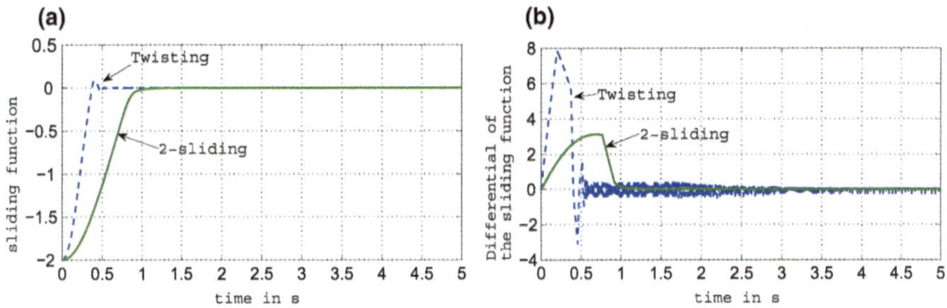

Fig. 1.6 Second order sliding mode. Case of a non perturbed system. **a** Evolution of the sliding function. **b** Evolution of the sliding function derivative

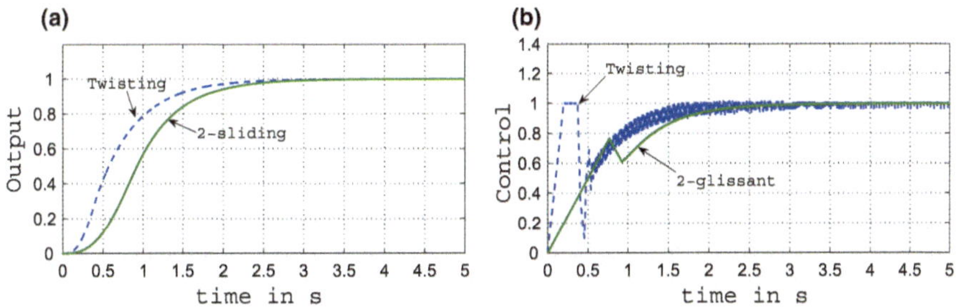

Fig. 1.7 Second order sliding mode. Case of an non perturbed system. **a** Evolution of the output. **b** Evolution of the control variable

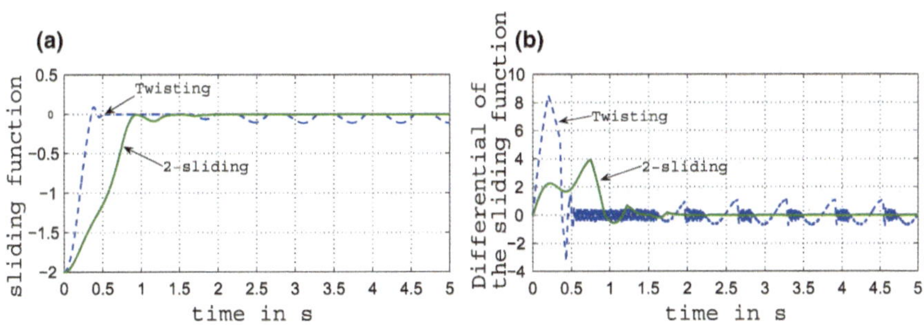

Fig. 1.8 Second order sliding mode. Case of a perturbed system. **a** Evolution of the sliding function. **b** Evolution of the sliding function derivative

According to these figures, it is notable that results of the 2-sliding mode controller present less oscillations compared to those achieved by the Twisting algorithm. However, for these algorithms, the convergence is performed in a finite time.

To test the robustness of these algorithms, a sinusoidal noise ($0.1\ \sin\ t$) has been added to the input. Simulation results presented in Fig. 1.8 show better robustness of the 2-sliding control law compared to those of the twisting type.

1.5 Conclusion

In this chapter, a reminder about different aspects of the variable structure control and the sliding mode control in the continuous domain for single input–single output systems has been presented. Sliding modes are detailed in the classical formalism (invariance conditions, different control forms and sliding functions are given), in the formalism of differential algebra (generalized sliding mode control) and the higher order sliding modes (essentially the second order sliding mode controller).

Acknowledgments This work has been supported by the Ministry of the Higher Education and Scientific Research in Tunisia.

References

Asada, H., & Slotine, J. J. E. (1986). *Robot Analysis and Control*. New York: Wiley, A WILEY - interscience publication.

Emelyanov, S. V. (1967). *Variable Structure Control Systems*. Moscow: Nauka.

Emelyanov, S. V., Korovin, S. K., & Levantvosky, L. V. (1986). Higher order sliding modes in the binary control systems. *Soviet Physics, Doklady*, *31*, 291–293.

Emelyanov, S. V., Utkin, V. I., Taran, V. A., Kostyleva, N. E., Shubladze, A. M., Ezerov, V. B., et al. (1970). *Theory of Variable Structure Control*. Moscow (in russian): Nauka.

Filippov, A. (1960). Equations différentielles à second membre discontinu. *Journal de mathématiques*, *51*(1), 99–128.

Fliess, M. (1990). Generalised controller canonical forms for linear and non-linear dynamics. *IEEE Transactions on Automatic Control*, *35*, 994–1001.

Fliess, M. & Messager, F. (1990). Vers une stabilisation non linéaire discontinue. *Proceedings of the 29th International Conference on Analysis and Optimization of Systems, Lecture Notes in Control and information*. Berlin: Springer. (editor)

Fliess, M. & Messager, F. (1991). *Methods of Non-linear Discontinuous Stabilisation*. C. I. Byrnes et A. Kurzhansky (Eds.). Birkhaüser.

Fridman, L., & Levant, A. (1999). *Higher Order Sliding Modes*. Sliding modes in Automatic Control: Int. School in Automatic Control of Lille.

Hamerlain, M. (1993). *Commande hiérarchisée à modèle de référence et à structure variable d'un robot manipulateur à muscles artificiels*. Ph.D. thesis, GARI/DGE/INSA, Toulouse.

Levant, A. (1993). Sliding order and sliding accurcy in sliding mode control. *International Journal of Control*, *58*(6), 1247–1263.

Levantovsky, L. V. (1985). Second order sliding algorithms: their realization. *Dynamics of Heterogeneous Systems*. Materials of the seminar (pp. 32–43), Moscow.

Levantovsky, L. V. (1986). Sliding modes with continuous control. *Proceedings of the All Union Scientific Practical Deminar on Application Experience of Distrubed Systems Novokuznezk* (Vol. 1, pp. 79–80) (in russian).

Lopez, P. & Nouri, A. S. (2006). *Théorie élémentaire et pratique de la commande par les régimes glissants*. Mathématiques et applications 55, SMAI, Springer.

Lyapunov, A. M. (1966). *Stability of Motion*. New York: Academic Press.

Medhafar, H., Damak, T., & Derbel, N. (2005). Direct adaptive fuzzy moving sliding mode controller design for robot manipulators. *International Journal of Computational Intelligence and Applications (IJCIA)*, *5*(1), 1–20.

Messager, F. (1992). *Sur la stabilisation discontinue des systèmes*. Ph.D. thesis, Orsay, Paris.

Mira, C. (1990). *Systèmes asservis non linéaires*. Paris: Hermès.
Neymark, Y. I. (1957). On sliding modes in relay control systems. *Automatiks i telemekhanika, 18*(1), 27–33.
Popovski, A. M. (1950). Linearization of sliding operation mode for a constant speed controller. *Automatiks i telemekhanika, 11*(3), 161–163.
Sira-Ramirez, H. (1988). Differential geometric methods in variable structure methods. *International Journal of Control, 48*(4), 1359–1390.
Sira-Ramirez, H. (1990). Structure at infinity, zero dynamics and normal forms of systems undergoing sliding motion. *International Journal of Systems Sciences, 21*(4), 665–674.
Sira-Ramirez, H., Ahmad, S., & Zribi, M. (1992). Dynamical feedback control of robotic manipulators with joint flexibility. *IEEE Transactions on Systems, Man and Cybernetics, 22*(4), 736–747.
Slotine, J. J. E., & Coestsee, J. A. (1986). Sliding controller synthesis for non-linear systems. *Int. J. Control, 43*(6), 1631–1651.
Slotine, J. J. E., & Li, W. (1991). *Applied Non-linear Control*. USA: Prentice-Hall international Inc.
Utkin, V. I. (1972). Equations of sliding mode in discontinuous systems. *Automation and Remote Control, 2*(2), 211–219.
Utkin, V. I. (1978). Discontinuous control systems: State of art in theory and application. Technical report, Institute of Control Sciences, Ed. MIR, Moscow, USSR.
Utkin, V. I. (1978). *Sliding Modes and their Application in Variable Structure Systems*. Moscow: Edition MIR.
Utkin, V. I. (1992). *Sliding Mode in Control Optimisation*. Berlin: Springer.
Vernhes, J. P. (1971). *Contribution à l'étude des systèmes à structure variable*. Ph.D. thesis, Université P. Sabatier.

Chapter 2
Conditions of Disturbances Rejection for Discrete First, Second Order and Repetitive Sliding Mode Controllers

Khadija Dehri, Majda Ltaief and Ahmed Said Nouri

Abstract Harmonic disturbance rejection is an important field of control theory and applications. In this paper a discrete first and second order sliding mode control for multivariable systems are investigated. The necessary conditions of harmonic disturbances rejection using first and second order sliding mode control laws are elaborated. In order to improve the performances of sliding mode control in periodic disturbances rejection, a discrete repetitive sliding mode control is presented. A necessary condition for the choice of the discontinuous terms in discrete repetitive sliding mode control is then developed. The different proposed control strategies have been tested on numerical simulation example. The obtained results are very satisfactory in terms of compensation of periodic disturbances using discrete repetitive sliding mode control.

Keywords Conditions of disturbances rejection · Discrete sliding mode control · Second order sliding mode control · Repetitive sliding mode control · Multivariable control systems · Rejection of periodic disturbances

2.1 Introduction

Due to the development and the progress of technology, many industrial processes are became more complex (multivariable, non linear, with parameter uncertainties and external disturbances, ...). Then, the representation of such systems by an exact mathematical model and the development of an adequate control law are extremely difficult.

K. Dehri (✉) · M. Ltaief · A.S. Nouri
Research Unit CONPRI, ENIG, University of Gabes, Gabes, Tunisia
e-mail: khadija.dehri@gmail.com

M. Ltaief
e-mail: majda.ltaief@enig.rnu.tn

A.S. Nouri
e-mail: ahmedsaid.nouri@enig.rnu.tn

N. Derbel et al. (eds.), *Applications of Sliding Mode Control*,
Studies in Systems, Decision and Control 79, DOI 10.1007/978-981-10-2374-3_2

To overcome this problem, various control strategies have been proposed in the last two decades. Sliding Mode Control (SMC) as a Variable Structure System (VSS) has been widely used in different engineering fields and carried out excellent performances in many industrial applications (Lopez and Nouri 2006; Yu and Kaynak 2009). The SMC is known by its robustness against uncertainties and external disturbances. The main idea behind SMC is to use a discontinuous control input to drive state trajectories towards desired sliding surface in finite time and maintain them on it. Nevertheless, the main drawback of the sliding mode control is the chattering phenomenon caused by the discontinuous part of the control law. The phenomenon of chattering is a harmful phenomenon and it often leads to undesirable results.

Due to the development of computer and the use of digital control, discrete sliding mode control (DSMC) has become more important research in many theoretical and practical control systems (Gao et al. 1995; Bartoszewicz 1998; Stoica 2008). However, the properties of continuous SMC is not available in the case of DSMC because of the finite sampling rate (Young et al. 1999). Then, the problem of chattering phenomenon is more difficult in discrete sliding mode control. Several methods have been proposed to reduce this problem. One solution is to replace the signum function by a smoother function in the boundary layer such as the saturation function. Another solution is to use a high order sliding mode control (HOSMC) (Mihoub et al. 2009; Romdhane et al. 2015; Cavallo and Natale 2004).

Another problem of the DSMC is its sensitivity to external disturbances. In a variety of industrial processes, these disturbances are, often, periodic signals (robotic, rotating machine tools, active noise control ...). In electrical networks, non-linear devices are the main source of harmonic disturbances. All power electronic converters can generate harmonic disturbances by injecting harmonic currents in the power system (Zhou and Wang 2003). Under the effect And turning fields, the couple can cause mechanical vibration in rotating machinery. For example, in the hard disk drives, the periodic disturbances are caused by the eccentricity on the track at the frequency of rotation of the disk (Chang et al. 2006). In the robot-assisted laparoscopic surgery of the digestive system, physiological movements caused by breathing and heartbeat can be considered as a periodic disturbances (Ginhoux 2003), etc.

In order to resolve this problem, several proposals have been made to modify the discrete classical sliding mode control. Bartoszewicz et al. (Ginhoux 2003) proposed a new non-switching reaching law for periodic review inventory systems. In (Mihoub et al. 2011), a multimodel discrete second order sliding mode control for non stationary system was formulated. Young et al. (Young et al. 1999) proposed a discrete sliding mode control with delayed disturbance compensator. In (Yan et al. 2013), a discrete sliding mode control with decoupled perturbation estimator based on computation time delay was developed. A discrete sliding mode control with a nonlinear observer is designed to estimate non measurable states and perturbation and applied to induction motors in (Castillo-Toledoa et al. 2008). Bandyopadhyay et al. (Bandyopadhyay and Fulwani 2009) proposed a nonlinear sliding surface and disturbance observer to design sliding mode control for discrete multiple-input multiple-output linear systems with matched perturbations.

In parallel a discrete adaptive sliding mode controller have been developed (Monsees 2002; Chan 1997).

Nevertheless, the use of disturbance observer can make control system more complex. Another approach is to combine the sliding mode control with repetitive control (Dehri et al. 2011, 2012a, b).

Repetitive control (RC) is one of the most common approach in dealing with the periodic disturbance rejection (Arimoto et al. 1984; Xuan et al. 2013). It has been investigated in the continuous and discrete time domain and in different engineering areas (robots (Xuan et al. 2007; Fateha et al. 2013), hard disk drives (Chang et al. 2006), pulse width modulation inverters (Zhou and Wang 2003), etc.). The main idea behind repetitive control is to remove errors that occur at the fundamental and harmonics frequency of the periodic signal. This control is based on the Internal Model Principle (IMP) (Francis and Wonham 1976) which consists in simply incorporating the model of the disturbance into the controller configuration. A periodic signal can be generated by a delay block with a positive feedback loop. However, the repetitive control is confronted with several problems: such as the problem of stability and the inability to take into account certain characteristics of processes (Doh and Ryoo 2006).

In order to resolve these problems, we have proposed to combine the repetitive control with the sliding mode control for discrete multivariable systems (Dehri et al. 2011, 2012a, b).

This work focuses on the synthesis of a necessary condition to reject periodic disturbances for multivariable systems using discrete first, second order and repetitive sliding mode controllers.

This paper is organized as follows. The synthesis of a necessary condition for rejection external disturbances using classical discrete multivariable sliding mode control is presented in Sect. 2.2. In Sect. 2.3, a necessary condition for rejection external disturbances in discrete second order sliding mode control for multivariable systems is developed. A necessary condition for rejection periodic disturbances of discrete repetitive sliding mode control is then proposed in Sect. 2.4. Section 2.5 gives the simulation results of the discrete first, second order and repetitive sliding mode controllers for a chemical reactor model subjected to periodic disturbances.

2.2 Conditions of Disturbances Rejection in Discrete Multivariable Sliding Mode Control

Consider a linear multivariable discrete-time system:

$$x(k+1) = Ax(k) + Bu(k) \tag{2.1}$$

where $x(k)$ and $u(k)$ are respectively the state and input vectors:

$$x(k) = \left[x_1(k) \dots x_n(k) \right]^\top$$

$$u(k) = \left[u_1(k) \dots u_m(k) \right]^\top$$

A and B are respectively $n \times n$ and $n \times m$ known constant matrices. We suppose that the matrix B is of full rank.

The sliding function is defined as:

$$S(k) = Cx(k) = [s_1(k) \cdots s_m(k)]^\top \tag{2.2}$$

where C is an (m, n) matrix chosen using the pole assignment method (Chang and Chen 2000).

The reaching law can be written as follows (Bartoszewicz 1998):

$$S(k+1) = \Phi S(k) - \begin{bmatrix} m_1 \text{sign}(s_1(k)) \\ m_2 \text{sign}(s_2(k)) \\ \vdots \\ m_m \text{sign}(s_m(k)) \end{bmatrix} \tag{2.3}$$

where Φ ($\Phi \in \Re^{m \times m}$) is a diagonal matrix such
$0 \leq \Phi_{i,\,i} < 1, \quad \forall\, i = 1 \dots m$ and m_i ($m_i \in \Re$) is a positive gain.
And sign is the signum function defined as:

$$\text{sign}(s_i(k)) = \begin{cases} -1 & \text{si} \;\; s_i(k) < 0 \\ 1 & \text{si} \;\; s_i(k) > 0 \end{cases}; \quad i \in \left[1 \dots m\right]$$

Using the considered reaching law, the sliding mode control law can be expressed as:

$$u(k) = (CB)^{-1}\left[-CAx(k) + \Phi S(k)\right] - (CB)^{-1} \begin{bmatrix} m_1 \text{sign}(s_1(k)) \\ \vdots \\ m_m \text{sign}(s_m(k)) \end{bmatrix} \tag{2.4}$$

To study the robustness of the control law, we suppose that the system is subject to external disturbances as follows:

$$x(k+1) = Ax(k) + Bu(k) + f(k) \tag{2.5}$$

where $f(k)$ is the external disturbances vector.

Suppose the following matching condition is satisfied:

$$f(k) = Bd(k); \quad d(k) = \begin{bmatrix} d_1(k) \\ d_2(k) \\ \vdots \\ d_m(k) \end{bmatrix} \tag{2.6}$$

Then, the system (2.5) is equivalent to:

$$x(k+1) = Ax(k) + B[u(k) + d(k)] \tag{2.7}$$

Conditions of Disturbances Rejection

By applying the sliding control law (2.4) to the system (2.5), the sliding functions vector is given by:

$$\begin{aligned} S(k+1) &= Cx(k+1) = CAx(k) + CBu(k) + CBd(k) \\ &= \Phi S(k) - \begin{bmatrix} m_1 \mathrm{sign}(s_1(k)) \\ m_2 \mathrm{sign}(s_2(k)) \\ \vdots \\ m_m \mathrm{sign}(s_m(k)) \end{bmatrix} + CBd(k) \end{aligned}$$

A necessary and sufficient condition for the existence of a quasi sliding mode is defined as (Sarpturk et al. 1987):

$$|s_i(k+1)| < |s_i(k)|, \quad \forall\ i = 1 \ldots m \tag{2.8}$$

The previous condition is equivalent to:

$$\begin{cases} (s_i(k+1) - s_i(k))\mathrm{sign}(s_i(k)) < 0, & \forall\ i = 1 \ldots m \\ (s_i(k+1) + s_i(k))\mathrm{sign}(s_i(k)) > 0, & \forall\ i = 1 \ldots m \end{cases}$$

Suppose that:

$$\bar{d}(k) = CBd(k) = \left[\bar{d}_1(k)\ \bar{d}_2(k)\ \ldots\ \bar{d}_m(k) \right]^\top$$

We have:

$$s_i(k+1) = \Phi_{i,\,i} s_i(k) - m_i \mathrm{sign}(s_i(k)) + \bar{d}_i(k), \quad \forall\ i = 1 \ldots m$$

Case $s_i(k) > 0$:
In this case, the conditions of existence of a quasi-sliding mode are:

$$\begin{cases} s_i(k+1) - s_i(k) < 0 \\ s_i(k+1) + s_i(k) > 0 \end{cases}$$

or

$$s_i(k+1) - s_i(k) = -(1 - \Phi_{i,i})s_i(k) - m_i + \bar{d}_i(k)$$

then:

$$s_i(k+1) - s_i(k) = (\Phi_{i,i} - 1)s_i(k) - m_i + \bar{d}_i(k) < 0 \Rightarrow m_i > (\Phi_{i,i} - 1)s_i(k) + \bar{d}_i(k)$$

We have:

$$s_i(k+1) + s_i(k) = (1 + \Phi_{i,i})s_i(k) - m_i + \bar{d}_i(k)$$

then:

$$s_i(k+1) + s_i(k) = (\Phi_{i,i} + 1)s_i(k) - m_i + \bar{d}_i(k) > 0 \Rightarrow m_i < (\Phi_{i,i} + 1)s_i(k) + \bar{d}_i(k)$$

Case $s_i(k) < 0$:
The conditions of existence of a quasi-sliding mode are:

$$\begin{cases} s_i(k+1) - s_i(k) > 0 \\ s_i(k+1) + s_i(k) < 0 \end{cases}$$

or

$$s_i(k+1) - s_i(k) = -(1 - \Phi_{i,i})s_i(k) + m_i + \bar{d}_i(k)$$

then

$$s_i(k+1) - s_i(k) = (\Phi_{i,i} - 1)s_i(k) + m_i + \bar{d}_i(k) > 0 \Rightarrow m_i > -(\Phi_{i,i} - 1)s_i(k) - \bar{d}_i(k)$$

We have:

$$s_i(k+1) + s_i(k) = (\Phi_{i,i} + 1)s_i(k) + m_i + \bar{d}_i(k) < 0 \Rightarrow m_i < -(\Phi_{i,i} + 1)s_i(k) - \bar{d}_i(k)$$

Theorem 1 *The discrete sliding mode control law defined in (2.4) allows the rejection of the external disturbances if and only if the gains m_i satisfy:*

$$(\Phi_{i,i} - 1)\,|s_i(k)| + \bar{d}_i(k)\,\text{sign}(s_i(k)) < m_i < (\Phi_{i,i} + 1)\,|s_i(k)| + \bar{d}_i(k)\,\text{sign}(s_i(k)) \tag{2.9}$$

According to this condition, it is noted that the rejection of external disturbances is possible only if we know exactly these disturbances. If the disturbances are not known, an estimation phase is required.

Also, the minimum quasi-sliding mode band $2\mu_i$ depends on the maximum norm of the disturbances:

$$2\mu_i > \frac{2m_i}{1 - \Phi_{i,i}}$$

Consequently, if these disturbances are relatively important, a large amplitude oscillations is appeared which can excite the high frequencies and damage the controlled system.

To resolve this problem, the discrete second order sliding mode control is described in the following paragraph.

2.3 Conditions of Disturbances Rejection in Discrete Multivariable Second Order Sliding Mode Control

Discrete second order sliding mode control (2-DSMC) is a major approach being used of higher order sliding mode control. It consists in force the the state to move on the sliding surface and to keep it first derivative null.

There are some works, in literature, which used 2-DSMC for multiple-input multiple-output (MIMO) systems because it is well known that the control problems of MIMO systems are very difficult (Chang 2002; Romdhane et al. 2015; Monsees 2002; Dehri et al. 2011, 2012a, b; Chang and Chen 2000).

In the case of the multivariable second order sliding mode control, the sliding functions vector is selected as follows (Mihoub et al. 2011):

$$\sigma(k) = S(k) + \beta S(k-1) = \left[\sigma_1(k) \ \ldots \ \sigma_m(k)\right]^\top \tag{2.10}$$

where β is an (m, m) diagonal matrix.

The equivalent control law is obtained if the following relation is verified:

$$\sigma(k+1) = \sigma(k) = 0$$

Then, the expression of the equivalent control law can be calculated as:

$$u_{eq}(k) = (CB)^{-1}\left[-\beta S(k) - CAx(k)\right]$$

The robustness is ensured by the addition of a discontinuous term $u_{dis}(k)$ such as:

$$\begin{aligned} u_{dis}(k) &= u_{dis}(k-1) - (CB)^{-1} T_e \begin{bmatrix} m_1^{'}\text{sign}(\sigma_1(k)) \\ \vdots \\ m_i^{'}\text{sign}(\sigma_i(k)) \\ \vdots \\ m_m^{'}\text{sign}(\sigma_m(k)) \end{bmatrix} \\ &= u_{dis}(k-1) - (CB)^{-1}\left[m_1^{''}\text{sign}(\sigma_1(k)) \ \ldots \ m_m^{''}\text{sign}(\sigma_m(k))\right]^\top \end{aligned}$$

with T_e is the sampling time.

The discrete second order sliding mode control is then expressed as:

$$u(k) = u_{eq}(k) + u_{dis}(k) \tag{2.11}$$

Conditions of Disturbances Rejection

Using (2.5) and (2.11), the sliding functions vector $S(k+1)$ can be written as:

$$S(k+1) = Cx(k+1) = -\beta S(k) + CBd(k) + CBu_{dis}(k)$$

Based on the last relation and (2.10), the sliding functions vector $\sigma(k+1)$ can be written as:

$$\begin{aligned}\sigma(k+1) &= \sigma(k) + CB(d(k) - d(k-1)) \\ &- \left[m_1^{''}\text{sign}(\sigma_1(k)) \; \ldots \; m_m^{''}\text{sign}(\sigma_m(k)) \right]^{\top}\end{aligned}$$

A necessary and sufficient condition for the existence of a quasi sliding mode in the case of discrete second order sliding mode control is defined as:

$$|\sigma_i(k+1)| < |\sigma_i(k)|, \quad \forall\ i = 1 \ldots m \tag{2.12}$$

The previous inequality is equivalent to the following two inequalities:

$$\begin{cases} (\sigma_i(k+1) - \sigma_i(k))\text{sign}(\sigma_i(k)) < 0 \\ (\sigma_i(k+1) + \sigma_i(k))\text{sign}(\sigma_i(k)) > 0 \end{cases}$$

Suppose that:

$$\tilde{d}(k) = CBd(k) - CBd(k-1) = \left[\tilde{d}_1(k)\ \tilde{d}_2(k) \ \ldots \ \tilde{d}_m(k) \right]^{\top}$$

We have:

$$\sigma_i(k+1) = \sigma_i(k) - m_i^{''}\text{sign}(\sigma_i(k)) + \tilde{d}_i(k), \quad \forall\ i = 1..m$$

Case $\sigma_i(k) > 0$:
The conditions of existence of a quasi-sliding mode becomes:

$$\begin{cases} \sigma_i(k+1) - \sigma_i(k) < 0 \\ \sigma_i(k+1) + \sigma_i(k) > 0 \end{cases}$$

We have

$$\sigma_i(k+1) - \sigma_i(k) = -m_i^{''} + \tilde{d}_i(k)$$

So:

$$\sigma_i(k+1) - \sigma_i(k) = -m_i^{''} + \tilde{d}_i(k) < 0 \Rightarrow m_i^{''} > \tilde{d}_i(k)$$

Using the following equality:

$$\sigma_i(k+1) + \sigma_i(k) = 2\sigma_i(k) - m_i^{''} + \tilde{d}_i(k)$$

We obtain:

$$\sigma_i(k+1) + \sigma_i(k) = 2\sigma_i(k) - m_i^{''} + \tilde{d}_i(k) > 0 \Rightarrow m_i^{''} < 2\sigma_i(k) + \tilde{d}_i(k)$$

Case $\sigma_i(k) < 0$:
The conditions of existence of a quasi-sliding mode are:

$$\begin{cases} \sigma_i(k+1) - \sigma_i(k) > 0 \\ \sigma_i(k+1) + \sigma_i(k) < 0 \end{cases}$$

which yields

$$\sigma_i(k+1) - \sigma_i(k) = m_i^{''} + \tilde{d}_i(k) > 0 \Rightarrow m_i^{''} > -\tilde{d}_i(k)$$

$$\sigma_i(k+1) + \sigma_i(k) = 2\sigma_i(k) + m_i^{''} + \tilde{d}_i(k) < 0 \Rightarrow m_i^{''} < -2\sigma_i(k) - \tilde{d}_i(k)$$

Theorem 2 *The discrete second order sliding mode control law defined in (2.11) allows the rejection of the external disturbances if and only if the gains $m_i^{''}$ satisfy:*

$$\tilde{d}_i(k)\,\mathrm{sign}(\sigma_i(k)) < m_i^{''} < 2\,|\sigma_i(k)| + \tilde{d}_i(k)\,\mathrm{sign}(\sigma_i(k)) \tag{2.13}$$

We can conclude that the discrete second order sliding mode control is able to reduce the chattering phenomenon and to reject constant disturbances without necessity of estimating them by the integration of the discontinuous term.

Also, the minimum quasi-sliding mode band depends on the maximum norm of the derivative of disturbances.

Conditions of Rejection of Periodic Disturbances

The disturbance vector is given as follows:

$$d(k) = \begin{bmatrix} \sum_{j=0}^{N_d} a_{1j}\cos(jwk) + b_{1j}\sin(jwk) \\ \vdots \\ \sum_{j=0}^{N_d} a_{ij}\cos(jwk) + b_{ij}\sin(jwk) \\ \vdots \\ \sum_{j=0}^{N_d} a_{mj}\cos(jwk) + b_{mj}\sin(jwk) \end{bmatrix} \tag{2.14}$$

The difference between $d(k)$ and $d(k-1)$ can be majored as:

$$|d(k) - d(k-1)| \leq wT_e \begin{bmatrix} \sum_{j=0}^{N_d} |a_{1j}| + |b_{1j}| \\ \vdots \\ \sum_{j=0}^{N_d} |a_{ij}| + |b_{ij}| \\ \vdots \\ \sum_{j=0}^{N_d} |a_{mj}| + |b_{mj}| \end{bmatrix} \leq wT_e d_{\max}$$

with $d_{\max} = \sup(d(k)) = \begin{bmatrix} d_{1\max} & \ldots & d_{m\max} \end{bmatrix}^{\top}$

The condition of periodic disturbances rejection in discrete second order sliding mode control is formulated as:

$$-wT_e d_{i\max} < m_i'' < 2|\sigma_i(k)| + wT_e d_{i\max} \tag{2.15}$$

The discrete sliding mode control can reject periodic disturbances if the period of these disturbances is very high.

A solution for this problem was proposed in (Dehri et al. 2011, 2012a, b) by combining the discrete sliding mode control and the repetitive control.

2.4 Conditions of Disturbances Rejection in Discrete Multivariable Repetitive Sliding Mode Control

In the presence of periodic disturbances, the discrete multivariable first and second order sliding mode control performances are decreased considerably. In order to overcome these problems, we propose to use the discrete multivariable repetitive sliding mode control.

We suppose that the disturbances vector $d(k)$ is periodic with the period N:

$$d(k) = \begin{bmatrix} d_1(k) \\ \vdots \\ d_i(k) \\ \vdots \\ d_m(k) \end{bmatrix} = \begin{bmatrix} d_1(k-N) \\ \vdots \\ d_i(k-N) \\ \vdots \\ d_m(k-N) \end{bmatrix} = d(k-N) \tag{2.16}$$

The difference between $s(k+1)$ and $s(k+1-N)$ can be calculated as:

$$s(k+1) - s(k+1-N) = \begin{bmatrix} CAx(k) + CBu(k) + CBd(k) \\ -CAx(k-N) - CBu(k-N) - CBd(k-N) \end{bmatrix}$$

Using the last relation and (2.3), the discrete repetitive sliding mode control law for multivariable systems subjected to periodic disturbances can be expressed as:

$$\begin{aligned} u(k) &= u(k-N) - (CB)^{-1}CA(x(k) - x(k-N)) \\ &\quad - (CB)^{-1}\begin{bmatrix} m_1 \operatorname{sign}(s_1(k)) \\ m_2 \operatorname{sign}(s_2(k)) \\ \vdots \\ m_i \operatorname{sign}(s_i(k)) \\ \vdots \\ m_m \operatorname{sign}(s_m(k)) \end{bmatrix} \\ &\quad + (CB)^{-1}(\Phi s(k) - s(k+1-N)) \end{aligned} \tag{2.17}$$

Conditions of Disturbance Rejection

Replacing the control law by its expression (2.17), the sliding functions vector is then given as follows:

$$\begin{aligned} &S(k+1) = Cx(k+1) = CAx(k) + CBu(k) + CBd(k) \\ &= CAx(k) + CBu(k-N) - CA(x(k) - x(k-N)) + \Phi S(k) - S(k+1-N) \\ &- \begin{bmatrix} m_1 \operatorname{sign}(s_1(k)) \\ m_2 \operatorname{sign}(s_2(k)) \\ \vdots \\ m_m \operatorname{sign}(s_m(k)) \end{bmatrix} + CBd(k) \\ &= \Phi S(k) - \begin{bmatrix} m_1 \operatorname{sign}(s_1(k)) \\ m_2 \operatorname{sign}(s_2(k)) \\ \vdots \\ m_m \operatorname{sign}(s_m(k)) \end{bmatrix} + CB(d(k) - d(k-N)) \end{aligned}$$

A necessary and sufficient condition for the existence of a quasi sliding mode is defined by (2.8).

Suppose that:

$$\underline{d}(k) = CB(d(k) - d(k-N)) = \begin{bmatrix} \underline{d}_1(k) \; \underline{d}_2(k) \ldots \underline{d}_m(k) \end{bmatrix}^{\top}$$

We have:

$$s_i(k+1) = \Phi_{i,\,i} s_i(k) - m_i \operatorname{sign}(s_i(k)) + \underline{d}_i(k), \quad \forall\, i = 1 \ldots m$$

Case $s_i(k) > 0$:
In this case, the conditions of existence of a quasi-sliding mode are:

$$\begin{cases} s_i(k+1) - s_i(k) < 0 \\ s_i(k+1) + s_i(k) > 0 \end{cases}$$

then:

$$s_i(k+1) - s_i(k) = (\Phi_{i,i} - 1)s_i(k) - m_i + \underline{d}_i(k) < 0 \Rightarrow m_i > (\Phi_{i,i} - 1)s_i(k) + \underline{d}_i(k)$$

$$s_i(k+1) + s_i(k) = (\Phi_{i,i} + 1)s_i(k) - m_i + \underline{d}_i(k) > 0 \Rightarrow m_i < (\Phi_{i,i} + 1)s_i(k) + \underline{d}_i(k)$$

Case $s_i(k) < 0$:
The conditions of existence of a quasi-sliding mode becomes

$$\begin{cases} s_i(k+1) - s_i(k) > 0 \\ s_i(k+1) + s_i(k) < 0 \end{cases}$$

Which gives:

$$s_i(k+1) - s_i(k) = (\Phi_{i,i} - 1)s_i(k) + m_i + \underline{d}_i(k) > 0 \Rightarrow m_i > -(\Phi_{i,i} - 1)s_i(k) - \underline{d}_i(k)$$

$$s_i(k+1) + s_i(k) = (\Phi_{i,i} + 1)s_i(k) + m_i + \underline{d}_i(k) < 0 \Rightarrow m_i < -(\Phi_{i,i} + 1)s_i(k) - \underline{d}_i(k)$$

Theorem 3 *The discrete multivariable repetitive sliding mode control law defined in (2.17) allows the rejection of the external disturbances if and only if the gains* m_i *satisfy:*

$$(\Phi_{i,i} - 1)\,|s_i(k)| + \underline{d}_i(k)\,\mathrm{sign}(s_i(k)) < m_i < (\Phi_{i,i} + 1)\,|s_i(k)| + \underline{d}_i(k)\,\mathrm{sign}(s_i(k)) \tag{2.18}$$

We remark that the discrete multivariable repetitive sliding mode control is able to reject periodic disturbances without necessity of estimating them.

Also, the minimum quasi-sliding mode band depends on the maximum norm of the difference of disturbances between instances k and $k - N$.

2.5 Simulation Results

Consider the mathematical model of a chemical reactor (Stoica 2008):

$$x(k+1) = Ax(k) + Bu(k) + Bd(k)$$

where:

$$A = \begin{bmatrix} 0.9580 & 0 & 0 & 0 \\ 0 & 0.9418 & 0 & 0 \\ 0 & 0 & 0.9048 & 0 \\ 0 & 0 & 0 & 0.9277 \end{bmatrix}; \quad B = \begin{bmatrix} 0.25 & 0 \\ 0.25 & 0 \\ 0 & 0.5 \\ 0 & 0.5 \end{bmatrix}$$

with

$$x(k) = \begin{bmatrix} x_1(k) \\ x_2(k) \\ x_3(k) \\ x_4(k) \end{bmatrix}; \quad u(k) = \begin{bmatrix} u_1(k) \\ u_2(k) \end{bmatrix}; \quad d(k) = \begin{bmatrix} d_1(k) \\ d_2(k) \end{bmatrix}$$

The periodic disturbances vector is chosen as:

$$\begin{cases} d_1(k) = 0.1 + 0.5 \sin\left(\frac{2\pi k}{N}\right) + 0.2 \cos\left(\frac{2\pi k}{N}\right) \\ d_2(k) = 0.2 + 0.4 \sin\left(\frac{2\pi k}{N}\right) + 0.3 \cos\left(\frac{2\pi k}{N}\right) \end{cases}; \quad N = 30$$

The retained synthesis parameters are:

$$\Phi_{1,1} = \Phi_{2,2} = 0.1; \qquad C = \begin{bmatrix} 286.7629 & -282.7629 & 16.9876 & -16.9876 \\ -12.6797 & 12.6797 & -96.1930 & 98.1930 \end{bmatrix}$$

The evolution of disturbances $d_1(k)$ and $d_2(k)$ is given in Fig. 2.1.

In this section, three kinds of controllers are considered: the first order discrete multivariable sliding mode control (DSMC), the second order discrete multivariable sliding mode control (2-DSMC) and the discrete multivariable repetitive sliding mode control (DRSMC).

Firstly, a first order discrete multivariable sliding mode control is used.

The simulation results are shown in Figs. 2.2, 2.3 and 2.4. Figure 2.2 gives the evolutions of the states. Figure 2.3 presents the evolutions of the control signals $u_1(k)$ and $u_2(k)$. The evolution of the sliding surfaces $s_1(k)$ and $s_2(k)$ is illustrated by Fig. 2.4.

It can be noted that all the states and sliding functions track the origin with periodic error due to the presence of external periodic disturbances. As a conclusion, the first order DSMC is not able to reject effectively the considered periodic disturbances.

This is proved by Theorem 1. Indeed, the selected gains are not satisfied the condition (2.9) without knowing perfectly the considered disturbances.

For classical discrete sliding mode control, it was noted:

- the presence of chattering phenomenon.
- the rejection of constant or harmonic disturbances is only possible when these disturbances are known or estimated.

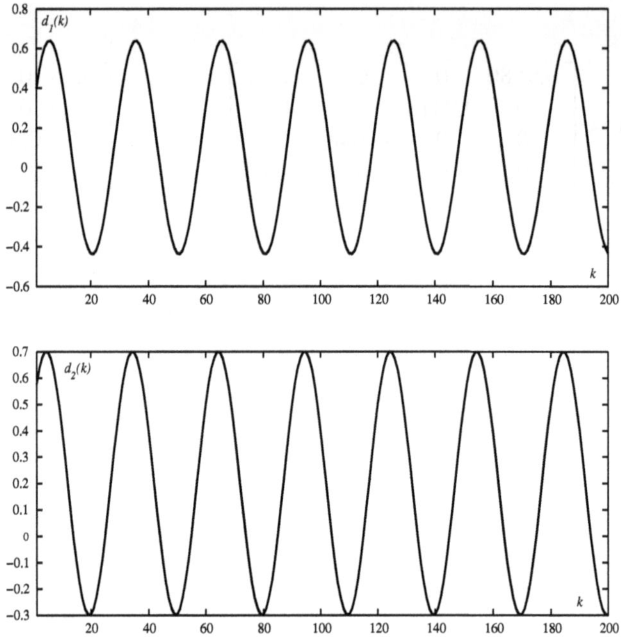

Fig. 2.1 Evolutions of periodic disturbances $d_1(k)$ and $d_2(k)$

Secondly, in order to ameliorate the performance of the first order discrete multivariable sliding mode control, we use the discrete multivariable second order sliding mode control.

The simulation results are shown in Figs. 2.5, 2.6 and 2.7. The evolution of the states is given in Fig. 2.5. Figure 2.6 shows the evolutions of the inputs $u_1(k)$ and $u_2(k)$. The evolution of sliding surfaces is presented in Fig. 2.7.

We observe that the presence of periodic disturbances can cause periodic errors at sliding surfaces. Also, it can be seen that a minor periodic errors are obtained using 2-DSMC as shown by comparing Figs. 2.2 and 2.5. This is due to the ability of discrete second order sliding mode control to reject constant terms of external disturbances.

Thirdly, the discrete multivariable repetitive sliding mode control (2.17), which is a combination between DSMC and repetitive approach, is used in order to reject periodic disturbances.

The simulation results of the system with the considered control law are shown in Figs. 2.8, 2.9 and 2.10. Figure 2.8 gives the evolutions of the states $x_1(k)$, $x_2(k)$, $x_3(k)$ and $x_4(k)$,. Figure 2.9 presents the evolutions of the inputs $u_1(k)$ and $u_2(k)$. The evolution of the sliding functions $s_1(k)$ and $s_2(k)$ is presented in Fig. 2.10. These figures proves that a relatively satisfactory performances are recorded in terms of rejecting periodic disturbances.

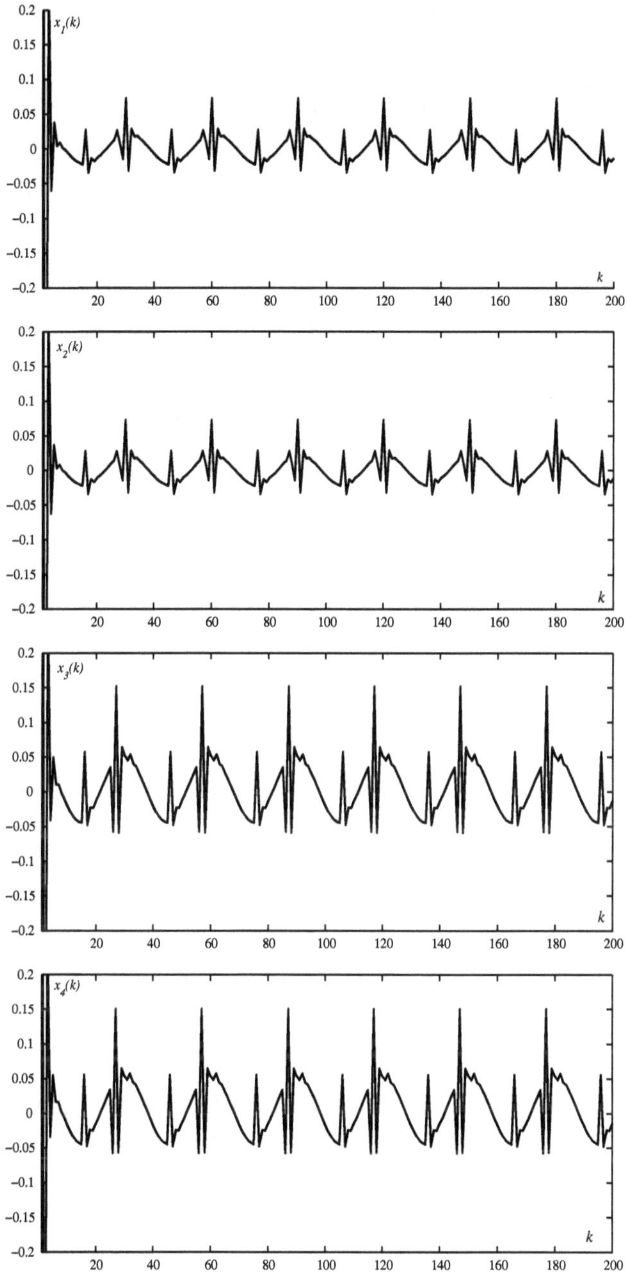

Fig. 2.2 Evolutions of the states $x_1(k)$, $x_2(k)$, $x_3(k)$ and $x_4(k)$ (DSMC)

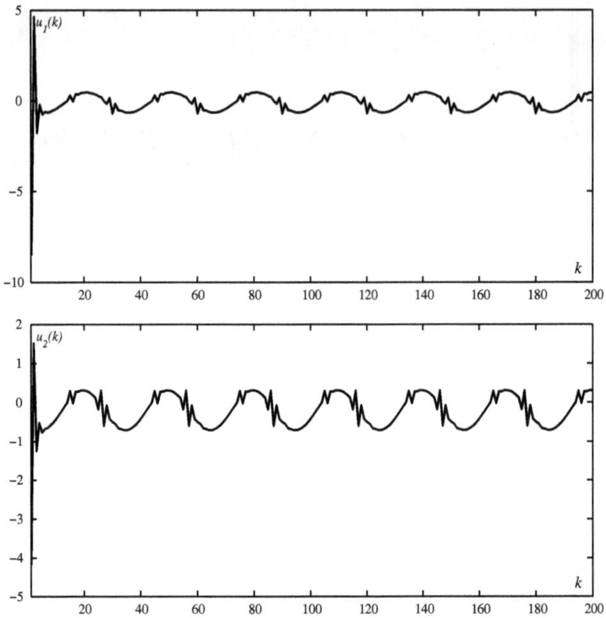

Fig. 2.3 Evolutions of the control signals $u_1(k)$ and $u_2(k)$ (DSMC)

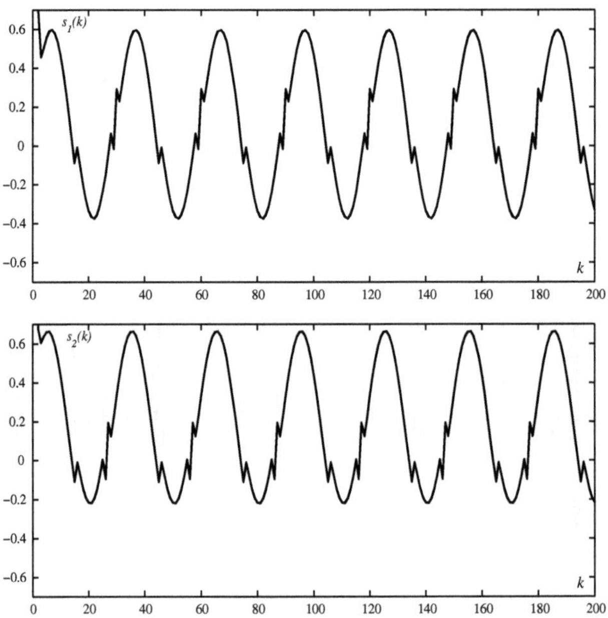

Fig. 2.4 Evolutions of the sliding functions $s_1(k)$ and $s_2(k)$ (DSMC)

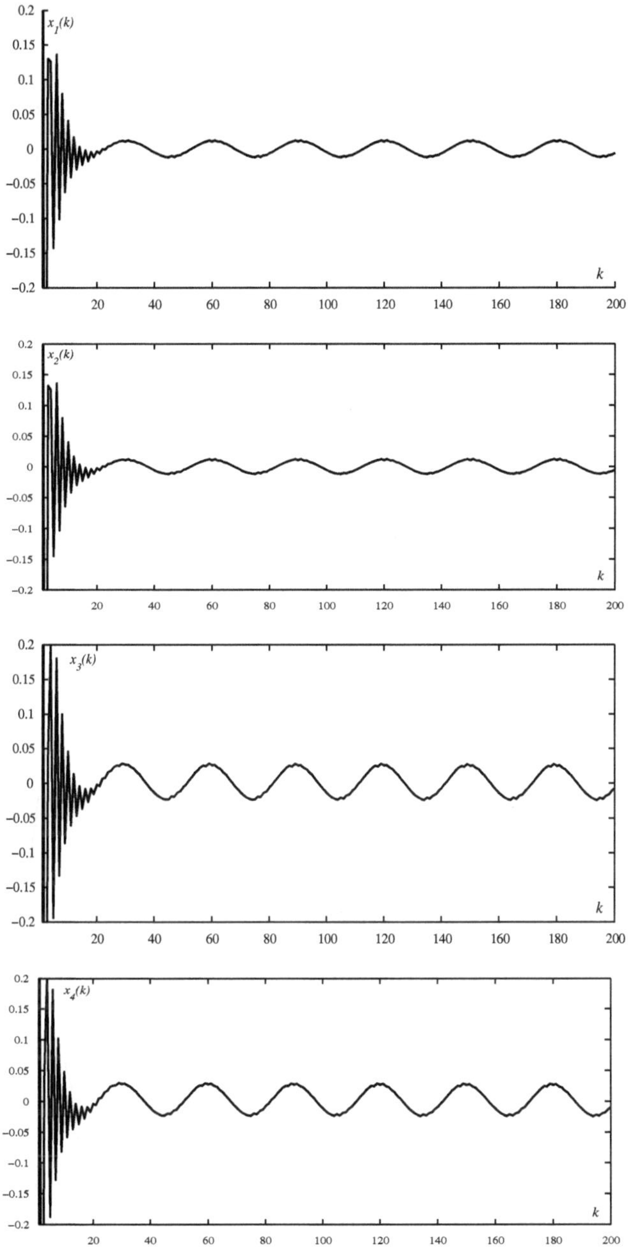

Fig. 2.5 Evolutions of the states $x_1(k)$, $x_2(k)$, $x_3(k)$ and $x_4(k)$(2-DSMC)

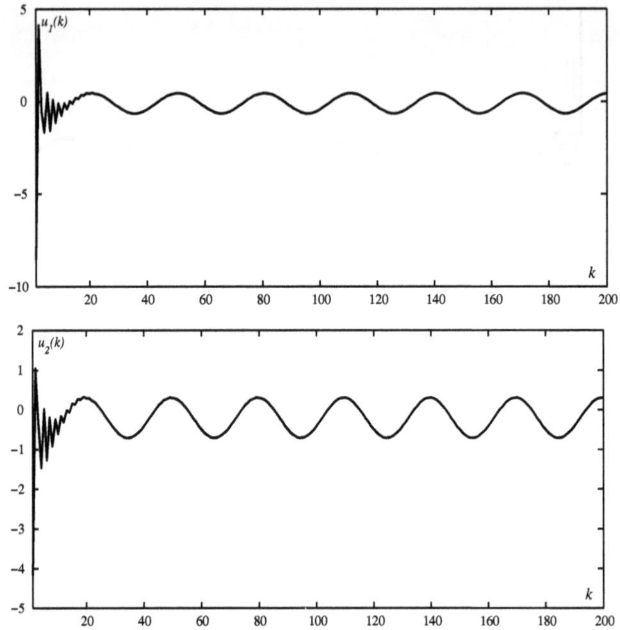

Fig. 2.6 Evolutions of the control signals $u_1(k)$ and $u_2(k)$ (2-DSMC)

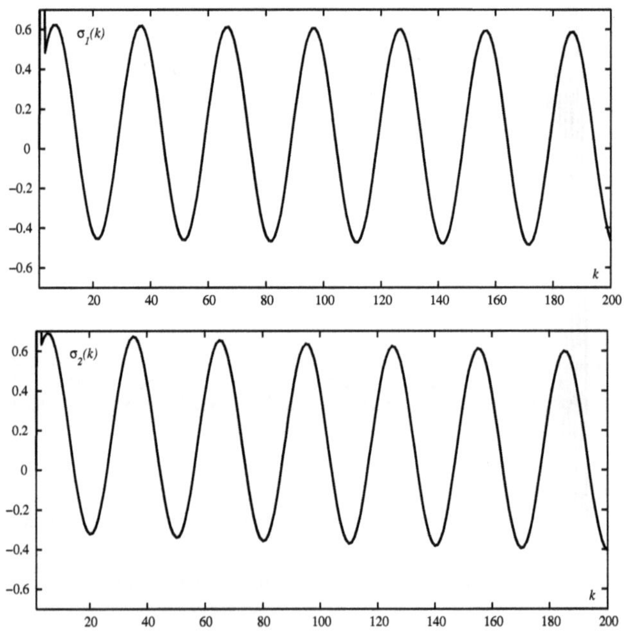

Fig. 2.7 Evolutions of the sliding functions $\sigma_1(k)$ and $\sigma_2(k)$ (2-DSMC)

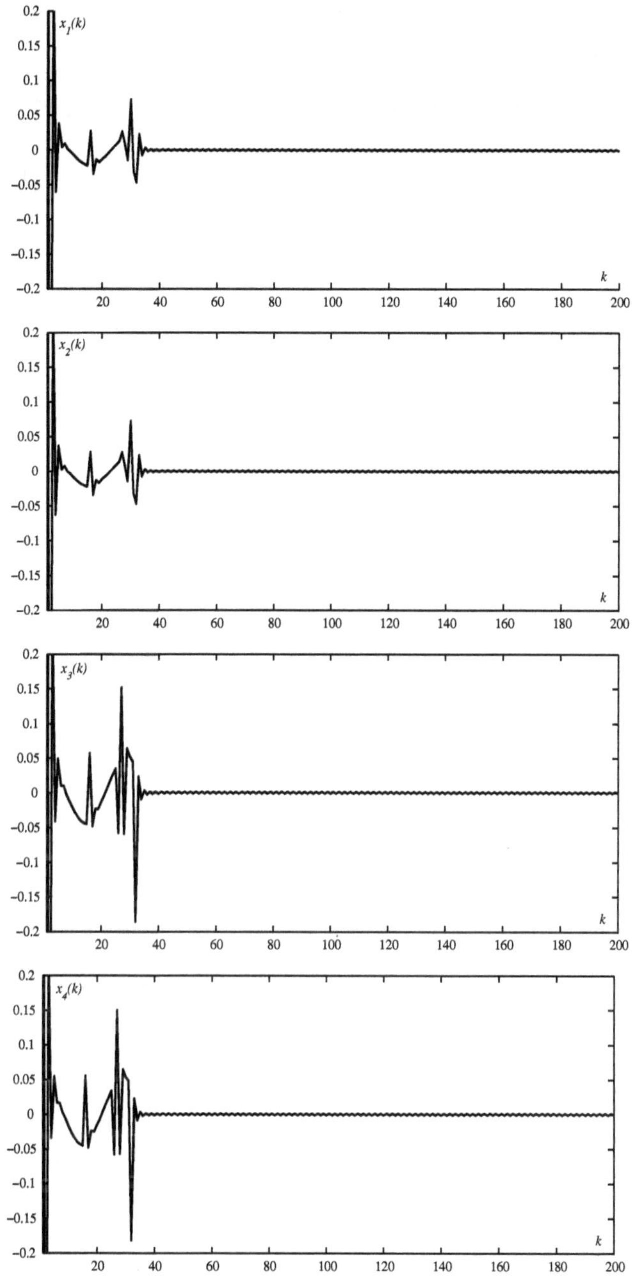

Fig. 2.8 Evolutions of the states $x_1(k)$, $x_2(k)$, $x_3(k)$ and $x_4(k)$ (DRSMC)

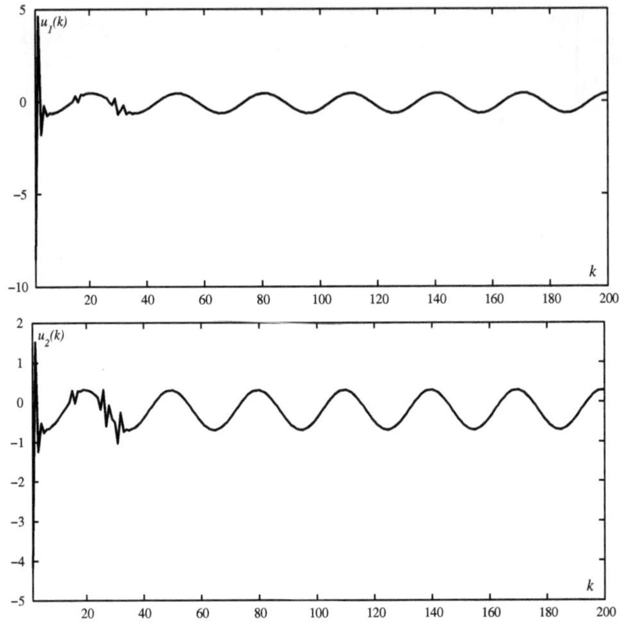

Fig. 2.9 Evolutions of the control signals $u_1(k)$ and $u_2(k)$ (DRSMC)

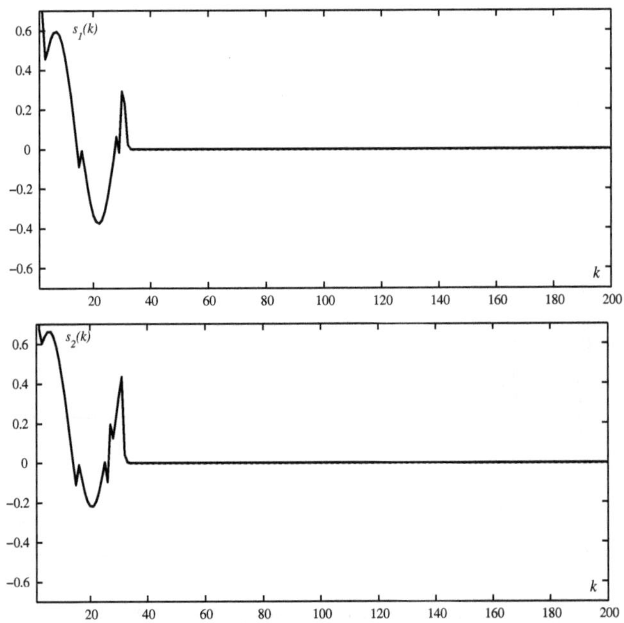

Fig. 2.10 Evolutions of the sliding functions $s_1(k)$ and $s_2(k)$ (DRSMC)

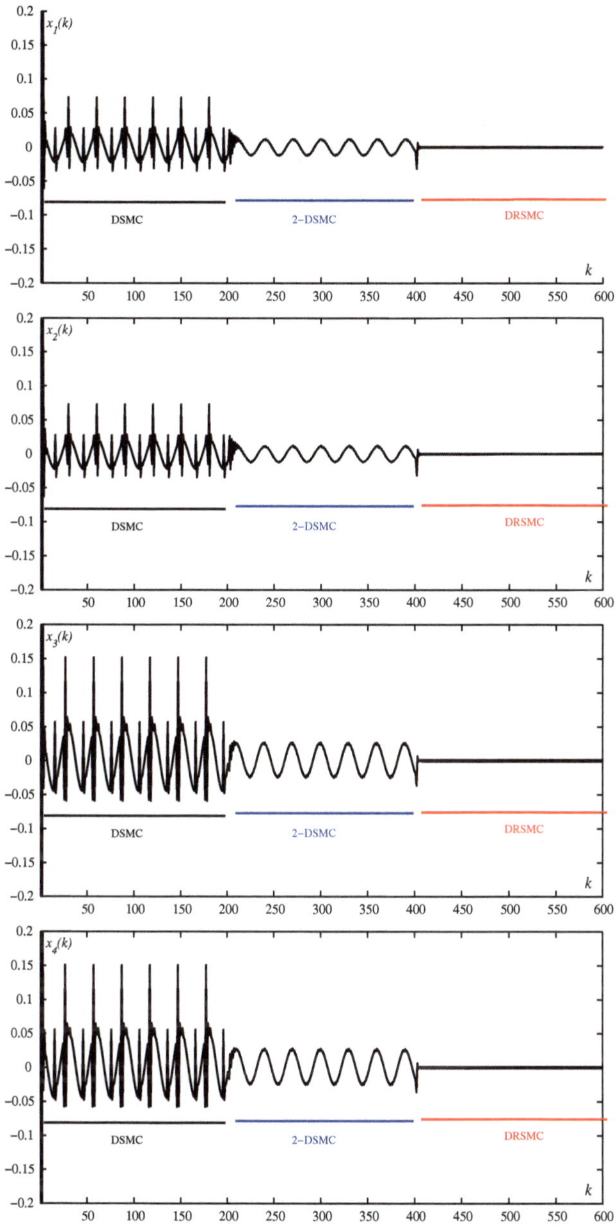

Fig. 2.11 Evolutions of the states $x_1(k)$, $x_2(k)$, $x_3(k)$ and $x_4(k)$

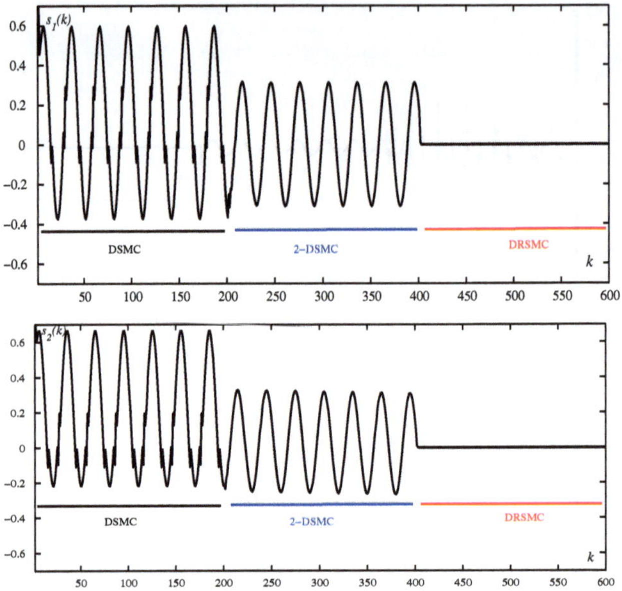

Fig. 2.12 Evolutions of the sliding functions $s_1(k)$ and $s_2(k)$

A comparison between the three kinds of sliding mode controllers: DSMC, 2-DSMC and DRSMC, as shown by Figs. 2.11 and 2.12, reveals that the use of DRSMC strategy reduce effectively the periodic disturbances.

2.6 Conclusion

In this paper, the problem of rejecting external disturbances in the discrete sliding mode control is studied. A solution to reject constant disturbances can be given by the discrete multivariable second order sliding mode control. To reject periodic disturbances, the discrete multivariable repetitive sliding mode control is proposed. Also, a necessary conditions of disturbances rejection in first, second order and repetitive sliding mode controllers has been developed for a discrete multivariable systems.

The effectiveness of the multivariable repetitive sliding mode control was validated through numerical simulation for a chemical reactor model. A comparison between the results obtained by the three kinds of sliding mode controllers showed a good rejection of periodic disturbances of the discrete multivariable repetitive sliding mode control.

Acknowledgments This work has been supported by the Ministry of the Higher Education and Scientific Research in Tunisia.

References

Arimoto, S., Kawamura, S., & Miyazaki, F. (1984). Bettering operation of dynamic systems by learning: a new control theory for servomechanism or mechatronics systems. *Proceedings of the 23rd Conference on Decision and Control*, pp. 1064–1069.

Bandyopadhyay, B., & Fulwani, D. (2009). High-performance tracking controller for discrete plant using nonlinear sliding surface. *IEEE Transactions on Indusrial Electronics*, *56*(9), 3628–3637.

Bartoszewicz, A. (1998). Discrete-time quasi-sliding-mode control strategies. *IEEE Transactions on Industrial Electronics*, *45*(4), 633–637.

Bartoszewicz, A., & Lesniewski, P. (2014). Reaching law approach to the sliding mode control of periodic review inventory systems. *IEEE Transactions on Automation Science and Engineering*, *11*(3), 810–817.

Castillo-Toledoa, B., Gennaro, S. D., Loukianov, A. G., & Rivera, J. (2008). Discrete time sliding mode control with application to induction motors. *Automatica*, *44*, 3036–3045.

Cavallo, A., & Natale, C. (2004). High-order sliding control of mechanical systems: theory and experiments. *Automatica*, *12*, 1139–1149.

Chan, C. Y. (1997). Discrete-time adaptive sliding mode control of a linear system in statespace form. *International Journal of Control*, *67*(6), 859–868.

Chang, J. L. (2002). Discrete sliding-mode control of mimo linear systems. *Asian Journal of Control*, *4*(2), 217–222.

Chang, J. L., & Chen, Y. P. (2000). Sliding vector design based on the pole assignment method. *Asian Journal of Control*, *2*, 10–15.

Chang, K., Shim, I., & Park, G. (2006). Adaptive repetitive control for an eccentricity compensation of optical disk drivers. *IEEE Transactions on Consumer Electronics*, *52*(2), 445–450.

Dehri, K., Ltaief, M., & Nouri, A.S. (2011). New discrete multivariable sliding mode control for multi-periodic disturbances rejection. *Proceedings of the International Multi-Conference on Systems, Signals and Devices*, Tunisia.

Dehri, K., Ltaief, M., Nouri, A. S. (2012a). Repetitive sliding mode control for nondecouplable systems: periodic disturbances rejection. *Proceedings of the 20th Mediterranean Conference on Control and Automation, MED'12*, July 3–6, Barcelona, Spain.

Dehri, K., Ltaief, M., & Nouri, A. S. (2012b). New discrete sliding mode control for nonlinear multivariable systems: multi-periodic disturbances rejection and stability analysis. *International Journal of Control Science and Engineering*, *2*(2), 7–15.

Doh, T.Y., & Ryoo, J.R. (2006). Robust stability condition of repetitive control systems and analysis on steady-state tracking errors. *Proceedings of the International Joint Conference on SICE-ICASE*, pp. 5169–5174.

Fateha, M. M., Tehranib, H. A., & Karbassic, S. M. (2013). Repetitive control of electrically driven robot manipulators. *International Journal of Systems Science*, *44*(4), 775–785.

Francis, B., & Wonham, W. (1976). The internal model principle of control theory. *Automatica*, *12*(5), 457–465.

Gao, W., Wang, Y., & Homaifa, A. (1995). Discrete-time variable structure control systems. *IEEE Transactions on Industrial Electronics*, *42*(2), 117–122.

Ginhoux, R. (2003). *Compensation des mouvements physiologiques en chirurgie robotisée par commande prédictive*. PhD thesis, Université Louis Pasteur Strasbourg.

Lopez, P., & Nouri, A.S. (2006). Théorie élémentaire et pratique de la commande par les régimes glissants. Berlin: Springer.

Mihoub, M., Nouri, A.S., Abdennour, R.B. (2009). Real-time application of discrete second order sliding mode control to a chemical reactor. *Control Engineering Practice*, *17*(9), 1089–1095.

Mihoub, M., Nouri, A.S., & Abdennour, R.B. (2011). *Multimodel discrete second order sliding mode control: stability analysis and real time application on a chemical reactor* pp. 473–490. In Tech.

Monsees, G. (2002). Discrete-time sliding mode control. *PHD Thesis, University of Gabes, Tunisia*.

Romdhane, H., Dehri, K., & Nouri, A. S. (2015). Second order sliding mode control for discrete decouplable multivariable systems via input-output models. *International Journal of Automation and Computing*, *12*(6), 630–638.

Sarpturk, S. Z., Istefanopulos, Y., & Kaynak, O. (1987). On the stability of discrete-time sliding mode control systems. *IEEE Transactions on Automatic Control*, *32*, 930–932.

Stoica, C. N. (2008). Robustification de lois de commande prédictives multivariables. *Thesis*.

Xuan, S. M., Xiong, H. X., & Yu, C. B. (2007). Repetitive Learning Control for Time-varying Robotic Systems: A Hybrid Learning Scheme. *Automatica*, *33*(11), 1189–1195.

Xuan, S. M., Jiang, Y. L., & Gang, H. H. (2013). Discrete Adaptive Repetitive Control: Convergence Analysis and Implementation. *Acta Automatica Sinica*, *39*(4), 400–7406.

Yan, T. H., He, B., Chen, X. D., & Xu, X. S. (2013). The discrete-time sliding mode control with computation time delay for repeatable run-out compensation of hard disk drives. *Mathematical Problems in Engineering*, *2013*(2), 1–13.

Young, K. D., Utkin, V. I., & Ozguner, U. (1999). A control engineer's guide to sliding mode control. *IEEE Transactions on Control Systems Technology*, *7*(3), 328–342.

Yu, X., & Kaynak, O. (2009). Sliding mode control with soft computing: a survey. *IEEE Transactions on Industrial Electronics*, *56*(9).

Zhou, K., & Wang, D. (2003). Digital repetitive controlled three-phase PWM rectifier. *IEEE Transactions on Power Electronics*, *18*(1), 309–316.

Chapter 3
Output Feedback Robust Exponential Higher Order Sliding Mode Control

Abdelhak Msaddek, Abderraouf Gaaloul and Faouzi M'sahli

Abstract The higher order sliding mode controller (HOSMC) is a robust control scheme used to overcome the chattering phenomenon which appears, generally, with standard sliding mode controller (SMC). In this chapter, a novel technique of HOSMC for uncertain nonlinear systems is presented. The proposed controller allows obtaining an exponential stability as well a finite time convergence to the sliding surface and guarantees robustness against uncertainties and external matched disturbances. Furthermore, the synthesis of the control law depends explicitly on the states of the system. But, in practice, most of systems admit one or more unknown states. Such problem represents a serious drawback when implementing the controller in real time. To solve this problem we incorporate a High Gain Observer (HGO) into the controller to estimate the missing states. These techniques of control and observation are applied to an induction motor system. Numerical simulations are developed to show the effectiveness of the resulting controller.

Keywords Nonlinear systems · Robustness · Higher order · Sliding mode control · High gain observer · Induction motor

3.1 Introduction

The sliding mode control (SMC) has demonstrated its effectiveness in main areas of application including robotics and electrical machinery (Defoort et al. 2007b; Ghanes 2005; Traore et al. 2012; Msaddek et al. 2014). This control approach is well known by its robustness against external matched disturbances, parametric variations,

A. Msaddek (✉) · A. Gaaloul · F. M'sahli
National School of Engineers of Monastir, University of Monastir,
Avenue Ibn Jazzar 5019, Monastir, Tunisia
e-mail: msaddekabdelhak@hotmail.fr

A. Gaaloul
e-mail: abderraouf.gaaloul@enim.rnu.tn

F. M'sahli
e-mail: faouzi.msahli@enim.rnu.tn

N. Derbel et al. (eds.), *Applications of Sliding Mode Control*,
Studies in Systems, Decision and Control 79, DOI 10.1007/978-981-10-2374-3_3

modeling uncertainties and nonlinearities (hysteresis, friction, etc.) that often characterize the real systems. Indeed, Utkin (1992) demonstrated the robustness property of SMC achieved using a high frequency switching to steer the states of a system into the sliding surface. However, the high-frequency switching leads, generally, to the appearing of an undesirable chattering phenomenon. This leads to the dissipation of a large quantity of energy in electric actuators and a rapid wear of mechanical actuators. This limits seriously the implementation of SMC in real time. To overcome this deficiency, several solutions have been proposed in the literature. Among them, the "sign" function is replaced by a smooth similar function such as the saturation function or the sigmoid function. Furthermore, some non conventional techniques such as fuzzy logic (FL) and neural networks were combined together with SMC (Roopoei and Zalghadri 2009; Tay et al. 2012; Golea et al. 2012; Jie et al. 2012; Nayak et al. 2013; Zhao et al. 2013; Msaddek et al. 2015b).

The most interesting way to get rid of the chattering phenomenon consists of enforcing a higher-order sliding mode (HOSM). The main objective of SMC of order ρ (called ρ-SMC) is to obtain a finite time convergence onto the manifold $S^{\rho} = \{s = \dot{s} = \ddot{s}... = s^{(\rho-1)}\}$, where s is the sliding variable. So, the philosophy of HOSMC consists of enforcing the sliding variable and its $\rho - 1$ first time derivatives to zero in finite time (Levant 1993). In Levant (2005), author proposed the so-called quasi-continuous SMC where the control input is chattering free except on the sliding surface.

Laghrouche et al. (2007) developed another approach of HOSMC based on the minimization of a quadratic criterion using the concept of SMC with integral action. This allows stabilizing in finite time a system of high order on the sliding surface. Besides, it permits to choose in advance the convergence time to the sliding surface. Although these algorithms are general, a priori accurate knowledge of the initial conditions of the system limits seriously the applicability of this approach. Recently, Defoort et al. (2006, 2007a, b, 2009), proposed a finite time HOSMC for a class of multivariable nonlinear systems. In these works, the drawbacks devoted in Laghrouche et al. (2007) are removed and a simple method for adjusting the synthesis parameters of the control law is presented. However, this controller contains necessarily a discontinuous part to reject the effect of disturbances and requires the knowledge of the maximum amplitude of the disturbances.

Mondal and Mahanta (2011) proposed a second order sliding mode controller based on a nonlinear sliding surface to control uncertain linear systems with matched uncertainty. Likewise, another approach of HOSMC presented in Ling et al. for uncertain nonlinear systems with relative degree three. The stability on the sliding surface is guaranteed and the chattering is reduced using these controllers. Nevertheless, these two approaches are applicable only for uncertain systems with relative degree two and three, respectively. In reality these strategies of control by HOSMC allow solving partially the problem of chattering phenomenon. However, this solution can not remove totally the discontinuous oscillations of the control signal especially when the sliding surface is reached.

In this work, we present a new technique of HOSMC (Msaddek et al. 2013, 2014) characterized by its exponential stability and its robustness against external matched disturbances, parametric variations and modeling uncertainties. Furthermore, the proposed approach is independent of the initial conditions and depends explicitly on the whole system state. However, in real application most of industrial processes admit one or more unknown states. This deficiency becomes a serious problem when implementing the controller in real time. In this paper, we present an efficient solution to overcome such drawback through the incorporation of a High Gain Observer presented in Farza et al. (2004) into the controller proposed in Msaddek et al. (2014).

The induction motor (IM) is widely used in industry, mainly due to its rigidness. Also, its maintenance is free operation, and relatively low cost. However, induction motors represent a theoretically challenging control problem since the dynamical system is nonlinear. The technique of vector control by indirect field oriented applied to induction motors permitted to have better performances comparable to DC motors. Nevertheless, it is very sensitive to parametric variations and external disturbances. To solve this problem, we apply the novel robust higher order sliding mode control combined with a high gain observer. Simulations results developed in this work show the effectiveness of the proposed output feedback controller.

This chapter is organized as follows. The next section is devoted to the presentation of the new HOSMC. In Sect. 3.3, we incorporate the HGO into the HOSMC. In Sect. 3.4, we present an application of the resulting controller to the model of the induction motor. Conclusions are reported in the last section of this work.

3.2 Higher Order Sliding Mode Control

3.2.1 Problem Formulation

Consider a nonlinear dynamical system described by

$$\begin{aligned} x &= f(x, t) + g(x, t)u + p(t) \\ s &= s(x, t) \end{aligned} \tag{3.1}$$

where $x = [x_1\ x_2\ \dots\ x_n]^\top \in X \subset \mathbb{R}^n$ is the state variable of the system with X an open set of $\mathbb{R}^n$ and the control input $u \in U \subset \mathbb{R}$ is a feature possibly discontinuous and bounded, depending on time and the system state, with U is an open set of $\mathbb{R}$, $f(x, t)$ and $g(x, t)$ are sufficiently differentiable vector fields.

Assumption 1: System (3.1) admits a $\rho \in \mathbb{N}$ constant and known relative degree with respect to the sliding variable $s(x, t)$. The system (3.1) can be written as follows (Laghrouche et al. 2007):

$$\begin{cases} \dot{z}_1 &= z_2 \\ \dot{z}_2 &= z_3 \\ &\vdots \\ \dot{z}_{\rho-1} &= z_\rho \\ \dot{z}_\rho &= \phi(x,t) + \varphi(x,t)u \end{cases} \tag{3.2}$$

where:

$$z = \begin{bmatrix} z_1 \\ z_2 \\ \vdots \\ z_{\rho-1} \\ z_\rho \end{bmatrix} = \begin{bmatrix} s \\ \dot{s} \\ \vdots \\ s^{(\rho-2)} \\ s^{(\rho-1)} \end{bmatrix}$$

with:

$$\begin{cases} \phi = \overline{\phi} + \delta_\phi \\ \varphi = \overline{\varphi} + \delta_\varphi \end{cases}$$

where $\overline{\phi}$ and $\overline{\varphi}$ are nominal known parts, δ_ϕ and δ_φ are unknown parts, including disturbances and uncertainties.

Assumption 2: The nominal part $\overline{\varphi}$ is assumed invertible.
The objective of the different techniques of higher order sliding mode control is to obtain a finite time convergence onto the manifold $S^\rho = \{s = \dot{s} = \cdots = s^{\rho-1} = 0\}$.

3.2.2 *Control Design*

Consider a chain of integrators, defined by

$$\begin{cases} \dot{z}_1 &= z_2 \\ \dot{z}_2 &= z_3 \\ &\vdots \\ \dot{z}_{\rho-1} &= z_\rho \\ \dot{z}_\rho &= z_{\rho+1} = w(z) \end{cases} \tag{3.3}$$

Theorem 1: (Msaddek et al. 2014) To stabilize exponentially the system (3.3) in presence of uncertainties on S^ρ, we propose the following control law.

$$w(z) = -\alpha \frac{a_1 z_1 + a_2 z_2 + \cdots a_\rho z_\rho}{b_1|z_1| + b_2|z_2| + \cdots b_\rho|z_\rho|} \tag{3.4}$$

with $\alpha > 0$, $a_i > 0$, $b_i > 0$, $(1 \le i \le \rho)$ and the polynomial $P(p) = a_1 + a_2 p + \cdots a_\rho p^{\rho-1}$ is Hurwitz.

Proof: Equation (3.4) can be written as follows:

$$a_1 s + a_2 \dot{s} + \cdots a_\rho s^\rho = f(t) \tag{3.5}$$

with

$$f(t) = -\dot{z}_\rho \frac{b_1|z_1| + b_2|z_2| + \cdots b_\rho|z_\rho|}{\alpha} \tag{3.6}$$

The Laplace Transform applied to (3.6) gives

$$\left(a_1 + a_2 p + \cdots a_\rho p^{\rho-1}\right) S(p) = F(p) \tag{3.7}$$

As $P(p) = a_1 + a_2 p + \cdots a_\rho p^{\rho-1}$ is Hurwitz, the solution of (3.5) is stable.

Proof: of the convergence of z to the zero vector of $\mathbb{R}^\rho$. Assume that the states of system are not in the manifold S^ρ. One has:

$$z_{\rho+1} = w(z) = -\alpha \, \frac{a_1 z_1 + a_2 z_2 + \cdots a_\rho z_\rho}{b_1|z_1| + b_2|z_2| + \cdots b_\rho|z_\rho|} \tag{3.8}$$

In other words, one obtains

$$\beta z_{\rho+1} = a_1 z_1 + a_2 z_2 + \cdots a_\rho z_\rho \tag{3.9}$$

with $\forall t > 0$:

$$\beta = -\frac{b_1|z_1| + b_2|z_2| + \cdots b_\rho|z_\rho|}{\alpha}$$

So, (3.9) is assumed a linear differential equation with second member. The equation without second member is given by:

$$a_1 z_1 + a_2 z_2 + \cdots a_\rho z_\rho = 0 \tag{3.10}$$

The roots of the polynomial $P(p)$ have a strictly negative real part. So, the polynomial which is given by

$$P(p) = a_1 + a_2 p + \cdots a_\rho p^{\rho-1} \tag{3.11}$$

can be rewritten as follows

$$P(p) = \prod_{i=1}^{r} (p - \lambda_i)^{\eta_i} \ , \quad \sum \eta_i = \rho - 1 \ , \quad \Re e(\lambda_i) < 0$$

with λ_i are the roots of the polynomial $P(p)$ with multiplicity degree η_i.

Therefore, the homogeneous solution of (3.9) is of the form

$$z_1(t) = \sum_{i=1}^{r} q_i(t)e^{\lambda_i t} \tag{3.12}$$

with $q_i(t)$ are polynomials of degree $\eta_i - 1$.

Also, one notes that the null function $z_1 = 0$ is a solution of the Eq. (3.9). Indeed if $z_1 = 0\ \forall t$, then $\beta = 0\ \forall t$. So, one can take it as a particular solution of (3.9). Consequently, the general solution of (3.9) is given by (3.12).

Coefficients of $q_i(t)$ are fixed using initial conditions. So the $z_i = z_1^{i-1}, (1 \le i \le \rho)$ converge exponentially to zero of $\mathbb{R}^\rho$ whatever the initial conditions are chosen.

Proof of robustness: Suppose that the control is affected by the disturbances as follows:

$$p_1(t)z_{\rho+1} + p_2(t) = -\alpha \frac{a_1 z_1 + a_2 z_2 + \cdots a_\rho z_\rho}{b_1|z_1| + b_2|z_2| + \cdots b_\rho|z_\rho|} \tag{3.13}$$

where $z_{\rho+1} = \dot{z}_\rho$, $p_1(t)$ and $p_2(t)$ are two bounded disturbances.

Equation (3.13) can be rewritten as follows

$$\beta\left[p_1(t)z_{\rho+1} + p_2(t)\right] = a_1 z_1 + a_2 z_2 + \cdots a_\rho z_\rho \tag{3.14}$$

The homogeneous solution of the Eq. (3.14) takes the form of (3.12), and $z_1 = 0$ is a particular solution of (3.14). So the general solution of (3.14) is given by (3.12).

So, $z_i = z_1^{i-1}$, $(1 \le i \le \rho)$ converge exponentially to zero of $\mathbb{R}^\rho$. Therefore, the proposed controller ensures the robustness against bounded disturbances. Now, consider system (3.2) which can be rewritten as follows (Defoort et al. 2007b):

$$\begin{cases} \dot{z}_1 &= z_2 \\ \dot{z}_2 &= z_3 \\ &\vdots \\ \dot{z}_{\rho-1} &= z_\rho \\ \dot{z}_\rho &= \nu(x,t) + [1 + \zeta(x,t)]w \end{cases} \tag{3.15}$$

with:

$$\begin{aligned} u &= \overline{\varphi}^{-1}(w - \overline{\phi}) \\ \nu &= \delta_\phi - \delta_\varphi \overline{\varphi}^{-1}\overline{\phi} \\ \zeta &= \delta_\varphi \overline{\varphi}^{-1} \end{aligned}$$

Functions $\nu(x,t)$ and $\zeta(x,t)$ and may include the uncertainties of the system.

Assumption 3: Functions $\nu(x,t)$ and $\zeta(x,t)$ are bounded. In addition, there is a positive function $a(x)$ and a positive constant b $(0 < b \le 1)$, such that:

$$\begin{cases} |\nu(x,t)| \leq a(x) \\ |\zeta(x,t)| \leq 1-b \end{cases} \tag{3.16}$$

Now consider the control law:

$$u = \overline{\varphi}^{-1}[w(z) - \overline{\phi}] \tag{3.17}$$

where $\overline{\varphi}$ and $\overline{\phi}$ are obtained according to (3.2) and is given by (3.4).

Theorem 2: (Msaddek et al. 2014) The controller (18) ensures a sliding mode of order ρ with respect to $s(x, t)$ provided that assumptions (3.2) and (3.3) are verified.

Proof: Using (3.13) in the equation system (3.15), one has:

$$\dot{z}_\rho = \nu(x,t) + [1 + \zeta(x,t)]w \tag{3.18}$$

If Assumption (3.3) is verified, one can write (3.18) as follows

$$\begin{aligned} w(z) &= [1 + \zeta(x,t)]^{-1}[\dot{z}_\rho - \nu(x,t)] \\ &= [1 + \zeta(x,t)]^{-1}\dot{z}_\rho - [1 + \zeta(x,t)]^{-1}\nu(x,t) \end{aligned} \tag{3.19}$$

Now, applying our approach (3.13) to (3.19) one obtains:

$$p_1(t)z_{\rho+1} + p_2(t) = -\alpha\,\frac{a_1z_1 + a_2z_2 + \cdots a_\rho z_\rho}{b_1|z_1| + b_2|z_2| + \cdots b_\rho|z_\rho|} \tag{3.20}$$

with:

$$p_1(t) = [1 + \zeta(x,t)]^{-1} \tag{3.21}$$

$$p_1(t) = -[1 + \zeta(x,t)]^{-1}\nu(x,t) \tag{3.22}$$

Now, using the proof of robustness presented above, one can conclude that the control law (3.17) allows stabilizing exponentially the uncertain system (3.1) on the sliding surface.

3.3 Output Feedback Control

The higher order sliding mode control previously presented is based on the assumption that all system states are known.

Nevertheless, in practice and for economic or technological reasons, this assumption is, generally, not true. So, to overcome this problem, we proposed an output feedback control which it is synthesized using the system outputs and unknown states are estimated using high gain observer.

Assumption 4: There exists a lipschitzian diffeomorphism that permit to put the system (3.1) under the following form used by Farza et al. (2004):

$$\begin{aligned} \dot{x} &= F(s,x)x + G(u,s,x) + \overline{\varepsilon}(t) \\ y &= \overline{C}x \end{aligned} \tag{3.23}$$

where the state:

$$x = \begin{bmatrix} x^1 \\ x^2 \\ \vdots \\ x^q \end{bmatrix}$$

with $x^k \in \mathbb{R}^{n_k}$, $k = 1, 2 \cdots q$ and $p = n_1 \geq n_2 \geq \cdots \geq n_q$, $\sum\limits_{k=1}^{q} n_k = n$, the input $u \in U$, U a compact subset of $\mathbb{R}^m$, the output $y \in \mathbb{R}^p$, $s(t)$ is a known signal:

$$\overline{\varepsilon} = \begin{bmatrix} 0 \\ \vdots \\ \varepsilon(t) \end{bmatrix}; \; \varepsilon = \begin{bmatrix} \varepsilon_1 \\ \vdots \\ \varepsilon_{n_q} \end{bmatrix}; \; \overline{C} = \begin{bmatrix} I_{n_1} & O_{n_1 \times n_2} & O_{n_1 \times n_3} & \cdots & O_{n_1 \times n_q} \end{bmatrix}$$

with each ε_i, $i = 1, \cdots, n_q$ being an unknown bounded function and I_{n_1} is the identity matrix and $O_{n_1 \times n_k}$ is the $n_1 \times n_k$ null matrix, $k = 2, \cdots q$,

$$F(s,x) = \begin{bmatrix} 0 & F_1(s,x^1) & 0 & \cdots & 0 \\ 0 & 0 & F_2(s,x^1,x^2) & \ddots & \vdots \\ \vdots & \vdots & 0 & \ddots & 0 \\ 0 & 0 & \vdots & 0 & F_{q-1}(s,x^1,\cdots,x^{q-1}) \\ 0 & 0 & 0 & \cdots & 0 \end{bmatrix}$$

is a block matrix with each F_k, $k = 1 \cdots q$ denoting a $n_k \times n_{k+1}$ rectangular matrix, and

$$G(u,s,x) = \begin{bmatrix} G_1(u,s,x^1) \\ G_2(u,s,x^1,x^2) \\ \vdots \\ G_q(u,s,x) \end{bmatrix}$$

The synthesis of the high gain observer necessitates some assumptions:

Assumption 5: There exist two positive constants α and β such that $\forall k = 1, \cdots q-1$, $\forall x \in \mathbb{R}^n$, $\forall t \geq 0$, one has:

$$0 < \alpha^2 I_{n_k+1} \leq F_k(s,x)^\top F_k(s,x) \leq \beta^2 I_{n_k+1}$$

Assumption 6: Function $\varepsilon(t)$ is bounded.

Assumption 7: Signals $s(t)$ and $\dot{s}(t)$ are bounded.

For the nonlinear system class (3.23), Farza et al. (2004) proposed a HGO given by the following form:

$$\dot{\widehat{x}} = F(s, \widehat{x})\widehat{x} + G(u, s, \widehat{x}) - \vartheta \Lambda^{+}(s, \widehat{x})\Delta_{\vartheta}^{-1} S^{-1} C^{\top} \overline{C}(\widehat{x} - x) \tag{3.24}$$

where $\widehat{x}$ is the system estimated states, $\Lambda^{+}(s, \widehat{x})$ is the left inverse of the matrix $\Lambda(s, \widehat{x})$ given by:

$$\Lambda^{+}(s, \widehat{x}) = \text{diag}\left[I_{n_1},\ F_1[s(t), \xi],\ F_1[s(t), \xi] \times F_2[s(t), \xi], \cdots, \prod_{i=1}^{q-1} F_i[s(t), \xi] \right] \tag{3.25}$$

Δ_{ϑ} is a diagonal matrix defined by:

$$\Delta_{\vartheta} = \text{diag}\left[I_{n_1}, \frac{1}{\vartheta} I_{n_1}, \cdots, \frac{1}{\vartheta^{q-1}} I_{n_1} \right] \tag{3.26}$$

The matrix S is the unique solution of the following algebraic Lyapunov equation:

$$S + A^{\top} S + SA - C^{\top} C = 0 \tag{3.27}$$

The solution of (3.27) is symmetric positive definite and it can be obtained by this matrix relation:

$$S^{-1} C^{\top} = \text{diag}\left[C_q^1 I_{n_1}, C_q^2 I_{n_1}, \cdots, C_q^q I_{n_1} \right]^{\top}$$

Matrices A and C are done by:

$$A = \begin{bmatrix} 0 & I_{n_1} & 0 & \cdots & 0 \\ 0 & 0 & I_{n_1} & \ddots & \vdots \\ \vdots & \vdots & \ddots & \ddots & 0 \\ 0 & 0 & 0 & \ddots & I_{n_1} \\ 0 & 0 & 0 & \cdots & 0 \end{bmatrix}, \quad C = \begin{bmatrix} I_{n_1} & O_{n_1} & \cdots O_{n_1} \end{bmatrix}$$

If we choose ϑ sufficiently big, the estimation error $e = (\widehat{x} - x)$ converges exponentially to zero. So, e is infinitely differentiable and ideally their derivatives converge exponentially to zero, in particular for $s = y - y_d$ and $y = h(x)$. When using the HGO to synthesize the control law one has:

$$\widehat{s} = \widehat{y} - y_d = h(\widehat{x}) - y_d$$

So, one obtains

$$\widehat{z} = \begin{bmatrix} \widehat{z}_1 \\ \widehat{z}_2 \\ \vdots \\ \widehat{z}_\rho \end{bmatrix} = \begin{bmatrix} \widehat{s} \\ \dot{\widehat{s}} \\ \vdots \\ \widehat{s}^{(\rho-1)} \end{bmatrix}$$

and

$$z_{\rho+1} = w(\widehat{z}) = -\alpha \frac{a_1\widehat{z}_1 + a_2\widehat{z}_2 + \cdots a_\rho\widehat{z}_\rho}{b_1|\widehat{z}_1| + b_2|\widehat{z}_2| + \cdots b_\rho|\widehat{z}_\rho|} \tag{3.28}$$

Authors in Farza et al. (2004) demonstrated that:

$$\lim_{\vartheta \longrightarrow +\infty} e = 0 \Longrightarrow \lim_{\vartheta \longrightarrow +\infty} \widehat{x} = x, \quad \lim_{\vartheta \longrightarrow +\infty} \widehat{z} = z$$

Then:

$$\lim_{\vartheta \longrightarrow +\infty} w(\widehat{z}) = w(z), \quad \lim_{\vartheta \longrightarrow +\infty} \phi(\widehat{x}, t) = \phi(x, t), \quad \lim_{\vartheta \longrightarrow +\infty} \varphi(\widehat{x}, t) = \varphi(x, t)$$

Thus, with a proper choice of ϑ, we can obtain a resulting control which may stabilize the real states of the system on the sliding surface. In other words, we must choose ϑ to get a convergence time of the observer faster than the convergence time of the control and the dynamics of the system. Thus, the resulting output feedback controller based on high gain observer for the nonlinear systems class (3.1) is given by the following expression:

$$u = \overline{\varphi}^{-1}(\widehat{x}, t)[w(z) - \overline{\phi}(\widehat{x}, t)] \tag{3.29}$$

3.4 Application to Induction Machine

3.4.1 Mathematical Model of the Induction Motor

The modeling of the induction machine (IM) described in the repository Park is given in the following system of equations (Alvarez-Salas 2002; Hirokazu et al. 2002; Mezouar et al. 2007; Traore et al. 2012; Msaddek et al. 2014).

$$\begin{cases} \dfrac{di_{sd}}{dt} = -\dfrac{L_{sr}^2 R_r + L_r^2 R_s}{\sigma L_s L_r^2} i_{sd} + \dfrac{L_{sr} R_r}{\sigma L_s L_r^2} \phi_{rd} + \omega_a i_{sq} + \dfrac{1}{\sigma L_s} u_{sd} \\ \dfrac{di_{sq}}{dt} = -\dfrac{L_{sr}^2 R_r + L_r^2 R_s}{\sigma L_s L_r^2} i_{sq} - \dfrac{p L_{sr}}{\sigma L_s L_r} \omega \phi_{rd} - \omega_a i_{sd} + \dfrac{1}{\sigma L_s} u_{sq} \\ \dfrac{d\phi_{rd}}{dt} = -\dfrac{R_r}{L_r} \phi_{rd} + \dfrac{L_{sr} R_r}{L_r} i_{sd} \\ \dfrac{d\omega}{dt} = \dfrac{T_{em}}{J} - \dfrac{T_r}{J} - \dfrac{f}{J} \omega \\ \dfrac{d\theta}{dt} = \omega \end{cases} \tag{3.30}$$

where R_r and R_s are rotor and stator resistances, L_r and L_s are rotor and stator inductances, L_{sr} is the mutual inductance, θ is the rotor position, ω is the rotor angular velocity, p is the number of pole pairs, J is the inertia of the rotor, f is the coefficient of viscous friction, T_r is the load torque, ϕ_{rd} is the rotor flux linkage, i_{sd} and i_{sq} stand for the $d-q$ axis currents, u_{sd} and u_{sq} are the $d-q$ axis voltages, $\sigma = 1 - \dfrac{L_{sr}^2}{L_s L_r}$ is the dispersion coefficient of Blondel, $T_{em} = \dfrac{p L_{sr}}{L_r} \phi_{rd} i_{sd}$ is the electromagnetic torque and $\omega_a = p\omega + \dfrac{L_{sr} R_r}{L_r \phi_{rd}} i_{sd}$.

The model of IM (3.30) can be written as follows:

$$\dot{x} = f(x, t) + g(x, t)u + p(t)$$

with:

$$x = \begin{bmatrix} i_{sd} \\ i_{sq} \\ \phi_{rd} \\ \omega \end{bmatrix}, \quad f(x, t) = \begin{bmatrix} -\dfrac{L_{sr}^2 R_r + L_r^2 R_s}{\sigma L_s L_r^2} i_{sd} + \dfrac{L_{sr} R_r}{\sigma L_s L_r^2} \phi_{rd} + \omega_a i_{sq} \\ -\dfrac{L_{sr}^2 R_r + L_r^2 R_s}{\sigma L_s L_r^2} i_{sq} - \dfrac{p L_{sr}}{\sigma L_s L_r} \omega \phi_{rd} - \omega_a i_{sd} \\ -\dfrac{R_r}{L_r} \phi_{rd} + \dfrac{L_{sr} R_r}{L_r} i_{sd} \\ \dfrac{T_{em}}{J} - \dfrac{f\omega}{J} \end{bmatrix}$$

and

$$g(x, t) = \begin{bmatrix} \dfrac{1}{\sigma L_s} & 0 \\ 0 & \dfrac{1}{\sigma L_s} \\ 0 & 0 \\ 0 & 0 \end{bmatrix}, \quad p(t) = \begin{bmatrix} 0 \\ 0 \\ 0 \\ -\dfrac{T_r}{J} \end{bmatrix}, \quad u = \begin{bmatrix} u_{sd} \\ u_{sq} \end{bmatrix}$$

So, this model (3.30) belongs to the nonlinear systems class (3.1) for which we can apply the proposed HOSMC.

3.4.2 Controllers Synthesis

The control objective is to enforce the rotor angular position θ and the rotor flux linkage ϕ_{rd} to track a desired trajectory θ_{ref} and $\phi_{rd,ref}$, respectively. So, we consider the following sliding variable

$$s = \begin{bmatrix} s_1 \\ s_2 \end{bmatrix} = \begin{bmatrix} \theta - \theta_{ref} \\ \phi_{rd} - \phi_{rd,ref} \end{bmatrix} \tag{3.31}$$

So, we have:

$$\begin{bmatrix} \dddot{s}_1 \\ \ddot{s}_2 \end{bmatrix} = \phi(x,t) + \varphi(x,t)u$$

where:

$$\begin{aligned} \overline{\phi}(x,t) &= \begin{bmatrix} \overline{\phi}_1(x,t) \\ \overline{\phi}_2(x,t) \end{bmatrix} \\ &= \begin{bmatrix} \alpha_{11} i_{sd}^2 + \alpha_{13} i_{sd}\phi_{rd} + \alpha_{23} i_{sq}\phi_{rd} + \alpha_{33}\phi_{rd}^2 + \alpha_4\omega - \dddot{\theta}_{ref} \\ \beta_1 i_{sd} + \beta_2 i_{sq} + \beta_3\phi_{rd} + \beta_4\omega - \ddot{\phi}_{rd,ref} \end{bmatrix} \end{aligned} \tag{3.32}$$

with:

$$\alpha_{11} = \frac{pL_{sr}^2 R_r}{JL_r^2}, \quad \alpha_{13} = -\frac{pL_{sr}}{JL_r}\left(\frac{R_r}{L_r} + \frac{f}{J} + \frac{L_{sr}^2 R_r + L_r^2 R_s}{\sigma L_s L_r^2}\right)$$

$$\alpha_{23} = \frac{pL_{sr}}{JL_r}\omega_a, \quad \alpha_{33} = \frac{pL_{sr}^2 R_r}{\sigma J L_s L_r^3}, \quad \alpha_4 = \left(\frac{f}{J}\right)^2$$

and:

$$\beta_1 = -\frac{R_r L_{sr}}{L_r^2}\left(R_r + \frac{R_r L_{sr}^2 + R_s L_r^2}{\sigma L_s L_r}\right), \quad \beta_2 = \frac{L_{sr} R_r}{L_r}\omega_a$$

$$\beta_3 = \left(\frac{R_r}{L_r}\right)^2 \left(1 + \frac{L_{sr}^2 R_r}{\sigma L_s L_r}\right), \quad \beta_4 = 0$$

where:

$$\overline{\varphi}(x,t) = \begin{bmatrix} 0 & \lambda_1 \\ \lambda_2 & 0 \end{bmatrix}, \quad \Delta_\varphi = \begin{bmatrix} 0 & 0 \\ 0 & 0 \end{bmatrix}, \quad \Delta_\phi = \begin{bmatrix} 0 \\ 0 \end{bmatrix}$$

with:

$$\lambda_1 = \frac{pL_{sr}}{\sigma L_s L_r J}\phi_{rd}, \quad \lambda_2 = \frac{L_{sr} R_r}{\sigma L_s L_r}$$

So, a third and a second order SMC are used, respectively, to control the rotor angular position θ and the rotor flux linkage ϕ_{rd}.

From the expression of the matrix $\overline{\varphi}(x,t)$, one can show that $u_1 = u_{sd}$ depends only on s_1 and its first derivatives. Besides, $u_2 = u_{sq}$ depends only on s_2, $\dot{s}_2$ and $\ddot{s}_2$, so the control law to applied to the IM is of the following expression.

$$u = \begin{bmatrix} u_{sd} \\ u_{sq} \end{bmatrix} = \overline{\varphi}^{-1}\left[w(z) - \overline{\phi}\right]$$

with

$$w(z) = \begin{bmatrix} w_1(z_1) \\ w_2(z_2) \end{bmatrix} = \begin{bmatrix} -\alpha_1 \dfrac{a_{11}z_{11} + a_{21}z_{21} + a_{31}z_{31}}{b_{11}|z_{11}| + b_{21}|z_{21}| + b_{31}|z_{31}|} \\ -\alpha_2 \dfrac{a_{12}z_{12} + a_{22}z_{22}}{b_{12}|z_{12}| + b_{22}|z_{22}|} \end{bmatrix} \tag{3.33}$$

where

$$w_1(z_1) = -9.5 \times 10^4 \, \frac{20z_{11} + 2000z_{21} + 0.5z_{31}}{3000|z_{11}| + 48.5|z_{21}| + |z_{31}|}$$

obtained for $\alpha_1 = 9.5 \times 10^4$, $a_{11} = 20$, $a_{21} = 2000$, $a_{31} = 0.5$, $b_{11} = 3000$, $b_{21} = 48.5$ and $b_{31} = 1$; and:

$$w_2(z_2) = -10^4 \, \frac{2.15z_{12} + z_{22}}{10|z_{12}| + |z_{22}|}$$

obtained for $\alpha_2 = 10^4$, $a_{12} = 2.15$, $a_{22} = 1$, $b_{12} = 10$, $b_{22} = 1$, where

$$z = \begin{bmatrix} z_1 \\ z_2 \end{bmatrix}, \quad z_1 = \begin{bmatrix} z_{11} \\ z_{21} \\ z_{31} \end{bmatrix} = \begin{bmatrix} s_1 \\ \dot{s}_1 \\ \ddot{s}_1 \end{bmatrix}, \quad z_2 = \begin{bmatrix} z_{12} \\ z_{22} \end{bmatrix} = \begin{bmatrix} s_2 \\ \dot{s}_2 \end{bmatrix}.$$

For this application, we assume $\dddot{\theta}_{ref} = \ddot{\omega}_{ref} = 0$ and $\ddot{\phi}_{rd,ref} = 0$.

To minimize the cost of the control implementation, we propose to insert a high gain observer which allows estimating the rotor flux linkage ϕ_{rd} and the rotor angular velocity ω. So the rotor angular position $\theta = \int \omega dt$. So, we can see that the model (3.30) of the IM can be written under the form (3.23) where:

$$\overline{C} = \begin{bmatrix} 1\,0\,0\,0 \\ 0\,1\,0\,0 \end{bmatrix}, \quad x_1 = \begin{bmatrix} i_{sd} \\ i_{sq} \end{bmatrix}, \quad x_2 = \phi_{rd}, \quad x_3 = \omega$$

$$F_1(s, x_1) = \begin{bmatrix} \omega_a \\ -\dfrac{L_{sr}^2 R_r + L_r^2 R_s}{\sigma L_s L_r^2} \end{bmatrix}, \quad F_2(s, x_1, x_2) = -\frac{R_r}{L_r}$$

Using the Eq. (3.24), we synthesized the following high gain observer for the IM (Msaddek et al. 2015a):

$$
\begin{aligned}
\frac{d\widehat{i}_{sd}}{dt} &= -\frac{L_{sr}^2 R_r + L_r^2 R_s}{\sigma L_s L_r^2}\widehat{i}_{sd} + \frac{L_{sr} R_r}{\sigma L_s L_r^2}\widehat{\phi}_{rd} + \widehat{\omega}_a \widehat{i}_{sq} + \frac{1}{\sigma L_s} u_{sd} - 3\vartheta(\widehat{i}_{sd} - i_{sd}) \\
\frac{d\widehat{i}_{sq}}{dt} &= -\frac{L_{sr}^2 R_r + L_r^2 R_s}{\sigma L_s L_r^2}\widehat{i}_{sq} - \frac{pL_{sr}}{\sigma L_s L_r}\widehat{\omega}\widehat{\phi}_{rd} - \widehat{\omega}_a \widehat{i}_{sd} + \frac{1}{\sigma L_s} u_{sq} - 3\vartheta(\widehat{i}_{sq} - i_{sq}) \\
\frac{d\widehat{\phi}_{rd}}{dt} &= \frac{R_r}{L_r}\left(L_{sr}\widehat{i}_{sd} - \widehat{\phi}_{rd}\right) + \frac{3\vartheta^2}{b}\left[-\widehat{\omega}_a(\widehat{i}_{sd} - i_{sd}) + \frac{L_{sr}^2 R_r + L_r^2 R_s}{\sigma L_s L_r^2}(\widehat{i}_{sq} - i_{sq})\right] \\
\frac{d\widehat{\omega}}{dt} &= \frac{\widehat{T}_{em}}{J} - \frac{T_r}{J} - \frac{f}{J}\widehat{\omega} + \vartheta^3 \frac{L_r}{bR_r}\left[\widehat{\omega}_a(\widehat{i}_{sd} - i_{sd}) - \frac{L_{sr}^2 R_r + L_r^2 R_s}{\sigma b L_s L_r^2}(\widehat{i}_{sq} - i_{sq})\right] \\
\frac{d\widehat{\theta}}{dt} &= \widehat{\omega}
\end{aligned}
$$

with:

$$
\widehat{T}_{em} = \frac{pL_{sr}}{L_r}\widehat{\phi}_{rd}\widehat{i}_{sd}
$$

$$
\widehat{\omega}_a = p\widehat{\omega} + \frac{L_{sr} R_r}{L_r \widehat{\phi}_{rd}}\widehat{i}_{sd}
$$

$$
\widehat{y} = \overline{C}\widehat{x} = \begin{bmatrix} \widehat{i}_{sd} \\ \widehat{i}_{sq} \end{bmatrix}
$$

The control law applied to the IM model (3.30) is given by:

$$
u = \begin{bmatrix} u_{sd} \\ u_{sq} \end{bmatrix} = \overline{\varphi}^{-1}(\widehat{x}, t)\left[w(\widehat{z}) - \overline{\phi}(\widehat{x}, t)\right]
$$

where $w(\widehat{z})$ is given by (3.31) and:

$$
\widehat{s} = \begin{bmatrix} \widehat{s}_1 \\ \widehat{s}_2 \end{bmatrix} = \begin{bmatrix} \widehat{\theta} - \theta_{ref} \\ \widehat{\phi}_{rd} - \phi_{rdref} \end{bmatrix}, \quad \widehat{z} = \begin{bmatrix} \widehat{z}_1 \\ \widehat{z}_2 \end{bmatrix}, \quad \widehat{z}_1 = \begin{bmatrix} \widehat{s}_1 \\ \dot{\widehat{s}}_1 \\ \ddot{\widehat{s}}_1 \end{bmatrix}, \quad \widehat{z}_2 = \begin{bmatrix} \widehat{s}_2 \\ \dot{\widehat{s}}_2 \end{bmatrix}.
$$

First, we consider the case where noises are absents. So the load torque $T_r(t) = 0$. Simulation results plotted on Figs. 3.1, 3.2 and 3.3 show that estimated states converge to the simulated states. Moreover, the control objective is fulfilled with high performances.

Now, we consider that the IM is affected by external disturbance. So, the load torque varies randomly as plotted on Fig. 3.4. Simulation results depicted on Figs. 3.5, 3.6 and 3.7 show the robustness of the proposed approach against external disturbances.

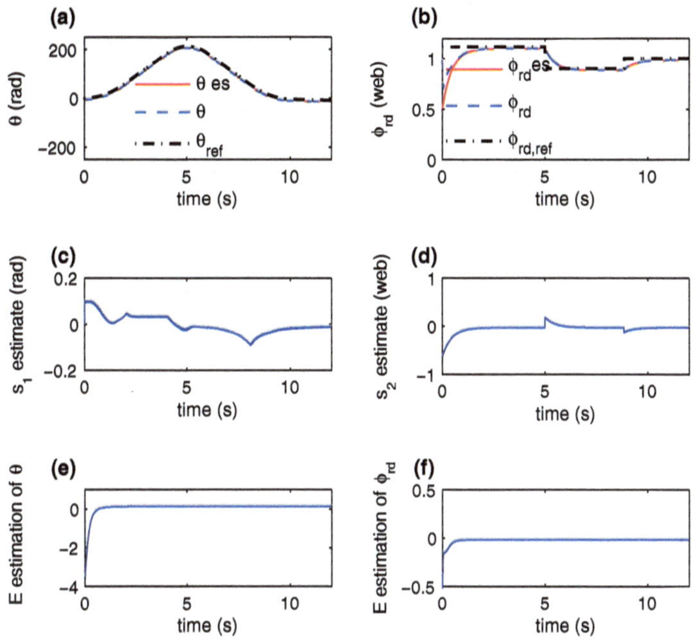

Fig. 3.1 Tracking performances without noises using the resulting controller, **a**: θ, $\widehat{\theta}$ and θ_{ref}, **b**: ϕ_{rd}, $\widehat{\phi}_{rd}$ and ϕ_{rdref}, **c**: $s_1 = \theta - \theta_{ref}$, **d**: $s_2 = \widehat{\phi}_{rd} - \phi_{rdref}$, **e**: $e_\theta = \widehat{\theta} - \theta$, **f**: $e_\phi = \widehat{\phi}_{rd} - \phi_{rd}$

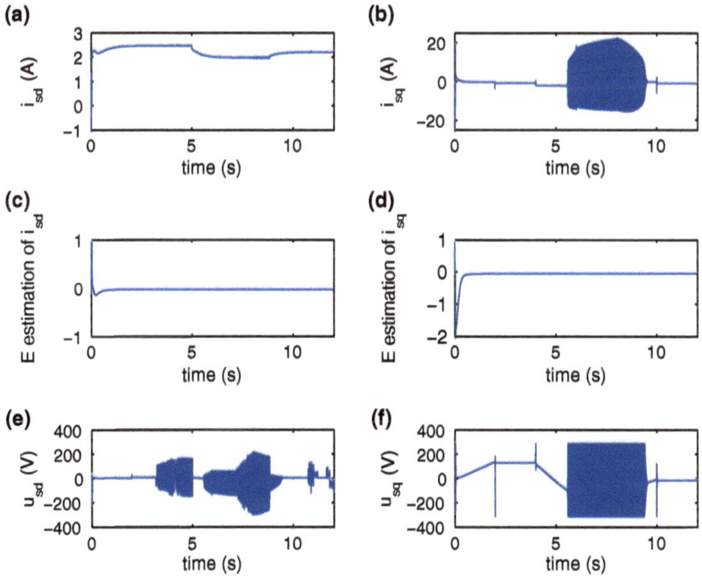

Fig. 3.2 Tracking performances without noises using the resulting controller, **a**: i_{sd}, **b**: i_{sq}, **c**: $e_{i_{sd}} = \widehat{i}_{sd} - i_{sd}$, **d**: $e_{i_{sq}} = \widehat{i}_{sq} - i_{sq}$, **e**: u_{sd}, **f**: u_{sq}

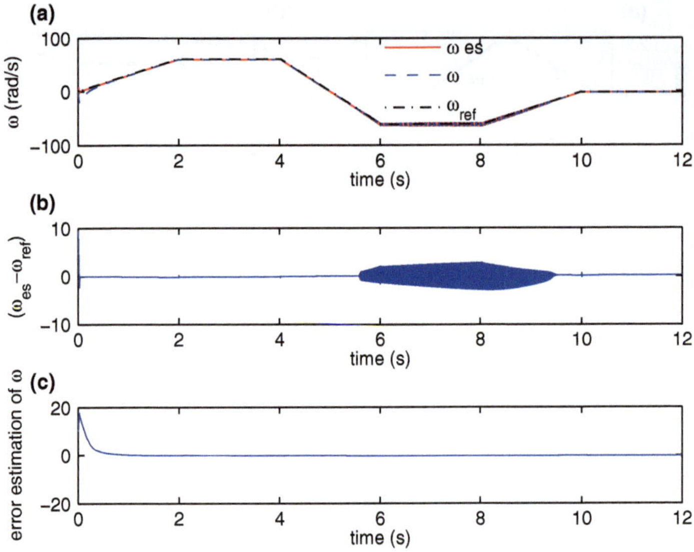

Fig. 3.3 Tracking performances without noises using the resulting controller, **a**: ω, $\widehat{\omega}$ and ω_{ref}, **b**: $\widehat{s}_1 = \widehat{\omega} - \omega_{ref}$, **c**: $\widehat{\omega} - \omega$

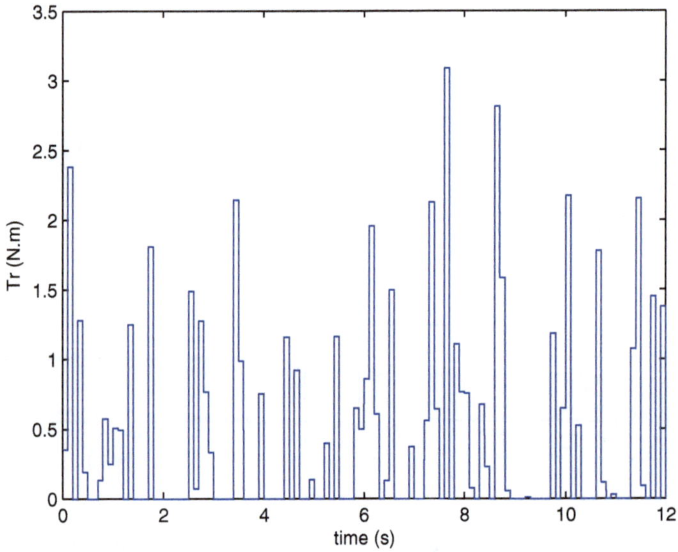

Fig. 3.4 Evolution of the Load torque T_r

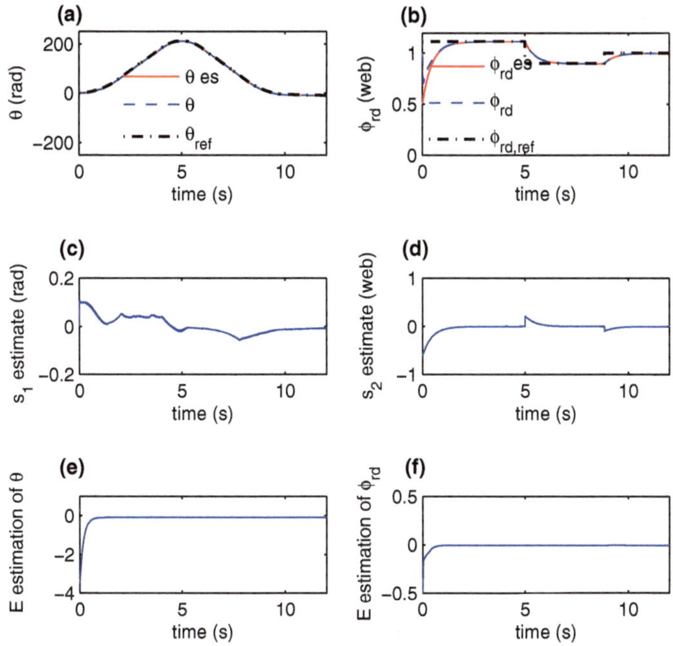

Fig. 3.5 Tracking performances under disturbances using the resulting controller, **a**: θ, $\widehat{\theta}$ and θ_{ref}, **b**: ϕ_{rd}, $\widehat{\phi}_{rd}$ and ϕ_{rdref}, **c**: $s_1 = \theta - \theta_{ref}$, **d**: $s_2 = \widehat{\phi}_{rd} - \phi_{rdref}$, **e**: $e_\theta = \widehat{\theta} - \theta$, **f**: $e_\phi = \widehat{\phi}_{rd} - \phi_{rd}$

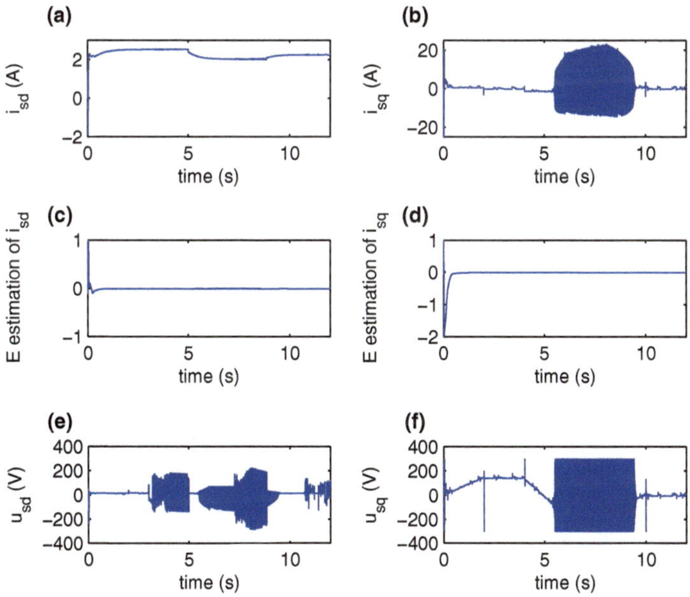

Fig. 3.6 Tracking performances under disturbances using the resulting controller, **a**: i_{sd}, **b**: i_{sq}, **c**: $e_{i_{sd}} = \widehat{i}_{sd} - i_{sd}$, **d**: $e_{i_{sq}} = \widehat{i}_{sq} - i_{sq}$, **e**: u_{sd}, **f**: u_{sq}

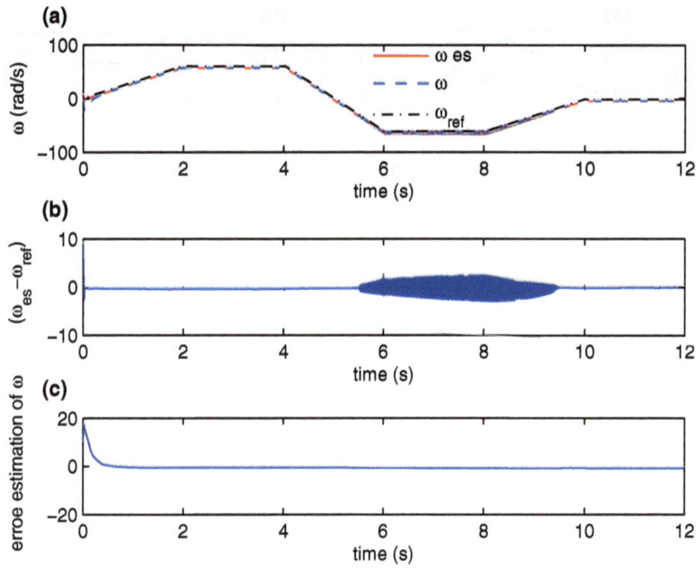

Fig. 3.7 Tracking performances under disturbances using the resulting controller, **a**: ω, $\widehat{\omega}$ and ω_{ref}, **b**: $\widehat{s}_1 = \widehat{\omega} - \omega_{ref}$, **c**: $\widehat{\omega} - \omega$

Table 3.1 Characteristics and parameters of the induction motor

Nominal power	1.5 kW	R_s	8 Ω
Voltage	220/380 V	R_r	4 Ω
Nominal current	6.4/3.7 A	L_s	0.47 H
Number of pole pair	$P = 2$	L_r	0.47 H
Frequency	50 Hz	L_{sr}	0.44 H
Rated speed	1410 rpm	J	0.04 kgm^2
Power factor	0.83	f	0.002 Nm/rad

3.5 Conclusion

In this chapter, we have proposed an output feedback controller for a class of nonlinear uncertain systems. This is achieved by the combination of higher order sliding mode controller together with a high gain observer to estimate the missing states of the system. The resulting controller has been applied to control the angular position of an induction motor. Simulation results show the performances and the robustness of such control strategy against external matched disturbances.

Nomenclature See Table 3.1.

References

Alvarez-Salas, R. (2002). Développement de loi de commande avec observateurs pour machine asynchrone. PhD Thesis, Institut National Polytechnique of Grenoble, France.

Defoort, M. (2007b). Contributions à la Planification et à la commande pour les robots mobiles Coopératifs. PhD Thesis, Ecole Centrale de Lille, France.

Defoort, M., Floquet, T. Kokosy, A., & Perruquetti, W. (2006). Finite-time control of a class of MIMO nonlinear systems using high order integral sliding mode control. International Workshop on Variable Structure Systems (VSS), Alghero, Italy.

Defoort, M., Palos, J., Floquet, T., Kokosy, A., & Perruquetti, W. (2007a). Practical stabilization and tracking of a wheeled mobile robot with integral sliding mode controller. *Proceedings of the IEEE International Conference on Decision and Control*, New Orleans, USA.

Defoort, M., Floquet, T., Kokosy, A., & Perruquetti, W. (2009). A novel higher order sliding mode control scheme. *Systems and control letters*, *58*, 102–108.

Farza, M., M'Saad, M., & Rossignol, L. (2004). Observer design for a class of MIMO nonlinear systems. *Automatica*, *40*, 135–143.

Ghanes, M. (2005). Observation et commande de la machine asynchrone sans capteur mécanique. PhD Thesis, Ecole Centrale de Nantes, France.

Golea, N., Debbache, G., & Golea, A. (2012). Neural network-based adaptive sliding mode control for uncertain non-linear MIMO systems. *International Journal of Modelling, Identification and Control*, *16*, 334–344.

Hirokazu, T., Giuseppe, G., & Hidetoshi, U. (2002). Consideration about problems and solutions of speed estimation method and parameter tuning for speed-senseless vector control of introduction motor drives. *IEEE Transaction on Industry Application*, *38*, 1282–1289.

Jie, S., Yong, Z., & Chengliang, Y. (2012). Longitudinal brake control of hybrid electric bus using adaptive fuzzy sliding mode control. *International Journal of Modelling, Identification and Control*, *15*, 147–155.

Laghrouche, S., Pleston, F., & Glumineau, A. (2007). Higher order sliding mode control based an integral sliding mode. *Automatica*, *43*, 531–537.

Levant, A. (1993). Sliding order and sliding accuracy in sliding mode control. *International Journal of Control*, *58*, 1247–1263.

Levant, A. (2005). Quasi-Continuous High-order Sliding-Mode controllers. *IEEE Transaction on Automatic Control*, *50*, 1812–1816.

Mezouar, A., Fellah, M. K., & Hadjeri, S. (2007). Adaptive sliding mode observer for induction motor using two-time-scale approach. *Electric Power Systems Research*, *77*, 604–618.

Mondal, S., & Mahanta, C. (2011). Nonlinear sliding surface based second order sliding mode controller for uncertain linear systems. *Communications in Nonlinear Science and Numerical Simulation*, *16*, 3760–3769.

Msaddek, A., Gaaloul, A., and M'sahli, F. (2013). A novel higher order sliding mode control: application to an induction motor. International conference on control, engineering and information technology (CEIT), Sousse, Tunisia.

Msaddek, A., Gaaloul, A., & M'sahli, F., (2014). Robust exponential stabilisation of a class of nonlinear systems using a novel technique of higher order sliding mode control. *International Journal of Modelling, Identification and Control*, *22*, 115–125.

Msaddek, A., Gaaloul, A. and M'sahli, F. (2015a). High Gain Observer Based Higher Order Sliding Mode Control: Application to an Induction Motor. *Proceedings of the 12th International Multi–Conference on Systems, Signals & Devices (SSD)*, Mahdia,Tunisia.

Msaddek, A., Gaaloul, A., & M'sahli, F., (2015b). Adaptive fuzzy supervision of the gain of the higher order sliding mode control. *International Journal of Automation and Control*, *9*, 228–246.

Nayak, N., Routray, S. K., & Rout, P. K. (2013). Non-linear control and stabilisation of VSC-HVDC transmission system based on Type-2 fuzzy sliding mode control. *International Journal of Automation and Control*, *7*, 1–20.

Roopoei, M., & Zalghadri, J. M. (2009). Chattering- free Fuzzy sliding mode control in MIMO Uncertain systems. *Nonlinear Analysis*, *71*, 4430–4437.

Tay, F., Man, Z., Cao, Z., Khoo, S., & Pin Tan, C. (2012). Sliding mode-like learning control for SISO complex systems with T-S fuzzy models. *International Journal of Modelling, Identification and Control*, *16*, 317–326.

Traore, D., De Leon, J., & Glumineau, A. (2012). Adaptive interconnected observer-based backstepping control design for sensorless induction motor. *Automatica*, *48*, 682–687.

Utkin, V. (1992). *Sliding Modes in Control and Optimization*. Berlin, Germany: Springer-Verlag.

Zhao, G., Li, H., & Song, Z. (2013). Adaptive dynamic fuzzy neural network-based decoupled sliding-mode controller with hybrid sliding surfaces. *International Journal of Automation and Control*, *7*, 183–201.

Chapter 4
Synthesis of an Optimal Sliding Function Using LMIs Approach for Time Delay Systems

Houda Romdhane, Khadija Dehri and Ahmed Said Nouri

Abstract During the reachability phase, the sliding mode control is sensitive to external disturbances and uncertainties. In this paper, we propose to determine coefficients of the sliding function using the technique of Linear Matrix Inequalities (LMIs) for single-input single-output time delay systems. This technique leads to an optimal choice of the sliding function to reduce the reachability phase. Using the proposed sliding function, a discrete second order sliding mode control is presented. The control law is based on an input–output model. Simulation results demonstrate that the proposed strategy leads to an optimal performance in terms of reduction of the reachability phase as well as the chattering phenomenon.

Keywords Discrete second order sliding mode control · Input time delay systems · Input–output model · Linear matrix inequalities · Reachability phase

4.1 Introduction

Sliding Mode Control (SMC) has been widely used in the literature. This success is due to its simplicity and robustness against both external disturbances and parametric uncertainties (Emelyanov 1967; Lopez and Nouri 2006; Perruquetti and Barbot 2002; Utkin 1992; Vecchio 2008). Sliding Mode Control consists of, firstly, bringing an arbitrary point to a specific surface called "sliding surface", which is the reachability phase. Secondly, ensuring the maintenance and the sliding along this surface until reaching the origin of the phase plane, which is the sliding phase.

H. Romdhane (✉) · K. Dehri · A.S. Nouri
Research Unit: Numerical Control of Industrial Processes (CONPRI),
National Engineering School of Gabes, University of Gabes, Gabes, Tunisia
e-mail: romdhanehouda@gmail.com

K. Dehri
e-mail: khadija.dehri@gmail.com

A.S. Nouri
e-mail: ahmedsaid.nouri@enig.rnu.tn

N. Derbel et al. (eds.), *Applications of Sliding Mode Control*,
Studies in Systems, Decision and Control 79, DOI 10.1007/978-981-10-2374-3_4

During the reachability phase, the sliding mode control is sensitive to external disturbances and uncertainties (Emelyanov 1967; Lopez and Nouri 2006; Utkin 1992). Hence, many researchers have been interested in solving this problem. One of the solutions was the choice of coefficients of the sliding function in an optimal way. Therefore, several methods are available to calculate these coefficients such as, the pole placement method (Dehri et al. 2011, 2012a, b), the Linear Matrix Inequalities (LMIs) approach (Abdnnabi and Nouri 2012a, b; Chaker et al. 2013; Lin et al. 2011; Silva and Edwards 2009; Silva et al. 2008, 2009; Wang and Yu 2013; Yadav and Singh 2012), etc.

The (LMIs) technique has become the most important technique in modern methods of automatic control (Boyd et al. 1994). Many problems can be formulated as convex optimization problems under LMI constraints. Such a formulation is very advantageous because it allows to solve optimization problems that appear difficult or even impossible to solve analytically (Gahinet et al. 1995; Sandberg 1965). An important result in convex optimization came from the introduction of the interior point methods initially developed by Karmarkar (1984) for linear programming. This method has been then extended in the space of definite positive matrix by Nesterov and Nemiroviskii (1993).

In the last few years, the LMIs technique has been used in the synthesis of the sliding mode control (Abdnnabi and Nouri 2012a, b; Chaker et al. 2013; Lin et al. 2011; Silva and Edwards 2009; Silva et al. 2008, 2009; Wang and Yu 2013; Yadav and Singh 2012). This technique allows an optimal determination of coefficients of the sliding function, so that it reduces the time of the reachability phase. Therefore, the sliding mode is reached quickly and the control law becomes robust since the initial instant.

In this context, a linear sliding function based on the LMIs has been proposed in Lin et al. (2011). An algorithm for designing a non linear sliding function vector for multivariables systems (MIMO) has been presented in Yadav and Singh (2012). The authors, in Chaker et al. (2013), proposed a method to select a sliding function and to synthesize a sliding mode control for a class of discrete-time linear saturated systems. A sliding mode control algorithm has been developed for a class of uncertain discrete network systems (Wang and Yu 2013). A discrete second order sliding mode control for the linear uncertain system with state delay has been presented in Abdnnabi and Nouri (2012a, b). In all of the above works, the sliding function based on LMIs has been developed with state space model.

However, in the literature, there are many systems which are described by their transfer functions i.e only their inputs and outputs are known. That's why, in our work, we are interested, in a first time, in synthesizing of a discrete second order sliding mode control via input output model (Romdhane et al. 2013a, b, 2014a, b) and in a second time, in the determination of sliding function's coefficients using the LMIs approach (Romdhane et al. 2015).

In this paper, firstly, we propose to synthesize a discrete second order sliding mode control via input output model for single-input single-output systems with input delay, then, to calculate sliding function's coefficients using LMIs approach in order to reduce the reachability phase. A comparison between the results obtained by the pole placement method with those obtained by the LMIs approach is given.

This paper is organized as follows. Section 4.2 presents a discrete second order sliding mode control for input time delay systems using the pole placement method. In Sect. 4.3, a discrete second order sliding mode control using an optimal sliding function is developed for such systems. This optimal sliding function is synthesized using the LMIs approach. Simulation results are given in Sect. 4.4. Section 4.5 concludes the paper.

4.2 Discrete Second Order Sliding Mode Control

Consider the discrete-time delay system described by the following model:

$$A(q^{-1})y(k) = q^{-d}B(q^{-1})u(k-1) \tag{4.1}$$

where $y(k)$, $u(k)$ and d ($d \geq 0$) are respectively the output, the input and the constant delay. $A(q^{-1})$ and $B(q^{-1})$ are two polynomials defined as:

$$\begin{aligned} A\left(q^{-1}\right) &= 1 + a_1 q^{-1} + \cdots + a_{n_A} q^{-n_A} \\ B\left(q^{-1}\right) &= b_0 + b_1 q^{-1} + \cdots + b_{n_B} q^{-n_B} \end{aligned}$$

In the case of second order sliding mode control, the sliding function is defined by:

$$\sigma(k) = S(k) + \beta S(k-1) \tag{4.2}$$

with

- β was chosen in the interval $]0,\ 1[$ in order to ensure the convergence of the sliding function vector $\sigma(k)$.
- $S(k)$ is the sliding function in the classical sliding mode control case. It is defined by:

$$S(k) = C(q^{-1})(y(k) - r(k)) \tag{4.3}$$

where:

- $C\left(q^{-1}\right)$ is a polynomial defined as follows:

$$C(q^{-1}) = 1 + c_1 q^{-1} + \cdots + c_{nc} q^{-nc}$$

- $r(k)$ is the desired trajectory.

In the case of time delay systems, the equivalent control can be calculated from:

$$\sigma(k+d+1) = \sigma(k+d) = 0 \tag{4.4}$$

The sliding function at instant $k+d+1$ is given by:

$$\begin{aligned}\sigma(k+d+1) &= S(k+d+1) + \beta S(k+d)\\ &= C\left(q^{-1}\right)\left(y(k+d+1) - r(k+d+1)\right) + \beta S(k+d)\end{aligned}$$

After calculation, the sliding function becomes:

$$\begin{aligned}\sigma(k+d+1) = E\left(q^{-1}\right) F\left(q^{-1}\right) B\left(q^{-1}\right) u(k) + G\left(q^{-1}\right) y(k+1)\\ -C\left(q^{-1}\right) r(k+d+1) + \beta S(k+d)\end{aligned} \tag{4.5}$$

with $F\left(q^{-1}\right)$ and $G\left(q^{-1}\right)$ are two polynomials solution of the diophantine polynomial equation:

$$C\left(q^{-1}\right) = A\left(q^{-1}\right) E\left(q^{-1}\right) F\left(q^{-1}\right) + q^{-d} G\left(q^{-1}\right) \tag{4.6}$$

where

$$\begin{aligned}F\left(q^{-1}\right) &= 1 + f_1 q^{-1} + \cdots + f_{n_F} q^{-n_F}; \quad n_F = d-1\\ G\left(q^{-1}\right) &= g_0 + g_1 q^{-1} + \cdots + g_{n_G} q^{-n_G}; \quad n_G = sup(n_c - 1, n_A)\\ E\left(q^{-1}\right) &= 1 - q^{-1}\end{aligned}$$

Therefore, the equivalent control law can be written as:

$$\begin{aligned}u_{eq}(k) = &-\left[E\left(q^{-1}\right) F\left(q^{-1}\right) B\left(q^{-1}\right)\right]^{-1} \beta S(k+d)\\ &-\left[E\left(q^{-1}\right) F\left(q^{-1}\right) B\left(q^{-1}\right)\right]^{-1} G\left(q^{-1}\right) y(k+1)\\ &+\left[E\left(q^{-1}\right) F\left(q^{-1}\right) B\left(q^{-1}\right)\right]^{-1} C\left(q^{-1}\right) r(k+d+1)\end{aligned} \tag{4.7}$$

The discontinuous term is defined as follows:

$$u_{dis}(k) = u_{dis}(k-1) - \left[E\left(q^{-1}\right) F\left(q^{-1}\right) B\left(q^{-1}\right)\right]^{-1} T_e M \,\text{sign}\left[\sigma(k)\right] \tag{4.8}$$

with T_e is the sampling rate, M is the discontinuous gain and sign is the signum function defined as:

$$\text{sign}\left[\sigma(k)\right] = \left|\begin{array}{lll} -1 & \text{if} & \sigma(k) < 0\\ 1 & \text{if} & \sigma(k) > 0\end{array}\right.$$

Then, the global discrete second order sliding mode control can be expressed by:

$$u(k) = u_{eq}(k) + u_{dis}(k) \tag{4.9}$$

4.3 Discrete Second Order Sliding Mode Control Using LMIs Approach

In this section, we consider the system previously used in Sect. 4.2.

An extended state space representation for the input–output model defined by (4.1) can be obtained by taking as the state vector present and past values of system's inputs and outputs (Granado et al. 2005) such that:

$$\begin{aligned} x(k) = [&y(k) \;\; y(k-1) \;\; \cdots \;\; y(k-n_A+1) \\ &u(k-d-1) \;\; u(k-d-2) \;\cdots u(k-d-n_B)]^\top \end{aligned} \tag{4.10}$$

Therefore, an equivalent state space representation is given by:

$$\begin{cases} x(k+1) = A_r\, x(k) + B_r\, u(k-d) \\ y(k) = F\, x(k) \end{cases} \tag{4.11}$$

where

$$A_r = \begin{pmatrix} -a_1 & -a_2 & \cdots & -a_{n_A-1} & -a_{n_A} & b_1 & \cdots & b_{n_B-1} & b_{n_B} \\ 1 & 0 & \cdots & 0 & 0 & 0 & \cdots & 0 & 0 \\ 0 & 1 & \cdots & 0 & 0 & 0 & \cdots & 0 & 0 \\ \vdots & \vdots & \ddots & \vdots & \vdots & \vdots & \ddots & \vdots & \vdots \\ 0 & 0 & \cdots & 1 & 0 & 0 & \cdots & 0 & 0 \\ 0 & 0 & \cdots & 0 & 0 & 0 & \cdots & 0 & 0 \\ 0 & 0 & \cdots & 0 & 0 & 1 & \cdots & 0 & 0 \\ \vdots & \vdots & \ddots & \vdots & \vdots & \vdots & \ddots & \vdots & \vdots \\ 0 & 0 & \cdots & 0 & 0 & 0 & \cdots & 1 & 0 \end{pmatrix}$$

$$B_r = [b_0 \;\; 0 \;\; 0 \;\; \cdots \;\; 0 \;\; 1 \;\; 0 \;\; \cdots \;\; 0]^\top$$

$$F = [1 \;\; 0 \;\; 0 \;\; \cdots \;\; 0 \;\; 0 \;\; 0 \;\; \cdots \;\; 0]$$

we assumed that the pair $(A_r,\ B_r)$ is controllable.

In order to simplify the development of the sliding mode control law, the system (4.11) can be written in a regular form. There exists a transformation matrix T_1 such that $T_1 B_r$ can be written as follows $T_1 B_r = \begin{bmatrix} B_{r_1} \\ B_{r_2} \end{bmatrix}$ with B_{r_2} is a non-zero constant.

To make the system in a regular form, we use the following transformation (Wang and Yu 2013; Yadav and Singh 2012):

$$z(k) = Tx(k) = \begin{pmatrix} I_{n-1} & -B_{r_1}B_{r_2}^{-1} \\ 0 & B_{r_2}^{-1} \end{pmatrix} x(k) \tag{4.12}$$

where $n = n_A + n_B$.

Therefore, system (4.11) becomes:

$$\begin{cases} z(k+1) = \tilde{A}z(k) + \tilde{B}u(k-d) \\ y(k) = \tilde{F}z(k) \end{cases} \tag{4.13}$$

with:

$$z(k) = \begin{bmatrix} z_1(k) \\ z_2(k) \end{bmatrix}; \quad \tilde{A} = TA_rT^{-1} = \begin{bmatrix} A_{11} & A_{12} \\ A_{21} & A_{22} \end{bmatrix}$$

$$\tilde{B} = TB_r = \begin{bmatrix} \left.\begin{matrix} 0 \\ \vdots \\ 0 \end{matrix}\right\} n-1 \\ 1 \end{bmatrix}; \quad \tilde{F} = FT^{-1}$$

Therefore, system (4.13) can be rewritten as follows:

$$\begin{cases} z_1(k+1) = A_{11}z_1(k) + A_{12}z_2(k) \\ z_2(k+1) = A_{21}z_1(k) + A_{22}z_2(k) + u(k-d) \\ y(k) = \tilde{F}z(k) \end{cases} \tag{4.14}$$

where $z_1(k) \in R^{n-1}$ and $z_2(k) \in R$.

In the case of state space model, the classical sliding function is defined as :

$$S(k) = \bar{C}z(k) = [C_r \quad 1]\,z(k) = C_rz_1(k) + z_2(k) \tag{4.15}$$

with $\bar{C} = [C_r \quad 1]$ and $C_r \in R^{1\times(n-1)}$.

In sliding mode, we have:

$$C_rz_1(k) + z_2(k) = 0 \;\Rightarrow\; z_2(k) = -C_rz_1(k) \tag{4.16}$$

The ideal sliding mode is governed by the following reduced system:

$$\begin{cases} z_1(k+1) = A_{11}z_1(k) + A_{12}z_2(k) \\ z_2(k) = -C_rz_1(k) \end{cases} \tag{4.17}$$

The reduced closed loop system can be written as follows:

$$z_1(k+1) = (A_{11} - A_{12}C_r)\, z_1(k) \tag{4.18}$$

The synthesis of a sliding mode control law thus results in determining a state feedback C_r which stabilizes the reduced system (4.18).

The determination of the state feedback C_r is essentially based on resolving the linear matrix inequality (LMI) cited in the following theorem.

Theorem 1 *The reduced system (4.18) is asymptotically stable if there exists a symmetric definite positive matrix $P = P^\top > 0$ and a matrix W such that the following LMI (4.19) is satisfied:*

$$\begin{pmatrix} -Q & * \\ A_{11}Q - A_{12}W & -Q \end{pmatrix} < 0 \tag{4.19}$$

where

- $Q = P^{-1}$
- $W = C_r P^{-1}$

$(*)$ *denotes the transposed elements in the symmetric position:* $(A_{11}Q - A_{12}W)^\top$

Proof Let's consider the following Lyapunov function:

$$v(k) = z_1^\top(k)\; P\, z_1(k)$$

with P is a symmetric definite positive matrix.

To ensure the convergence of system (4.18), it is necessary to satisfy the condition $\Delta v(k) = v(k+1) - v(k) < 0$.

$$\begin{aligned} \Delta v(k) &= z_1^\top(k+1)\, P z_1(k+1) - z_1^\top(k)\, P z_1(k) \\ &= ((A_{11} - A_{12}C_r)\, z_1(k))^\top P\, ((A_{11} - A_{12}C_r)\, z_1(k)) - z_1^\top(k)\, P z_1(k) \\ &= z_1^\top(k) \left((A_{11} - A_{12}C_r)^\top P\, (A_{11} - A_{12}C_r) - P\right) z_1(k) \end{aligned}$$

Then, the reduced system (4.18) is asymptotically stable if and only if:

$$(A_{11} - A_{12}C_r)^\top P\, (A_{11} - A_{12}C_r) - P < 0 \tag{4.20}$$

Using the Schur complement, the inequality (4.20) is equivalent to:

$$\begin{pmatrix} -P & * \\ A_{11} - A_{12}C_r & -P^{-1} \end{pmatrix} < 0 \tag{4.21}$$

(∗) denotes the transposed elements in the symmetric position: $(A_{11} - A_{12}C_r)^{\top}$

To express the last inequality as a LMI, we perform a pre- and post-multiplication of inequality (4.21) by $diag\{P^{-1}, I\}$ and we take $Q = P^{-1}$ and $W = C_r P^{-1}$, therefore, inequality (4.21) becomes:

$$\begin{pmatrix} -Q & * \\ A_{11}Q - A_{12}W & -Q \end{pmatrix} < 0 \tag{4.22}$$

Hence, Theorem 1 is verified and the reduced system (4.18) is asymptotically stable.

Once, we obtain the coefficients of the classical sliding function $S(k)$, we can express the sliding function $\sigma(k)$ such that in (4.2) recalled bellow:

$$\sigma(k) = S(k) + \beta S(k-1)$$

with $S(k)$ is the sliding function defined in (4.15).
Therefore, the equivalent control law $u_{eq}(k)$, obtained from the condition $\sigma(k+d+1) = 0$, can be written as follows:

$$u_{eq}(k) = \left(\bar{C}\tilde{B}\right)^{-1}\left(-\beta S(k+d) - \bar{C}\tilde{A}z(k+d)\right) \tag{4.23}$$

where $\left(\bar{C}\tilde{B}\right)$ is assumed to be non-singular matrix.

The discontinuous term can be expressed as:

$$u_{dis}(k) = u_{dis}(k-1) - \left(\bar{C}\tilde{B}\right)^{-1} T_e M \operatorname{sign}(\sigma(k)) \tag{4.24}$$

with T_e is the sampling rate.

Then, the global discrete second order sliding mode controller is given by:

$$u(k) = u_{eq}(k) + u_{dis}(k) \tag{4.25}$$

4.4 Simulation Example

In this section, a real discrete-time delay system is used in order to prove the effectiveness of the proposed method. The system consists of two masses and a spring as shown in Fig. 4.1.

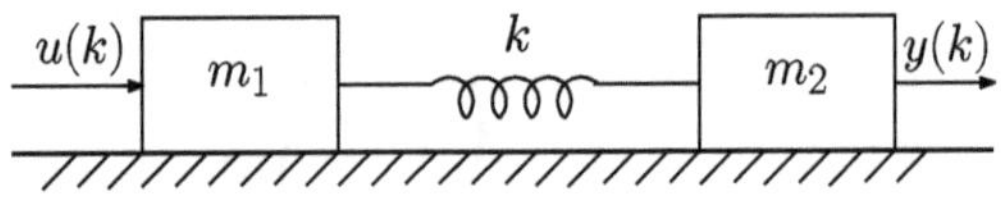

Fig. 4.1 Two-masses-spring system

By taking $k = 1$, the discrete-time input–output representation of the considered system with the sampling rate $T_e = 0.1$ is given by:

$$A\left(q^{-1}\right) y\left(k\right) = q^{-d} B\left(q^{-1}\right) u\left(k-1\right)$$

with:

- $A\left(q^{-1}\right) = 1 - 4q^{-1} + 6.2q^{-2} - 4.4q^{-3} + 1.2q^{-4}$
- $B\left(q^{-1}\right) = 0.0001$
- $d = 3$

The desired trajectory $r(k)$ is chosen equal to zero.

The synthesis parameters are chosen as:

$$\beta = 0.1\,, \qquad M = 0.001$$

In the next subsections, we use for the determination of sliding function's coefficients, in the first time, the pole placement method and, in the second time, the LMIs approach. Then, we apply the proposed discrete second order sliding mode control to the considered system. Finally, we compare the obtained results.

4.4.1 Synthesis of Sliding Function's Coefficients Using Pole Placement Method

In this case, the polynomial $C\left(q^{-1}\right)$ is chosen as:

$$C\left(q^{-1}\right) = 1 + 0.2q^{-1} - 0.24q^{-2}$$

Simulation results are shown in Figs. 4.2, 4.3 and 4.4. Figure 4.2 shows the evolution of the output $y(k)$, Fig. 4.3 presents the evolution of the controller $u(k)$ and Fig. 4.4 illustrates the evolution of the sliding surface $\sigma(k)$.

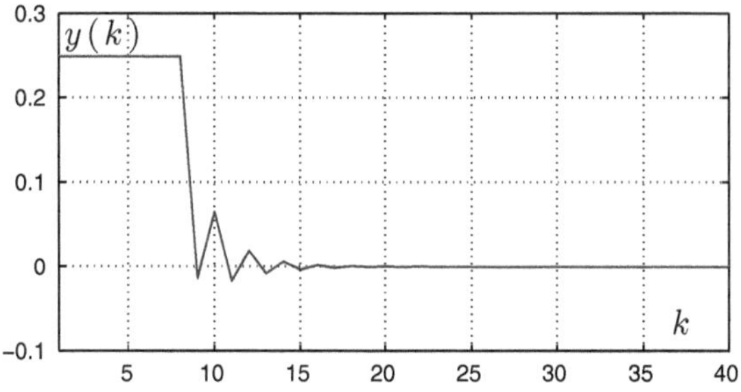

Fig. 4.2 Evolution of the output $y(k)$ (2-DSMC with pole placement method)

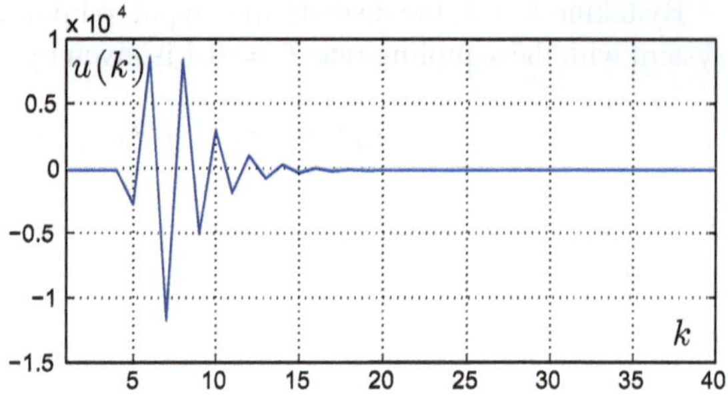

Fig. 4.3 Evolution of control input $u(k)$ (2-DSMC with pole placement method)

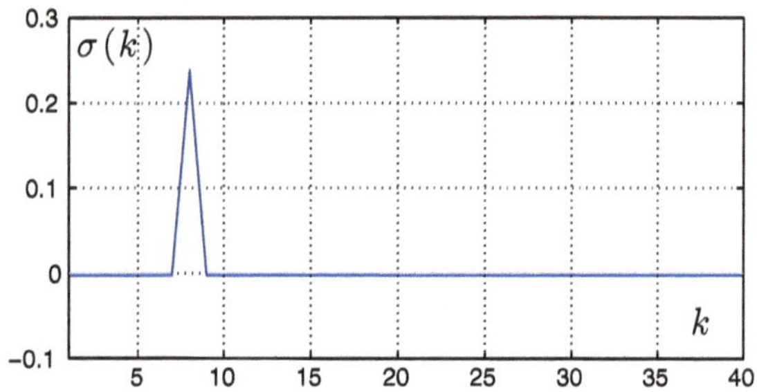

Fig. 4.4 Evolution of the sliding function $\sigma(k)$ (2-DSMC with pole placement method)

It is clear that the sliding mode control is achieved after a few moments. During this time, the robustness of the sliding mode control law is not guaranteed. For this reason, we propose to synthesize coefficients of the sliding function with LMIs.

4.4.2 Synthesis of Sliding Function's Coefficients Using LMIs Approach

The resolution of the LMI gives the following parameters:

$$P = \begin{bmatrix} 26.0021 & 0.0105 & -0.0102 \\ 0.0105 & 38.7353 & -0.0026 \\ -0.0102 & -0.0026 & 61.4095 \end{bmatrix}$$

$$\bar{C} = [-3.3342 \quad 5.1683 \quad -3.6678 \quad 1]$$

Simulation results of the discrete second order sliding mode control using LMI approach are shown in Figs. 4.5, 4.6 and 4.7. Figure 4.5 presents the evolution of the output $y(k)$, Fig. 4.6 illustrates the evolution of the controller $u(k)$ and Fig. 4.7 shows the evolution of the sliding surface $\sigma(k)$.

We observe that the LMIs approach gives optimal values of sliding function's coefficients allowing the reduction of the reachability phase duration.

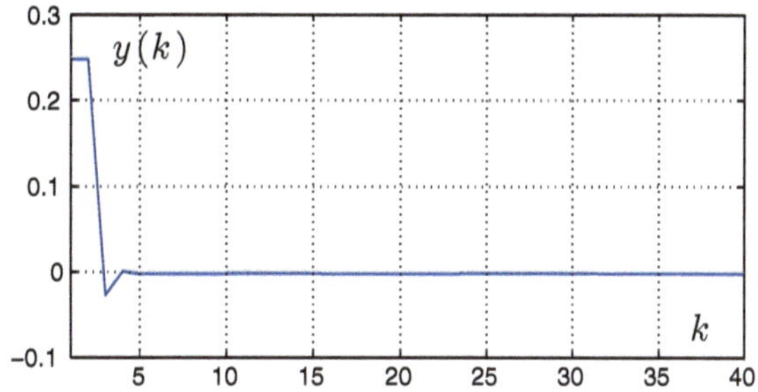

Fig. 4.5 Evolution of the output $y(k)$ (2-DSMC with LMI)

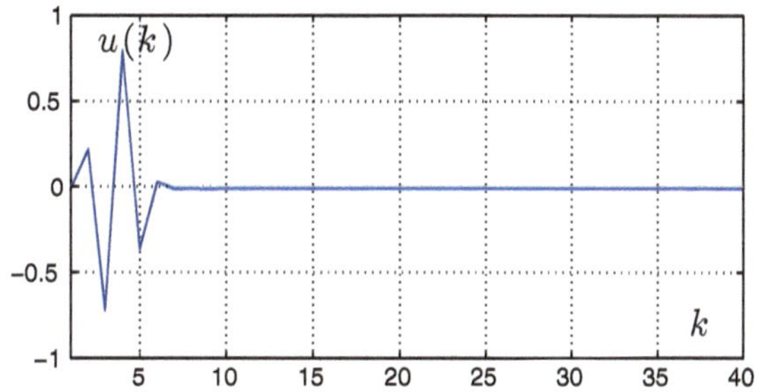

Fig. 4.6 Evolution of control input $u(k)$ (2-DSMC with LMI)

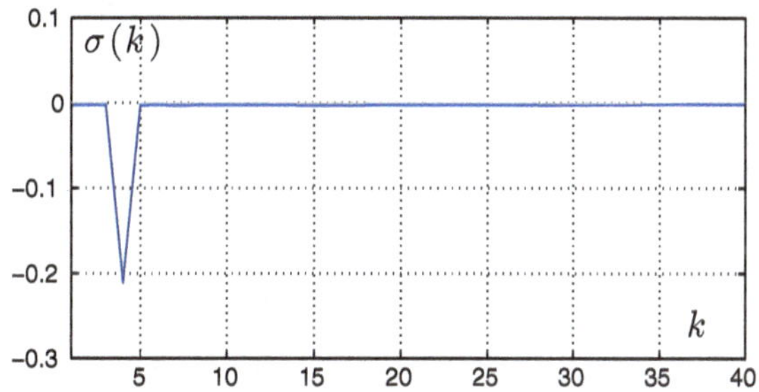

Fig. 4.7 Evolution of the sliding function $\sigma(k)$ (2-DSMC with LMI)

In order to prove the effectiveness of the proposed method, a comparison between simulation results obtained using the LMI technique with those obtained by the pole placement method is given in Figs. 4.8 and 4.9.

We note, according to the simulation results, that when we choose coefficients of the sliding function using the pole placement method, the system converges to zero but the sliding mode is achieved after a few moments. During this time, the robustness of the sliding mode control law is not guaranteed. However, if we choose coefficients of the sliding surface using the LMIs technique, we obtain optimal values of these coefficients allowing the reduction of the duration of the reachability phase and ensuring the convergence of the output in finite time.

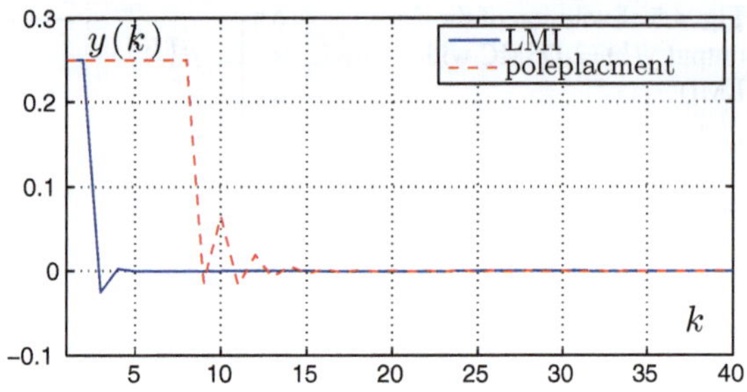

Fig. 4.8 Evolution of the output $y(k)$ (2-DSMC)

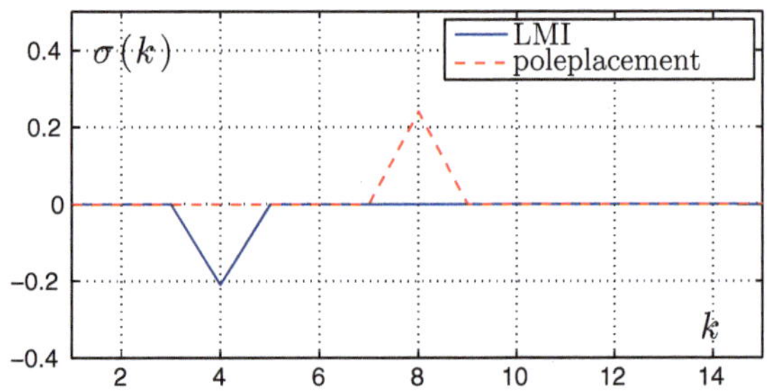

Fig. 4.9 Evolution of the sliding function $\sigma(k)$ (2-DSMC)

4.5 Conclusion

In order to reduce the reachability phase duration, we proposed to determine sliding function's coefficients using the LMIs approach for time delay systems. This technique gives optimal values of these coefficients. To prove the effectiveness of the proposed strategy, we compared the results obtained using the LMIs approach with those given by the pole placement method. Good performances are obtained in terms of the reduction of the reachability phase.

Acknowledgments This work has been supported by the Ministry of Higher Education and Scientific Research in Tunisia.

References

Abdnnabi, N., & Nouri, A. S. (2012a). A new sliding surface for discrete second order sliding mode control for time delay systems with time varying uncertainties. *International Journal of Computer Applications*, *52*(21), 16–22.

Abdnnabi, N., & Nouri, A. S. (2012b). A new sliding surface for discrete second order sliding mode control of time delay systems. In *9th International Multi-Conference on Systems, Signals and Devices (SSD'12), Chemnitz, Germany*.

Boyd, S., Ghaoui, L. E., & Feron, E. (1994). *Linear matrix inequalities in system and control theory*. Philadelphia: SIAM.
Chaker, Z., Anis, S., & Garcia, G. (2013). Discrete-time sliding mode control of a class of saturated systems. *World Applied Sciences Journal, 28*(10), 1355–1360.
Dehri, K., Ltaief, M., & Nouri, A. S. (2011). New discrete multivariable sliding mode control for multi-periodic disturbances rejection. In *8th International Multi-Conference on Systems, Signals and Devices (SSD'11), Sousse, Tunisia*.
Dehri, K., Ltaief, M., & Nouri, A. S. (2012a). Discrete second order sliding mode control for nonlinear multivariable systems. In *MELECON'12, Hammamet, Tunisie* (pp. 387–390).
Dehri, K., Ltaief, M., & Nouri, A. S. (2012b). Repetitive sliding mode control for nondecouplable multivariable system: Periodic disturbances rejection. In *20th Mediterranean Conference on Control and Automation (MED), Barcelona, Spain* (pp. 1194–1199).
Emelyanov, S. V. (1967). *Variable structure control systems*. Moscow: Nauka.
Gahinet, P., Nemivski, A., Laud, A. J., & Chilali, M. (1995). *LMI control toolbox for use with matlab*. Natick: The Mathworks Inc.
Granado, E., Colmenares, W., Bernussou, J., & Garcia, G. (2005). LMI based robust output feedback MPC. In *44th IEEE Conference on Decision and Control, and the European Control Conference*.
Karmarkar, N. (1984). A new polynomial-time algorithm for linear programming. *Combinatorica, 4*, 373–395.
Lin, Z., Xia, Y., Shi, P., & Wu, H. (2011). Robust sliding mode control for uncertain linear discret systems independent of time-delay. *International Journal of Innovative Computing, Information and Control, 7*(2), 869–880.
Lopez, P., & Nouri, A. S. (2006). *Théorie élémentaire et pratique de la commande par régimes glissants*. New York: Springer.
Nesterov, Y., & Nemiroviskii, A. (1993). *Interior-point polynomial algorithms in convex programming*. Philadelphia: SIAM.
Perruquetti, W., & Barbot, J. (2002). *Sliding mode control in engineering*. New York: Marcel Dekker.
Romdhane, H., Dehri, K., & Nouri, A. (2013a). Discrete second order sliding mode control for input-output model discrete second order sliding mode control for input-output model. In *International Conference on Control, Engineering and Information Technology (CEIT'13)* (Vol. 4, pp. 73–77).
Romdhane, H., Dehri, K., & Nouri, A. S. (2013b). Discrete second order sliding mode control for multivariable systems via input-output models. In *14th International Conference on Science and Techniques of Automatic Control and Computer Engineering (STA'2013), Sousse, Tunisia* (pp. 468–473).
Romdhane, H., Dehri, K., & Nouri, A. S. (2014a). Discrete second order sliding mode control for nondecouplable multivariable systems via input-output model. In *22nd Mediterranean Conference on Control and Automation(MED), University of Palermo, Italy*.
Romdhane, H., Dehri, K., & Nouri, A. S. (2014b). Stability analysis of discrete input output second order sliding mode control. *International Journal of Modelling, Identification and Control, 22*(2), 159–169.
Romdhane, H., Dehri, K., & Nouri, A. S. (2015). Discrete input-output second order sliding mode control using LMIs approach. In *12th International Multi-Conference on Systems, Signals and Devices (SSD'15), Mahdia, Tunisia*.
Sandberg, I. W. (1965). On the boundedness of solutions of nonlinear integral equations. *Bell System Technical Journal, 44*, 439–453.
Silva, J. M. A.-D., & Edwards, C. (2009). Compensator-based sliding mode output feedback simultaneous stabilisation controller design via LMIs for uncertain systems. In *48th IEEE Conference on Decision and Control and 28th Chinese Control Conference, Shanghai, P.R. China*.
Silva, J. M. A.-D., Spurgeon, S. K., & Edwards, C. (2008). Sliding mode output feedback controller design using linear matrix inequalities with application to aircraft systems. In *Proceedings of the 10th International Workshop on VSS*.

Silva, J. M. A.-D., Edwards, C., & Spurgeon, S. K. (2009). Linear matrix inequility based dynamic output feedback sliding mode control for uncertain plants. In *American Control Conference, Hyatt Regency Riverfront, St. Louis, MO, USA*.
Utkin, V. I. (1992). *Sliding modes in control and optimization*. New York: Springer.
Vecchio, C. (2008). Sliding mode control: Theoretical developments and applications to uncertain mechanical systems. Ph.D. thesis, University degli Studi of Pavia.
Wang, H., & Yu, C. (2013). Design of congestion control scheme for uncertain discrete network systems. *International Journal of Computers, Communication and Control*, *8*(6), 901–906.
Yadav, N. K., & Singh, R. K. (2012). Robust discrete-time nonlinear sliding mode controller with plant uncertainties. *International Journal of Engineering, Science and Technology*, *4*(1), 38–45.

Chapter 5
Robust Flight Control of an Underactuated Quadrotor via Sliding Modes

Chih-Chen Yih

Abstract An underactuated quadrotor has four actuators and six degrees of freedom to be controlled. Nevertheless, by deliberately controlling the velocities of the four propellers, the underactuated quadrotor can track the desired position trajectory and maintain the correct attitude during flight. To improve the robustness and performance of the underactuated quadrotor system, we propose two sliding mode control to deal with the parametric variations and the external disturbances. We first establish the quadrotor model in terms of the translational and rotational dynamics along with the disturbances and the model uncertainties. By specifying the desired pitch and roll angle as the virtual control, we then design dual sliding modes: one for the translational and the other for the rotational dynamics. Our Lyapunov-based stability analysis shows that the proposed control schemes can guarantee the asymptotical stability of the error dynamics for the position and attitude control of the underactuated quadrotor. Numerical simulations also indicate that the sliding mode control can effectively follow the desired trajectory and maintain the proper attitude in the presence of parametric variations and external disturbances.

Keywords Quadrotor · Sliding mode · Underactuated systems · Robust control

Nomenclature

B	$\equiv$ quadrotor body
P_i	$\equiv$ propeller i
F_w	$\equiv$ inertia world frame
F_B	$\equiv$ body frame
F_{P_i}	$\equiv$ ith propeller frame
R_T	$\equiv$ transform matrix from body angular rates to Euler ones
p	$\equiv$ position of B in F_ω

C.-C. Yih (✉)
Department of Mechanical and Automation Engineering, I-Shou University, 84001 Kaohsiung, Taiwan
e-mail: ccyih@isu.edu.tw

N. Derbel et al. (eds.), *Applications of Sliding Mode Control*,
Studies in Systems, Decision and Control 79, DOI 10.1007/978-981-10-2374-3_5

q	$\equiv$ Euler angle of B in F_ω
$^{\omega}R_B$	$\equiv$ rotation matrix from F_B to F_ω
$^{B}R_{P_i}$	$\equiv$ rotation matrix from F_{P_i} to F_B
ω_i	$\equiv$ ith propeller spinning velocity about Z_{P_i}
ω_{P_i}	$\equiv$ angular velocity in the ith propeller frame
T_{P_i}	$\equiv$ force in the ith propeller frame
ω_B	$\equiv$ angular velocity of B in F_B
τ_B	$\equiv$ torque in F_B
τ_{P_i}	$\equiv$ torque in F_{P_i}
τ_{di}	$\equiv$ ith propeller air drag torque about Z_{P_i}
T_i	$\equiv$ ith propeller thrust along Z_{P_i}
τ_{ω_i}	$\equiv$ motor torque along Z_{P_i}
m	$\equiv$ total quadrotor mass
I_{P_i}	$\equiv$ inertia of the ith propeller P_i
I_B	$\equiv$ inertia of the quadrotor body B
k_f	$\equiv$ propeller thrust coefficient
k_m	$\equiv$ propeller drag coefficient
L	$\equiv$ distance of F_{P_i} to F_B
g	$\equiv$ gravity constant

5.1 Introduction

In recent years, small quadrotors have attracted much attention in the field of research on unmanned aerial vehicles (UAV). Due to the simple mechanical structure, researchers have been using microprocessors and micro-electro-mechanical sensors to stabilize the attitude and the altitude of the quadrotor. With the aid of GPS or the lightweight camera, quadrotor UAVs can easily hover and delicately fly in indoor or outdoor environment. The control of the quadrotor is accomplished by varying the pitch or speed of one pair or two pairs of symmetrically positioned propellers to generate the lifting force or the turning torque (Mahony et al. 2012; Mellinger et al. 2012; Xilun and Yushu 2013; Hyon et al. 2012; Hoffmann et al. 2007; Nonami et al. 2010).

An underactuated quadrotor has four actuators and six degrees of freedom to be controlled. By deliberately controlling the velocities of the four propellers, the underactuated quadrotor can track the desired position trajectory and maintain the correct attitude during flight. Nevertheless, the robust flight control is essential for a quadrotor subject to uncertainties such as varying payloads, external disturbances, model inaccuracies, and actuator dynamics.

Researchers have been proposing several approaches to the problem of quadrotor control. These approaches can be roughly categorized according to their different control design: PID control (Alexis et al. 2011a; Bouabdallah et al. 2004), feedback linearization (Mokhtari et al. 2006b; Lee et al. 2009; Das et al. 2009b), optimal control (Satici et al. 2013; Raffo et al. 2010; Alexis et al. 2012), back-stepping (Das et al. 2009a; Madani and Benallegue 2006; Bouabdallah and Siegwart 2005; Young–Cheol and Hyo–Sung 2015), sliding mode control (Xu and Özgüner 2006; Derafa et al. 2012; Besnard et al. 2012), robust control (Cabecinhas et al. 2015; Alexis et al. 2011b; Islam et al. 2015), and fuzzy/neural control (Dierks and Jagannathan 2010; Byung et al. 2013; Rabhi et al. 2011). Due to its capability to deal with uncertainties, the sliding mode control has become one of the most researched nonlinear control methods (Utkin 1992; Shtessel et al. 2014). For example, Xu and Özgüner (2006) presented a sliding mode control based on a simplified dynamic model for the stabilization of a quadrotor. The quadrotor system is divided into a fully-actuated subsystem and an underactuated subsystem. With a specific coordinate transform, one can arrange the underactuated subsystem as a cascaded subsystem, and design a sliding mode control accordingly. They also use a continuous approximation to reduce chattering associated with the discontinuous sign function.

To alleviate the undesired chattering and to maintain the robust performance of the first-order sliding mode control, researchers have proposed second order sliding mode control schemes such as the super twisting control algorithm (Levant 1993, 2001, 2007; Utkin 2003). This algorithm can improve robustness to modeling inaccuracies and external disturbances, and avoid the chattering induced by the conventional sliding mode control. Research works in Moreno and Osorio (2012), Pico et al. (2013), Utkin (2003) have shown, by Lyapunov stability analysis, the stability and finite-time convergence of the super twist control algorithm for single-variable systems.

Derafa et al. (2012) used the super twisting algorithm for the attitude tracking control of a quadrotor UAV. Their theoretical and experimental work demonstrated the effectiveness and robustness of the proposed attitude control under bounded external disturbances. Several control schemes combined with sliding mode disturbance observers (Mokhtari et al. 2006a; Beballegue et al. 2008) have been proposed for quadrotor control. For instance, Besnard et al. (2012) designed an observer-based sliding mode control to deal with model uncertainties and wind disturbances. Their proposed sliding mode observer is based on the super twist algorithm, and a low-pass filter is used to smooth the discontinuous term in the sliding mode control. In addition, the authors considered the feedback loop for the speed control of the four rotors.

This chapter exploits the structure of the quadrotor dynamics and uses the concept of virtual control to recast the underactuated as a fully-actuated one and design the sliding mode control accordingly. We propose two sliding mode control to improve the robustness and performance of the underactuated quadrotor system in the presence of the parametric variations and the external disturbances. We first establish the translational and rotational dynamics of the quadrotor. Using the desired pitch and roll angle as the virtual control, we propose the sliding mode attitude and position control scheme. Our Lyapunov-based stability analysis shows that the proposed control schemes can guarantee the asymptotical stability of the error dynamics for the position and attitude control of the quadrotor. From numerical simulation results, we can conclude that the proposed sliding mode control can effectively track the desired trajectory and retain the proper attitude under parametric variations and external disturbances.

This chapter is organized as follows: Sect. 5.2 discusses the quadrotor dynamics and properties. Section 5.3 presents the proposed sliding mode and the stability analysis. In Sect. 5.4, the proposed sliding mode control scheme is applied to a quadrotor for numerical simulation. Section 5.5 presents some conclusions.

5.2 Dynamic Model

In this section, we formulate the dynamic model from Newton-Euler equations. First, we derive the nonlinear model of the quadrotor including rotational and translational dynamics. Then, we present some properties associated with the rotational dynamics to facilitate the stability analysis in the later section.

5.2.1 Dynamics

The rotational dynamics of the quadrotor shown in Fig. 5.1 is:

$$\tau_B = I_B \dot{\omega}_B + \omega_B \times I_B \omega_B + \sum_{i=1}^{4} {}^{B}R_{P_i} \tau_{Pi} \tag{5.1}$$

where:

$$\tau_{Pi} = I_{Pi} \dot{\omega}_{Pi} + \omega_{Pi} \times I_{Pi} \omega_{Pi} - \tau_{di} \tag{5.2}$$

$$\tau_{di} = \left[0\,0\ \ -k_m \omega_i^2 \text{sign}(\omega_i)\right]^T \tag{5.3}$$

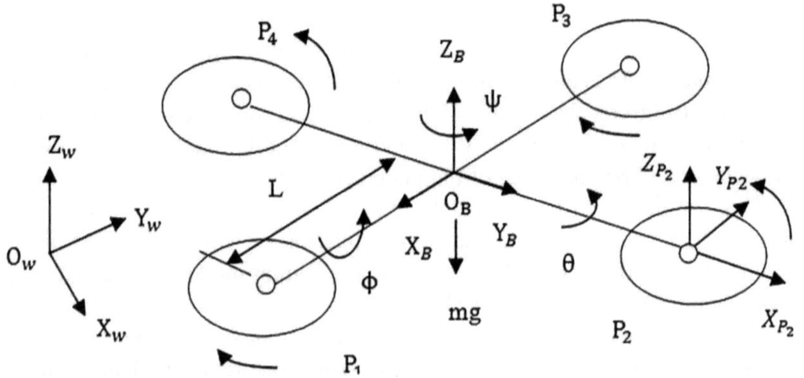

Fig. 5.1 The model of the quadrotor

We define:

$$\operatorname{sign}(x) = \begin{vmatrix} 1 & \text{if } x > 0 \\ -1 & \text{if } x < 0 \end{vmatrix}$$

$\omega_1 < 0, \omega_3 < 0$ (clockwise along the positive Z_{Pi} axis)
$\omega_2 > 0, \omega_4 > 0$ (counter-clockwise along the positive Z_{Pi} axis)

The angular velocity in the propeller frame is

$$\omega_{Pi} = {}^{B}R_{P_i}\omega_B + [0\;0\;\omega_i]^T \tag{5.4}$$

and the toque in body frame is:

$$\tau_B = \sum_{i=1}^{4} {}^{B}O_{P_i} \times {}^{B}R_{P_i} T_{P_i} \tag{5.5}$$

and the force in the propeller frame is:

$$T_{P_i} = \left[0\;0\;k_f\omega_i^2\right]^T \tag{5.6}$$

The transformation between body angular rates to the Euler rates is:

$$\dot{q} = R_T\omega_B \tag{5.7}$$

Taking derivative of (5.7) and ignoring I_{Pi} in (5.2), we have:

$$\ddot{q} = \dot{R}_T R_T^{-1}\dot{q} + R_T I_B^{-1}\left(\tau_B - \omega_B \times I_B\omega_B + \sum_{i=1}^{4} {}^{B}R_{P_i}\tau_{di}\right) \tag{5.8}$$

We define:

$$\tau = \sum_{i=1}^{4} {}^{B}O_{P_i} \times {}^{B}R_{P_i} T_{P_i} + \sum_{i=1}^{4} {}^{B}R_{P_i}\tau_{d_i} \tag{5.9}$$

and

$$\Psi = R_T^{-1} \tag{5.10}$$

It follows from (5.7) and (5.8) that:

$$\omega_B = \Psi\dot{q} \tag{5.11}$$

$$\begin{aligned}\Psi^T I_B \Psi\ddot{q} &= -\Psi^T I_B\dot{\Psi}\dot{q} - \Psi^T\omega_B \times I_B\omega_B + \Psi^T\tau \\ &= -\left[\Psi^T I_B\dot{\Psi} + \Psi^T(\Psi\dot{q} \times I_B\Psi)\right]\dot{q} + \Psi^T\tau\end{aligned} \tag{5.12}$$

Adding the disturbance torque, we obtain:

$$M(q)\ddot{q} + C(q,\dot{q})\dot{q} = \Psi^T(\tau + \tau_d) \tag{5.13}$$

where $q = [\phi\,\theta\,\psi]^T \in \mathbb{R}^3$ is the attitude vector of the roll, the pitch, and the yaw angle. $M(q) \in R^{3\times 3}$ is the symmetric and positive definite inertia matrix, $C(q,\dot{q})\dot{q} \in \mathbb{R}^3$ is the vector of centrifugal and Coriolis forces and $\tau = [\tau_\phi\ \tau_\theta\ \tau_\psi] \in \mathbb{R}^3$ is the vector of torques. Ψ is the transformation matrix and τ_d is the disturbance torque. In (5.13), we have:

$$M(q) = \Psi^T I_B \Psi \tag{5.14}$$

and:

$$C(q,\dot{q}) = [\Psi^T I_B \dot{\Psi} + \Psi^T(\Psi\dot{q} \times I_B\Psi)] \tag{5.15}$$

The matrix of $\Psi(q)$ is:

$$\Psi(q) = \begin{bmatrix} 1 & 0 & -s_\theta \\ 0 & c_\phi & c_\theta s_\phi \\ 0 & -s_\phi & c_\theta c_\phi \end{bmatrix} \tag{5.16}$$

Denoting $s_{q_i} = \sin q_i$ and $c_{q_i} = \cos q_i$, we have the inertia matrix M:

$$\begin{aligned}
&M_{11} = I_x,\ M_{12} = 0,\ M_{13} = -I_x s_\theta \\
&M_{22} = I_y c_\phi^2 + I_z s_\phi^2,\ M_{23} = c_\theta c_\phi s_\phi (I_y - I_z) \\
&M_{33} = I_x s_\theta^2 + I_y c_\theta^2 s_\phi^2 + I_z c_\theta^2 c_\phi^2
\end{aligned}$$

The matrix of $C(q,\dot{q})$ can be expressed as:

$$\begin{aligned}
&C_{11} = 0,\ C_{12} = -I_x\dot{\psi}c_\theta + (I_y - I_z)(\dot{\theta}s_\phi c_\phi + \dot{\psi}c_\theta s_\phi^2 - \dot{\psi}c_\theta c_\phi^2) \\
&C_{13} = -I_y\dot{\psi}c_\theta^2 s_\phi c_\phi + I_z\dot{\psi}c_\theta^2 s_\phi c_\phi \\
&C_{21} = I_x\dot{\psi}c_\theta - (I_y - I_z)(\dot{\theta}s_\phi c_\phi + \dot{\psi}c_\vartheta s_\phi^2 - \dot{\psi}c_\theta c_\phi^2) \\
&C_{22} = (I_z - I_y)\dot{\phi}s_\phi c_\phi \\
&C_{23} = -I_x\dot{\psi}s_\theta c_\theta + I_y\dot{\psi}s_\theta c_\theta s_\phi^2 + I_z\dot{\psi}s_\theta c_\theta c_\phi^2 n \\
&C_{31} = -I_x\dot{\theta}c_\theta + (I_y - I_z)\dot{\psi}c_\theta^2 s_\phi c_\phi \\
&C_{32} = I_x\dot{\psi}s_\theta c_\theta + (I_z - I_y)[\dot{\theta}s_\theta s_\phi c_\phi + \dot{\phi}c_\theta(s_\phi^2 - c_\phi^2)] - (I_y s_\phi^2 + I_z c_\phi^2)s_\theta c_\theta \\
&C_{33} = (I_x - I_y s_\phi^2 - I_z c_\phi^2)\dot{\theta}s_\theta c_\theta + (I_y - I_z)\dot{\phi}c_\theta^2 s_\phi c_\phi
\end{aligned}$$

Now, consider the translational dynamics of the quadrotor:

$$m\ddot{p} = [0\,0\,-mg]^T + {}^W R_B \sum_{i=1}^{4} {}^B R_{P_i} T_{P_i} \tag{5.17}$$

$${}^B R_{P_i} = R_Z(\beta_i) \tag{5.18}$$

with:

$$R_Z(\beta_i) = \begin{bmatrix} \cos\beta_i & -\sin\beta_i & 0 \\ \sin\beta_i & \cos\beta_i & 0 \\ 0 & 0 & 1 \end{bmatrix}, \quad \beta_i = (i-1)\frac{\pi}{2}, \quad i = 1, 2, 3, 4.$$

Considering the disturbance force, we have the translational dynamics:

$$m\ddot{p} = \begin{bmatrix} c_\phi s_\theta c_\psi + s_\phi s_\psi \\ c_\phi s_\theta s_\psi - s_\phi c_\psi \\ c_\phi c_\theta \end{bmatrix} f - \begin{bmatrix} 0 \\ 0 \\ mg \end{bmatrix} + u_d \tag{5.19}$$

where m is the mass of the quadrotor and $p = [x \ y \ z]^T \in \mathbb{R}^3$. $f \in \mathbb{R}$ is the lifting force, g is the gravity constant and $u_d \in \mathbb{R}^3$ is the disturbance force.

The vector of the force and the torque in (5.13) and (5.19) can be expressed by:

$$\begin{bmatrix} f \\ \tau_\phi \\ \tau_\theta \\ \tau_\psi \end{bmatrix} = \begin{bmatrix} k_f & k_f & k_f & k_f \\ 0 & Lk_f & 0 & -Lk_f \\ -Lk_f & 0 & Lk_f & 0 \\ k_m & -k_m & k_m & -k_m \end{bmatrix} \begin{bmatrix} \omega_1^2 \\ \omega_2^2 \\ \omega_3^2 \\ \omega_4^2 \end{bmatrix} \tag{5.20}$$

where ω_i is the ith propeller spinning velocity.

5.2.2 Properties

We can state several fundamental properties of the attitude dynamic equation of motion. Let us use $\lambda_M(A)$ and $\lambda_m(A)$ for the largest and smallest eigenvalue of a matrix A. We denote the Euclidean norm for an $n \times 1$ vector x by $||x|| = \sqrt{x^T x}$.
Property 2.1: The inertia matrix is symmetric, positive definite, and is bounded by:

$$0 < \lambda_m(M) < ||M(q)|| < \lambda_M(M)$$

Property 2.2: The matrix $\dot{M}(q) - 2C(q, \dot{q})$ is skew symmetric.

5.3 Sliding Mode Flight Control

A sliding mode control refines the dynamics of a nonlinear system by applying a discontinuous control signal to force the system to slide along an intersection of the system's nominal operation. In accordance to the current location in the state space, a sliding mode control inherently possesses a variable structure because it switches from one continuous structure to another. In this section, we first present a conventional sliding mode control that exploits the structure of the rotational dynamics and a chattering suppression scheme to reduce the switching gain. Then we proposed a second order sliding mode control for the translational dynamics to avoid the chattering problem.

5.3.1 Sliding Mode Attitude Control

Denoting $\widehat{M}$ and $\widehat{C}$ as the nominal matrix of M and C, we define $\widetilde{M} = M - \widehat{M}$ and $\widetilde{C} = C - \widehat{C}$. We can express the rotational dynamics as follows:

$$\widehat{M}(q)\ddot{q} + \widehat{C}(q, \dot{q})\dot{q} = \Psi^T(\tau + \tau_d) + h_1(q, \dot{q}, \ddot{q}) \tag{5.21}$$

with:

$$h_1(q, \dot{q}, \ddot{q}) = -\widetilde{M}(q)\ddot{q} - \widetilde{C}(q, \dot{q})\dot{q}$$

Define

$$\dot{q}_r = \dot{q}_d - \Lambda_1(q - q_d) \tag{5.22}$$

and:

$$s_1 = \dot{q} - \dot{q}_r = \dot{q} - \dot{q}_d + \Lambda_1(q - q_d) \tag{5.23}$$

where Λ_1 is a positive and diagonal matrix and $\dot{q}_d$ is the desired velocity. Taking the derivative of s_1, we have

$$\dot{s}_1 = \ddot{q} - \ddot{q}_r = \ddot{q} - \ddot{q}_d + \Lambda_1(\dot{q} - \dot{q}_d) \tag{5.24}$$

Setting $\dot{s}_1$ to be zero, we can obtain the equivalent control for sliding mode motion. Define:

$$s_1 = [s_{11}\ s_{12}\ s_{13}]^T \tag{5.25}$$

$$\text{sign}(s_1) = [\text{sign}(s_{11})\ \text{sign}(s_{12})\ \text{sign}(s_{13})]^T \tag{5.26}$$

Now, we propose the control:

$$\tau = \Psi^{-T}\left[\widehat{M}\ddot{q}_r + \widehat{C}\dot{q}_r - K_1\text{sign}(s_1) - K_3 s_1 - K_4\widetilde{q}\right] \tag{5.27}$$

where K_1, K_3 and K_4 are diagonal matrices.

Define the Lyapunov function:

$$V_1 = \frac{1}{2}s_1^T M(q) s_1 + \frac{1}{2}\widetilde{q}^T K_4 \widetilde{q} \tag{5.28}$$

The derivative of the Lyapunov function is:

$$\begin{aligned}\dot{V}_1 &= s_1^T M(q)\dot{s}_1 + \frac{1}{2}s_1^T \dot{M}(q)s_1 + \tilde{q}^T K_4\dot{\tilde{q}} \\ &= -s_1^T \widetilde{M}\ddot{q}_r - s_1^T \widetilde{C}\dot{q}_r - s_1^T K_1 \mathrm{sign}(s_1) - s_1^T K_3 s_1 - \tilde{q}^T \Lambda_1 K_4 \tilde{q} \end{aligned} \quad (5.29)$$

If the switching gain satisfies the following condition:

$$\lambda_m(K_1) \geq \left(||\widetilde{M}||\,||\ddot{q}_r|| + ||\widetilde{C}||\,||\dot{q}_r|| + ||\tau_d||\right) + \gamma_1 \quad (5.30)$$

from (5.29), we have:

$$\dot{V}_1 \leq -\gamma_1 s_1^T \mathrm{sign}(s_1) - s_1^T K_3 s_1 - \tilde{q}^T K_4 \tilde{q} < 0 \text{ (if } s_1 \neq 0) \quad (5.31)$$

where γ_1 is a positive constant.

As a result, the controller carries the system towards the sliding surface $s_1 = 0$. To avoid the system chattering that may excite unmodeled dynamics, in practice, one can use the continuous term s_1 with a boundary layer (Slotine 1991) to replace the discontinuous term $\mathrm{sign}(s_1)$. Recently developed higher-order sliding mode (Bartolini et al. 2003; Emelyanov et al. 1996; Bartolini et al. 1998) and adaptive sliding mode (Utkin 2003; Plestan et al. 2010; Utkin and Poznyac 2013) can also be employed to suppress the chattering phenomenon. In this chapter, we use the gain-adaptation method developed by Lee and Utkin (2007) to decrease the switching gain that is a function of the equivalent control.

For example, $K_1 = \mathrm{diag}([k_1\ k_2\ k_3])$ is replaced by the adaptation law:

$$k_i(t) = k_{ci}|\eta_i| + k_{mi} \quad (5.32)$$

where $k_{ci} > 0$, $k_{mi} > 0$ and η_i is obtained by filtering the $\mathrm{sign}(s_{1i})$ using a low-pass filter:

$$\zeta_i\dot{\eta}_i + \eta_i = \mathrm{sign}(s_{1i}),\ \ \eta_i(0) = 0 \quad (5.33)$$

where ζ_i is a positive constant.

5.3.2 Sliding Mode Position Control

The transitional dynamics can be expressed by

$$\ddot{p} = \frac{1}{m}u + \mu \quad (5.34)$$

where the virtual control u is:

$$u = \begin{bmatrix} u_x \\ u_y \\ u_z \end{bmatrix} = \begin{bmatrix} c_{\phi d} s_{\theta d} c_{\psi d} + s_{\phi d} s_{\psi d} \\ c_{\phi d} s_{\theta d} s_{\psi d} + s_{\phi d} c_{\psi d} \\ c_{\psi d} c_{\theta d} \end{bmatrix} f - \begin{bmatrix} 0 \\ 0 \\ mg \end{bmatrix} \tag{5.35}$$

and the uncertain term μ is:

$$\mu = \begin{bmatrix} c_{\phi} s_{\theta} c_{\psi} + s_{\phi} s_{\psi} \\ c_{\phi} s_{\theta} s_{\psi} + s_{\phi} c_{\psi} \\ c_{\psi} c_{\theta} \end{bmatrix} \frac{f}{m} - \begin{bmatrix} c_{\phi d} s_{\theta d} c_{\psi d} + s_{\phi d} s_{\psi d} \\ c_{\phi d} s_{\theta d} s_{\psi d} + s_{\phi d} c_{\psi d} \\ c_{\psi d} c_{\theta d} \end{bmatrix} \frac{f}{m} + \frac{u_d}{m} \tag{5.36}$$

The nominal transitional dynamics can be expressed as

$$\ddot{p} = \frac{1}{\widehat{m}} u + \widehat{\mu} \tag{5.37}$$

where $\widehat{m}$ denotes the nominal mass and $\widehat{\mu} = 0$ for the nominal model.

Define the sliding surface $s_2 \in \mathbb{R}^3$:

$$s_2 = \dot{p} - \dot{p}_d + \Lambda_2(p - p_d) \tag{5.38}$$

where Λ_2 is a positive and diagonal matrix and $\dot{p}_d$ is the desired velocity.

Taking the derivative of s_2 and using (5.37), we obtain:

$$\dot{s}_2 = \frac{1}{\widehat{m}} u - \ddot{p}_d + \Lambda_2(\dot{p} - \dot{p}_d) \tag{5.39}$$

The second derivative of s_2 is:

$$\ddot{s}_2 = \frac{1}{\widehat{m}} \dot{u} - \dddot{p}_d + \Lambda_2[\dot{s}_2 - \Lambda_2(\dot{p} - \dot{p}_d)] \tag{5.40}$$

Define the Lyapunov function:

$$V_2 = \frac{1}{2} s_2^T \Lambda_3 s_2 + \frac{1}{2} \dot{s}_2^T \dot{s}_2 + \alpha \mathrm{sign} s_2 \tag{5.41}$$

where: $\alpha = [\alpha_1 \ \alpha_2 \ \alpha_3]$, $\alpha_i > 0$ for $i = 1, 2, 3$.

The derivative of the Lyapunov function is:

$$\dot{V}_2 = \dot{s}_2^T \left(\ddot{s}_2 + \Lambda_3 s_2 + \Lambda \mathrm{sign}(s_2) \right) \tag{5.42}$$

where Λ_4 is a diagonal matrix with diagonal elements $[\alpha_1 \ \alpha_2 \ \alpha_3]$. It follows that if we choose:

$$\ddot{s}_2 + \Lambda_3 s_2 + \Lambda_4 \mathrm{sign}(s_2) = -K_2 \dot{s}_2 \tag{5.43}$$

then:

$$\dot{V}_2 = -\dot{s}_2^T K_2 \dot{s}_2 \tag{5.44}$$

Using (5.40) and (5.43), we have:

$$\dot{u} = \widehat{m}\Big(- (K_2 + \Lambda_2)\dot{s}_2 - \Lambda_3 s_2 - \Lambda_4 sign(s_2) + \dddot{p}_d + \Lambda_2^2(\dot{p} - \dot{p}_d)\Big) \tag{5.45}$$

Substituting s_2 into (5.44) yields the nominal virtual control:

$$\begin{aligned}\dot{u} = -(K_2 + \Lambda_2)u + \widehat{m}\Big[&(K_2 + \Lambda_2)\ddot{p}_d + \dddot{p}_d - K_2\Lambda_2(\dot{p} - \dot{p}_d)\\ &-\Lambda_3 s_2 - \Lambda_4 \mathrm{sign}(s_2)\Big]\end{aligned} \tag{5.46}$$

It follows from (5.43) that $\dot{s}_2 \longrightarrow 0$, $\ddot{s}_2 \longrightarrow 0$. From (42), we have

$$\Lambda_3 s_2 = -\Lambda_4 \mathrm{sign}(s_2) \tag{5.47}$$

which ensures that:

$$s_2 = 0 \tag{5.48}$$

It can be shown through the robustness analysis that the nominal control proposed in (5.46) can deal with bounded uncertain terms in (5.34), given the switching gain Λ_4 is large enough. The criteria to select Λ_4 can be determined analytically through the robustness analysis. Nevertheless, notice that the analysis only provides a sufficient condition; therefore, in practice, one can always determine this switching gain numerically.

The real control input can be computed as follows:

$$f = \sqrt{u_x^2 + u_y^2 + (u_z + mg)^2} \tag{5.49}$$

The desired roll and pitch angle are:

$$\phi_d = \sin^{-1} \frac{u_x S_{\psi d} - u_y C_{\psi d}}{\sqrt{u_x^2 + u_y^2 + (u_z + mg)^2}} \tag{5.50}$$

$$\theta_d = \tan^{-1} \frac{u_x C_{\psi d} + u_y S_{\psi d}}{u_z + mg} \tag{5.51}$$

Notice that the maximum desired roll and pitch angles ϕ_d and θ_d are limited for a quadrotor. In the extreme case of $u_x = 0$, $u_y = 0$ and $u_z = -mg$, we can see that the force f is zero from (5.49) and therefore the desired roll and pitch angles in (5.50) and (5.51) are undefined and meaningless due to the lack of lifting force ($f = 0$).

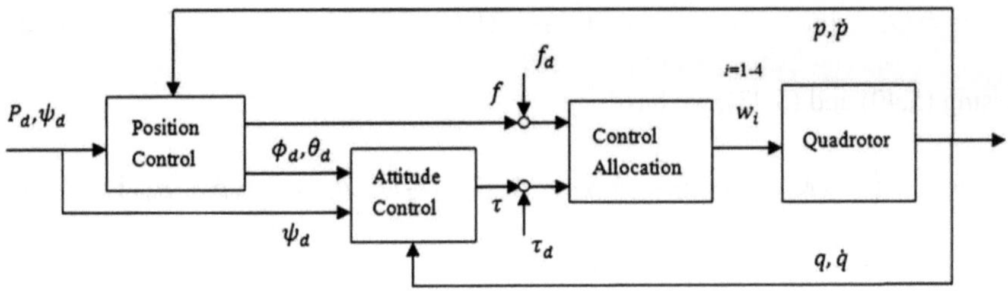

Fig. 5.2 The quadrotor control scheme

The vector of the squared speed for each rotor is:

$$\begin{bmatrix} \omega_1^2 \\ \omega_2^2 \\ \omega_3^2 \\ \omega_4^2 \end{bmatrix} = \begin{bmatrix} k_f & k_f & k_f & k_f \\ 0 & Lk_f & 0 & -Lk_f \\ -Lk_f & 0 & Lk_f & 0 \\ k_m & -k_m & k_m & -k_m \end{bmatrix}^{-1} \begin{bmatrix} f \\ \tau_\phi \\ \tau_\theta \\ \tau_\psi \end{bmatrix} \tag{5.52}$$

The attitude and the position control scheme of the quadrotor is shown in Fig. 5.2.

5.4 Numerical Simulation

In order to illustrate the design of the proposed scheme, we give an example of a quadrotor to numerically evaluate the robustness under the presence of the parameter change and the external disturbance.

5.4.1 Simulation Parameters

The following parameters of the quadrotor for the simulation are used: $m = 2\,\text{kg}$, $L = 0.275\,\text{m}$, $I_x = 0.025\,\text{kgm}^2$, $I_y = 0.025\,\text{kgm}^2$, $I_z = 0.040\,\text{kgm}^2$, $k_f = 9.3 \times 10^{-5}$ Ns^2, $k_m = 1.1 \times 10^{-6}\,\text{Ns}^2$, $g = 9.81\,\text{m/s}^2$.

We use the following initial conditions: $[x(0)\ y(0)\ z(0)] = [0\ 0\ 0]$ and the final desired positions (expressed in m) $[x_d(t_f)\ y_d(t_f)\ z_d(t_f)] = [20\ 10\ 5]$ where $t_f = 10\text{s}$.

The desired yaw angle is $\psi_d = 0$.

The desired position trajectories (expressed in m) are:

$$x_d(t) = 0.2t^3 - 0.03t^4 + 0.0012t^5$$
$$y_d(t) = 0.1t^3 - 0.015t^4 + 0.0006t^5$$
$$z_d(t) = 0.05t^3 - 0.0075t^4 + 0.0003t^5$$

Now, the design parameters are chosen as follows: $\Lambda_1 = 100I_2$, $\Lambda_2 = 100I_3$, $\Lambda_3 = 100I_3, \Lambda_4 = 1.0I_3, K_1 = 2I_3, K_2 = 10I_3, K_3 = 40I_3, K_4 = 100I_3, k_{ci} = 4I_3$, $k_{mi} = 0.1I_3$ $(i = 1, 2, 3)$.

5.4.2 *Simulation Results*

The simulation results of the proposed sliding mode attitude and position control are shown in Figs. 5.3, 5.4, 5.5, 5.6, 5.7 and 5.8. Figures 5.8 and 5.4 present the attitude and position trajectory of the quadrotor under the parametric variations (increasing the body weight to 125 %). The propelling forces generated by the four rotors of the quadrotor are shown in Fig. 5.5. From the simulation results in Fig. 5.3, 5.4 and 5.5, we can see that the proposed sliding mode control can successfully control the quadrotor from the initial position, via the desired trajectory, and to the final destination while maintaining the desired attitude.

As the stability analysis in the previous section demonstrates the robustness with respect to the parametric uncertainties and external disturbances, we now further evaluate the effects of the external disturbances on the quadrotor. We use the disturbance force and torque exerting on each attitude angle and each axis of movement ($2\sin 4t$ N and $0.1\sin 4t$ Nm) for the simulation. As shown in Figs. 5.6, 5.7 and 5.8, we can see that the disturbance has little effects on the sliding mode control due to its capability to reject disturbance and to pull the state variables back to the

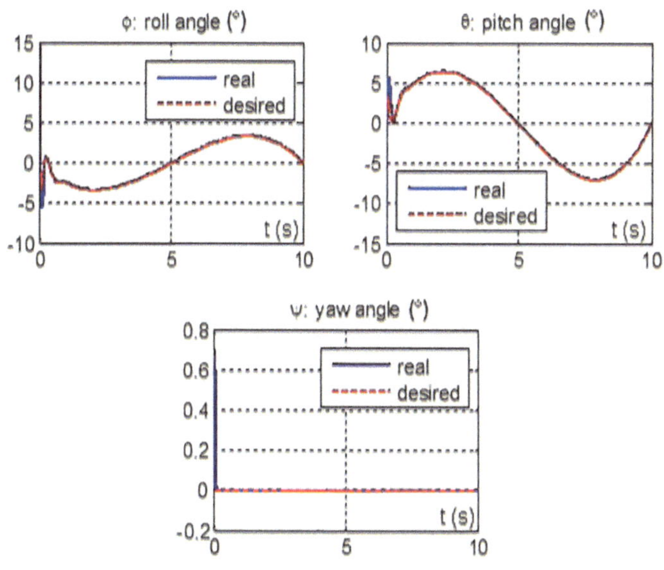

Fig. 5.3 The attitude trajectory under parametric variations

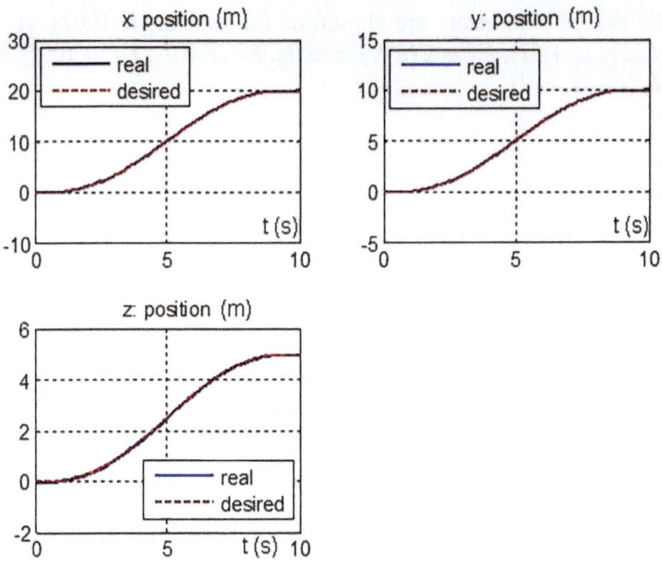

Fig. 5.4 The position trajectory under parametric variations

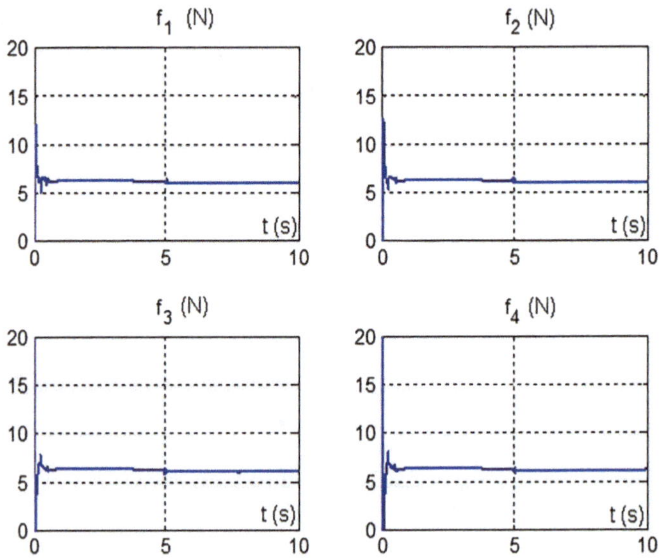

Fig. 5.5 The propelling force under parametric variations

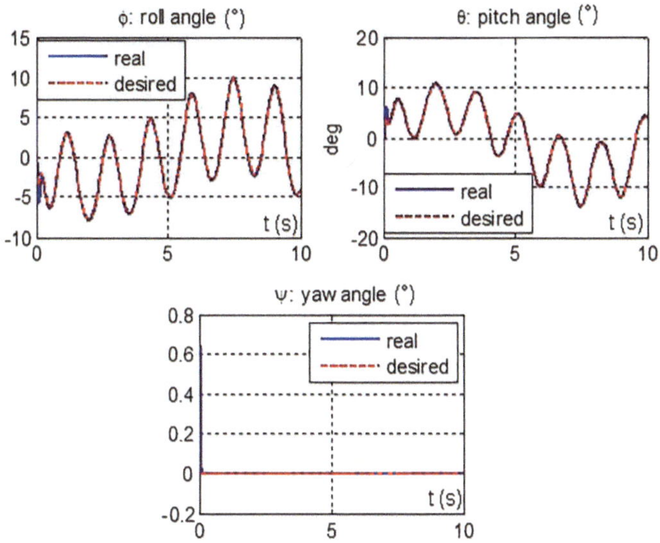

Fig. 5.6 The attitude trajectory under parametric variations and disturbances

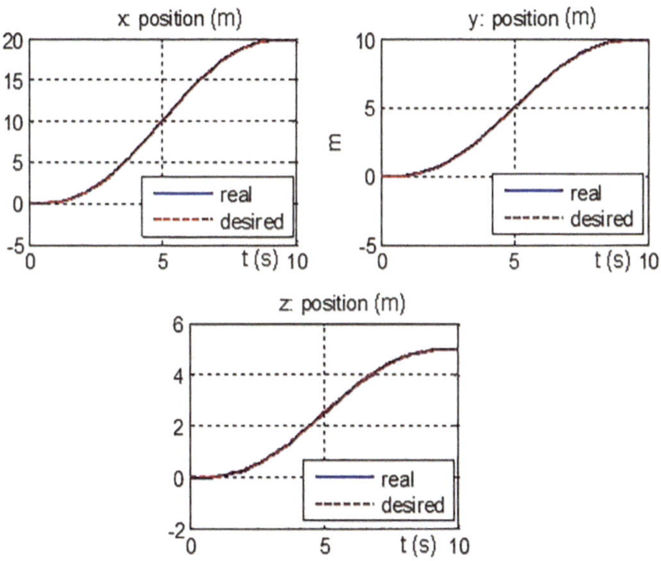

Fig. 5.7 The position trajectory under parametric variations and disturbances

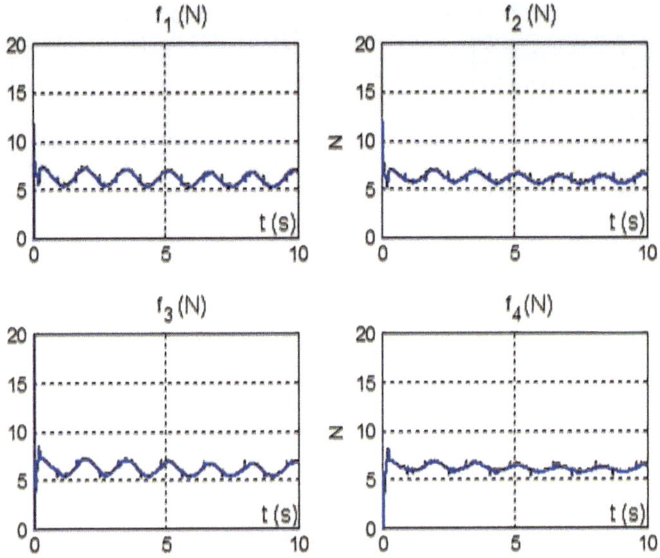

Fig. 5.8 The propelling force under parametric variations and disturbances

sliding surface. The improvement of robustness offered by sliding mode control is expected because sliding mode has nonlinear terms to counter the effects of varying parameters and bounded disturbances.

5.5 Conclusions

In this chapter, we propose two sliding mode control to deal with the parametric variations and the disturbances and to improve the robustness and performance of the quadrotor. Using Lyapunov-based stability analysis, we have shown that the proposed control schemes can guarantee the asymptotical stability of the error dynamics for the position and attitude control of the underactuated quadrotor.

From the simulation results, we demonstrate that conventional first-order sliding mode, coupled with a scheme to reduce the switching gain and to suppress the chattering, can successfully solve the control problem for an underactuated quadrotor. Moreover, we show the alternative control design of a second-order sliding mode via the Lyapunov stability analysis to get rid of the chattering. In the future, we will work on the experimental evaluation of the proposed sliding mode control and integrate the control schemes in the fault-tolerant flight control system.

References

Alexis, K., Nikolakopoulos, G., & Tzes, A. (2011a). Autonomous quadrotor position and attitude PID/PIDD control in GPS-denied environments. *The International Review of Automatic Control, 3*, 421–430.

Alexis, K., Nikolakopoulos, G., & Tzes, A. (2011b). Switching model predictive attitude control for a quadrotor helicopter subject to atmospheric disturbances. *Control Engineering Practice, 19*, 1195–1207.

Alexis, K., Nikolakopoulos, G., & Tzes, A. (2012). Model predictive quadrotor control: Attitude, altitude and position experimental studies. *IET Control Theory & Applications, 6*, 1812–1827.

Bartolini, G., Ferrara, A., & Usai, E. (1998). Chattering avoidance by second order sliding modes control. *IEEE Transactions on Automatic Control, 43*, 241–247.

Bartolini, G., Pisano, A., Punta, E., & Usai, E. (2003). A survey of applications of second–order sliding mode control to mechanical systems. *International Journal of Control, 76*, 875–892.

Beballegue, A., Mokhtari, A., & Fridman, L. (2008). High-order sliding-mode observer for a quadrotor UAV. *International Journal of Robust and Nonlinear Control, 18*, 427–440.

Besnard, L., Shtessel, Y. B., & Landrum, B. (2012). Quadrotor vehicle control via sliding mode controller driven by sliding mode disturbance observer. *Journal of the Franklin Institute, 349*, 658–684.

Bouabdallah, S., & Siegwart, R. (2005). Backstepping and sliding–mode techniques applied to an indoor micro quadrotor. *International Conference on Robotics and Automation, Proceedings, ICRA 2005* (pp. 2247–2252).

Bouabdallah, S., Noth, A., & Siegwart, R. (2004). PID vs LQ control techniques applied to an indoor micro quadrotor. *International Conference on Intelligent Robots and Systems, Proceedings, IROS, 2004*(3), 2451–2456.

Byung-Yoon, L., Hae-In, L., & Min-Jea, T. (2013). Analysis of adaptive control using on-line neural networks for a quadrotor UAV. *13th International Conference on in Control, Automation and Systems, Proceedings, ICCAS 2013* (pp. 1840–1844).

Cabecinhas, D., Cunha, R., & Silvestre, C. (2015). A globally stabilizing path following controller for rotorcraft with wind disturbance rejection. *IEEE Transactions on Control Systems Technology, 23*, 708–714.

Das, A., Lewis, F., & Subbarao, K. (2009a). Backstepping approach for controlling a quadrotor using lagrange form dynamics. *Journal of Intelligent and Robotic Systems, 56*, 127–151.

Das, A., Subbarao, K., & Lewis, F. (2009b). Dynamic inversion with zero-dynamics stabilisation for quadrotor control. *IET Control Theory & Applications, 3*, 303–314.

Derafa, L., Benallegue, A., & Fridman, L. (2012). Super twisting control algorithm for the attitude tracking of a four rotors UAV. *Journal of the Franklin Institute, 349*, 685–699.

Dierks, T., & Jagannathan, S. (2010). Output feedback control of a quadrotor UAV using neural networks. *IEEE Transactions on Neural Networks, 21*, 50–66.

Emelyanov, S., Korovin, S., & Levant, A. (1996). High–order sliding modes in control systems. *Computational Mathematics and Modeling, 7*, 294–318.

Hoffmann, G., Huang, H., Waslander, S., & Tomlin, C. (2007). Quadrotor helicopter flight dynamics and control: Theory and experiment. *Proceedings of the AIAA Guidance, Navigation, Control Conference* (pp. 1 –20).

Hyon, L., Jaemann, P., Daewon, L., & Kim, H. J. (2012). Build your own quadrotor: Open-source projects on unmanned aerial vehicles. *IEEE Robotics & Automation Magazine, 19*, 33–45.

Islam, S., Liu, P. X., & El Saddik, A. (2015). Robust control of four-rotor unmanned aerial vehicle with disturbance uncertainty. *IEEE Transactions on Industrial Electronics, 62*(1563–1571), 2015.

Lee, H., & Utkin, V. I. (2007). Chattering suppression methods in sliding mode control systems. *Annual Reviews in Control, 31*, 179–188.

Lee, D., Kim, H. J., & Sastry, S. (2009). Feedback linearization vs. adaptive sliding mode control for a quadrotor helicopter. *International Journal of Control, Automation and Systems, 7*, 419–428.

Levant, A. (1993). Sliding order and sliding accuracy in sliding mode control. *International Journal of Control*, *58*, 1247–1263.
Levant, A. (2001). Universal SISO sliding-mode controllers with finite-time convergence. *IEEE Transactions on Automatic Control*, *49*, 1447–1451.
Levant, A. (2007). Principles of 2-sliding mode design. *Automatica*, *43*, 576–586.
Madani, T., & Benallegue, A. (2006). *Backstepping Control for a Quadrotor Helicopter, International Conference on Intelligent Robots and Systems, Proceedings, IEEE/RSJ 2006* (pp. 3255–3260).
Mahony, R., Kumar, V., & Corke, P. (2012). Multirotor aerial vehicles: Modeling, estimation, and control of quadrotor. *IEEE Robotics & Automation Magazine*, *19*, 20–32.
Mellinger, D., Michael, N., & Kumar, V. (2012). Trajectory generation and control for precise aggressive maneuvers with quadrotors. *The International Journal of Robotics Research*, *31*, 664–674.
Moreno, A., & Osorio, M. (2012). Strict Lyapunov functions for the super-twisting algorithm. *IEEE Transactions on Automatic Control*, *57*, 1035–1040.
Mokhtari, A., Benallegue, A., & Orlov, Y. (2006a). Exact linearization and sliding mode observer for a quadrotor unmanned aerial vehicle. *International Journal of Robotics and Automation*, *21*, 39–49.
Mokhtari, A., M'sirdi, N. K., Meghriche, K., & Belaidi, A., (2006b). Feedback linearization and linear observer for a quadrotor unmanned aerial vehicle. *Advanced Robotics*, *20*, 71–91.
Nonami, K., Kendoul, F., Suzuki, S., Wang, W., & Nakazawa, D. (2010). *Autonomous flying robots*. Japan: Springer.
Pico, J., Pico-Marco, E., Vignoni, A., & De Battista, H. (2013). Stability preserving maps for finite-time convergence: Super-twisting sliding-mode algorithm. *Automatica*, *49*, 534–539.
Plestan, F., Shtessel, Y., Bregeault, V., & Poznyak, A. (2010). New methodologies for adaptive sliding mode control. *International Journal of Control*, *83*, 1907–1919.
Rabhi, A., Chadli, M., & Pegard, C. (2011). Robust fuzzy control for stabilization of a quadrotor. *15th International Conference on Advanced Robotics, Proceedings, ICAR 2011* (pp. 471–475).
Raffo, G. V., Ortega, M. G., & Rubio, F. R. (2010). An integral predictive/nonlinear control structure for a quadrotor helicopter. *Automatica*, *46*, 29–39.
Satici, A. C., Poonawala, H., & Spong, M. W. (2013). Robust Optimal Control of Quadrotor UAVs. *IEEE Access*, *1*, 79–93.
Shtessel, Y., Edwards, C., Fridman, L., & Levant, A. (2014). *Sliding mode control and observation*. New York: Springer.
Slotine, J. J. E., & Li, W. (1991). *Applied nonlinear control*. New Jersey: Prentice Hall.
Utkin, V. (1992). *Sliding modes in control and optimization*. Berlin: Springer.
Utkin, V. (2003). On convergence time and disturbance rejection of super-twisting control. *IEEE Transactions on Automatic Control*, *58*, 2013–2017.
Utkin, V. I., & Poznyak, A. S. (2013). Adaptive sliding mode control with application to super-twist algorithm: Equivalent control method. *Automatica*, *49*, 39–47.
Xilun, D., & Yushu, Y. (2013). Motion planning and stabilization control of a multipropeller multifunction aerial robot. *IEEE/ASME Transactions on Mechatronics*, *18*, 645–656.
Xu, R., & Özgüner, Ü. (2006). Sliding mode control of a quadrotor helicopter. *45th IEEE Conference on Decision and Control, Proceedings, CDC 2006* (pp. 4957–4962).
Young-Cheol, C., & Hyo-Sung, A. (2015). Nonlinear control of quadrotor for point tracking: Actual implementation and experimental tests. *IEEE/ASME Transactions on Mechatronics*, *20*, 1179–1192.

Chapter 6
Sliding Mode Control of an Inverted Pendulum

Olfa Jedda, Jalel Ghabi and Ali Douik

Abstract This paper presents solutions to attenuate the chattering phenomenon raised by the classic sliding mode control. The first solution consists in approximating the discontinuity in the control law, origin of chatter effect, by using a continuous function. Another solution is to use the second order sliding mode control. Subsequently, these different algorithms will be applied to an inverted pendulum in order to achieve dynamic output tracking. Simulation results are presented to illustrate the efficiency of these algorithms.

Keywords Sliding mode control · Chattering phenomenon · Saturation function · Twisting algorithm · Super-twisting algorithm · Inverted pendulum

6.1 Introduction

Since the late 1970s, sliding mode control (SMC) has attracted a significant interest from the control research community, and this is due to its insensitivity to model parametric uncertainties and external disturbances (Bandyopadhyay et al. 2009; Utkin 1977; Young et al. 1999). Indeed, the SMC design consists of two basic steps: selecting a stable sliding surface on the basis of control objectives and desired properties of the closed loop system, and synthesizing a discontinuous control law in such a way that the state trajectory of the system reaches the sliding surface in finite time and then remains there (Bandyopadhyay et al. 2009; Bregeaut 2010). Yet, in practice and in the presence of switching imperfections, the control cannot switch at a very

---

O. Jedda (✉) · J. Ghabi
National Engineering School of Monastir, University of Monastir, Monastir, Tunisia
e-mail: olfa_jedda@outlook.com

J. Ghabi
e-mail: jalel.ghabi@yahoo.fr

A. Douik
National Engineering School of Sousse, University of Sousse, Sousse, Tunisia
e-mail: Ali.douik@enim.rnu.tn

N. Derbel et al. (eds.), *Applications of Sliding Mode Control*,
Studies in Systems, Decision and Control 79, DOI 10.1007/978-981-10-2374-3_6

high frequency, and then the discontinuity in the control law yields to the so-called chattering phenomenon.

This phenomenon can actuate disregarded high-frequency dynamics, degrade system performances and even cause mechanical damages (Perruquetti and Barbot 2002). In this regard, several approaches have been proposed in the literature in order to overcome this problem such as the one proposed by Slotine and Sastry (1983), which consists in substituting the signum function, origin of the discontinuity in the control law, by a smooth function like saturation function (Hung et al. 1993; Perruquetti and Barbot 2002; Slotine 1984; Slotine and Li 1991). Actually, the state trajectory will evolve inside a thin boundary layer neighboring the switching surface and then the chatter effect will be attenuated close to this surface.

Another approach, introduced by Levant (2003), the higher-order sliding mode control (HOSMC), consists in constraining the state trajectory to reach in finite time the sliding set defined by:

$$S^r = \left\{x \in \mathbb{R}^n : s = \dot{s} = \ddot{s} = \ldots = s^{(r-1)} = 0\right\} \tag{6.1}$$

In addition to the chatter elimination, HOSMC ensures a better accuracy and resolves the problem of the restriction to a relative degree one, while preserving the main features of the standard sliding mode (Levant 2003; Perruquetti and Barbot 2002). In this study, only second-order sliding mode control (SOSMC) and specially twisting and super-twisting algorithms (Emelyanov et al. 1996; Levant 1993; Perruquetti and Barbot 2002; Shtessel et al. 2014) will be demonstrated for the control of the inverted pendulum as a nonlinear and unstable system.

In Sect. 6.2, we represent the first order sliding mode theory. Section 6.3 is devoted to second order sliding mode algorithms. In Sect. 6.4, Simulations results for both first and second order sliding mode algorithms applied to the inverted pendulum will be shown. Finally, concluding remarks will be given in Sect. 6.5.

6.2 First Order Sliding Mode Control

Well known for its high accuracy and robustness against parametric variations and external disturbances, the sliding mode control consists, by means of a discontinuous control, in constraining the system states to reach, in finite time, and then to evolve onto a sliding manifold defined by:

$$S = \left\{x \in \mathbb{R}^n : s = 0\right\} \tag{6.2}$$

Hence, the design of SMC involves two major steps: the first is to select a stable surface depending upon desired system dynamics, and the second is to synthesize a control law so that the system state trajectory evolves onto the chosen surface (Bregeaut 2010; Levant 2003).

Consider a single input nonlinear system as

$$\dot{x} = f(x) + b(x)u \tag{6.3}$$

where $x \in \mathbb{R}^n$ is the state vector and $u \in \mathbb{R}$ is the control input. The control law of the sliding mode controller is expressed as follows:

$$u = u_{eq} + u_d \tag{6.4}$$

with u_{eq} is the equivalent control and u_d is the discontinuous control. Using the equivalent control approach (DeCarlo et al. 1988; Gao and Hung 1992; Utkin 1992), u_{eq} is obtained by setting $\dot{s} = 0$. It ensures the maintain of the state trajectory onto the switching surface $s = 0$ during the sliding mode (Bandyopadhyay et al. 2009; Hung et al. 1993; Perruquetti and Barbot 2002).

For the system (6.3), the first time derivative of the sliding variable $s = s(x)$ is given by:

$$\dot{s} = \frac{\partial s}{\partial x}\left(f(x) + b(x)u\right) \tag{6.5}$$

Assuming that $\left(\frac{\partial s}{\partial x} b(x)\right)$ is invertible, the equivalent control law is expressed by:

$$u_{eq} = -\left(\frac{\partial s}{\partial x} b(x)\right)^{-1} \frac{\partial s}{\partial x} f(x) \tag{6.6}$$

Yet, a η-reachability condition defined in (Bandyopadhyay et al. 2009; Edwards and Spurgeon 1998; Perruquetti and Barbot 2002) as follows:

$$s\dot{s} \leq -\eta |s|, \quad \eta > 0 \tag{6.7}$$

must be met in order to ensure a finite time convergence to the sliding surface. Hence, the classic sliding mode control that satisfies the above condition is given by Riachy (2008):

$$u = -\left(\frac{\partial s}{\partial x} b(x)\right)^{-1} \left(\frac{\partial s}{\partial x} f(x) + k \,\mathrm{sign}(s)\right) \tag{6.8}$$

where k is a positive constant that verifies

$$s\dot{s} = s(-k \,\mathrm{sign}(s)) = -k|s| \leq -\eta |s| \Leftrightarrow k \geq \eta \tag{6.9}$$

and sign is the signum function defined by

$$\mathrm{sign}(s) = \left| \begin{array}{ll} 1 & \text{if } s > 0 \\ 0 & \text{if } s = 0 \\ -1 & \text{if } s < 0 \end{array} \right. \tag{6.10}$$

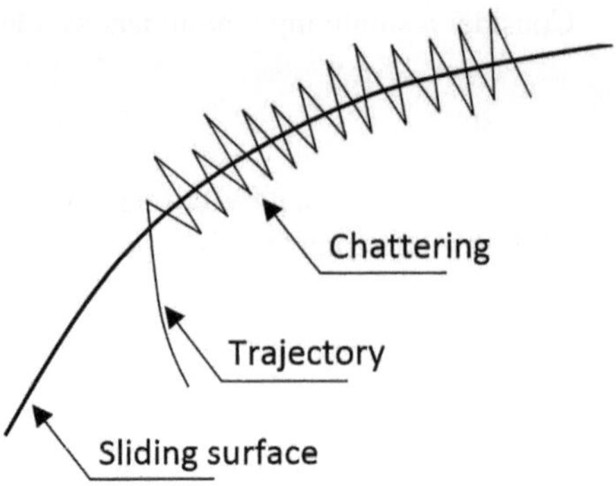

Fig. 6.1 Chattering phenomenon

Nevertheless, the discontinuity of the signum function in the vicinity of the sliding surface $s = 0$ involves an infinite frequency commutation of the control law which cannot exist in practice because of switching imperfections. This leads to the so-called chattering phenomenon (Fig. 6.1) which can degrade performances of the controlled system, excite disregarded high-frequency dynamics and even deteriorate the control member.

To attenuate the chatter effect, several approaches were proposed in literature. The main idea of the first approach is to substitute signum function by a continuous one such as saturation function defined by:

$$\mathrm{sat}(s, \varphi) = \left| \begin{array}{ll} \dfrac{s}{\varphi} & \text{if } \left|\dfrac{s}{\varphi}\right| \leq 1 \\ \mathrm{sign}(s) & \text{if } \left|\dfrac{s}{\varphi}\right| > 1 \end{array} \right. \tag{6.11}$$

The continuous function will ensure the convergence of system state trajectory to a thin boundary layer in the vicinity of the sliding surface. Consequently, the chatter effect will be reduced to the detriment of optimum accuracy and robustness of sliding mode.

The above restrictions were removed with the higher order sliding mode control which satisfies a finite time convergence of not only the sliding variable to zero, but also of a finite number of its time derivatives. In other words, for a relative degree $r > 1$, i.e. $\frac{\partial}{\partial u}s^{(i)} = 0\,(i = [|1, r-1|])$ and $\frac{\partial}{\partial u}s^{(r)} \neq 0$, it is necessary to ensure, for stability reasons, the following equalities:

$$s = \dot{s} = \ddot{s} = \cdots = s^{(r-1)} = 0 \tag{6.12}$$

Hence, HOSMC removes another drawback of classic sliding mode: the restriction to a relative degree one. However, this will require an additional information about the $(r-1)$ time derivatives of s. In what follows, only second order sliding mode algorithms will be treated.

6.3 Second Order Sliding Mode Control

The second order sliding mode controller consists on forcing the system state trajectories to evolve in finite time onto the second order sliding set defined by:

$$S^2 = \left\{x \in \mathbb{R}^n : s = \dot{s} = 0\right\} \tag{6.13}$$

Return to the system

$$\begin{aligned} \dot{x} &= f(x) + b(x)\,u \\ s &= s(x) \end{aligned} \tag{6.14}$$

with $x \in X = \{x : |x_i| \leq x_{i\,\max}, i \in [|1, n|]\}$ and $u \in U = \{u : |u| \leq u_{\max}\}$.

Let the relative degree be equal to one, differentiating the sliding variable s twice yields to the following relation:

$$\ddot{s} = \frac{\partial \dot{s}}{\partial x}\left(f(x) + b(x)\,u\right) + \frac{\partial \dot{s}}{\partial u}\dot{u} = \alpha(x) + \beta(x)\,\dot{u} \tag{6.15}$$

It is assumed that if $|s(x)| < s_0$ then there are positive constants Φ, Γ_m and Γ_M such that

$$\begin{aligned} |\alpha(x)| &< \Phi \\ 0 < \Gamma_m \leq \beta(x) &\leq \Gamma_M \end{aligned} \tag{6.16}$$

Subsequently, Only Twisting and Super-Twisting algorithms will be studied.

6.3.1 Twisting Algorithm

The twisting algorithm, one of the first known second order sliding mode algorithms, ensures a finite time convergence of the state trajectory to the origin of the phase plane $(s, \dot{s})$ after executing a certain number of rotations around it (Fig. 6.2), and this is due to the commutation of the control between two values.

For a relative degree 1, the control algorithm is defined by the following control law:

$$\dot{u}_{Tw} = \left| \begin{array}{ll} -u & \text{if } |u| > u_{max} \\ -k_m \text{sign}(s) & \text{if } s\dot{s} \leq 0 \text{ and } |u| \leq u_{max} \\ -k_M \text{sign}(s) & \text{if } s\dot{s} > 0 \text{ and } |u| \leq u_{max} \end{array} \right. \tag{6.17}$$

The sufficient conditions that ensure a finite time convergence to the second order sliding set are:

$$k_M > k_m > 0,\ k_m > \frac{4\Gamma_M}{s_0},\ k_m > \frac{\Phi}{\Gamma_m},\ \Gamma_m k_M - \Phi > \Gamma_M k_m + \Phi \tag{6.18}$$

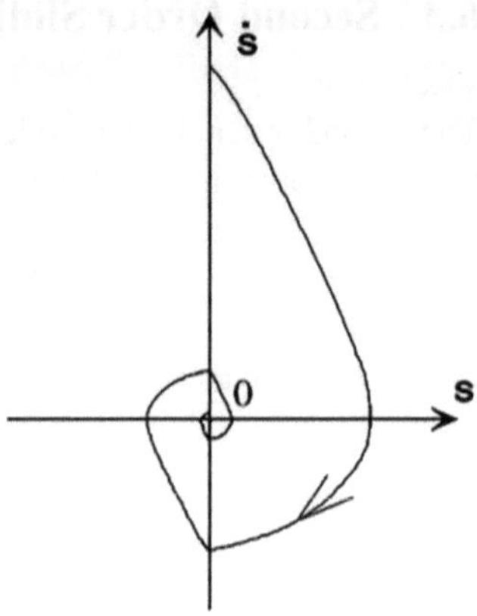

Fig. 6.2 Twisting algorithm phase trajectory

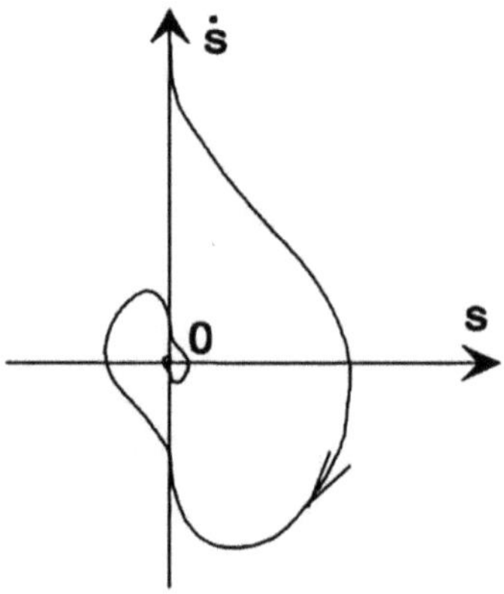

Fig. 6.3 Super-twisting algorithm phase trajectory

6.3.2 *Super-Twisting Algorithm*

Unlike twisting algorithm, this algorithm is only able to stabilize in finite time (Fig. 6.3) systems whose relative degree is equal to one and then it does not require information about the time derivative of the sliding variable.

The control law is given by:

$$\begin{aligned} u_{ST} &= -\alpha |s|^{\rho} \text{sign}\,(s) + u_1 \\ \dot{u}_1 &= -w \text{ sign}\,(s) \end{aligned} \tag{6.19}$$

and the corresponding sufficient conditions are:

$$w > \frac{\Phi}{\Gamma_m}, \quad \alpha^2 \geq \frac{4\Phi}{{\Gamma_m}^2} \frac{\Gamma_M\,(w + \Phi)}{\Gamma_m\,(w - \Phi)}, \quad 0 < \rho \leq 0.5 \tag{6.20}$$

Choosing $\rho = 0.5$ ensures the achievement of real second-order sliding mode.

6.4 Application to an Inverted Pendulum

The inverted pendulum is an interesting classic system which consists of a pendulum attached by a rotation joint to a cart. This cart, driven by a DC motor, can move along a horizontal guide rail in order to maintain the pendulum in its vertical balance. This system has two degrees of freedom whose generalized coordinates are: x for the horizontal cart movement and θ for the pendulum rotation (Fig. 6.4).

Let $x = \begin{bmatrix} x_1 \ x_2 \end{bmatrix}^\top = \begin{bmatrix} \theta \ \dot{\theta} \end{bmatrix}^\top$ be the state vector and u be the force applied to the cart, the dynamic equations of the inverted pendulum are:

$$\begin{cases} \dot{x}_1 = x_2 \\ \dot{x}_2 = f(x_1, x_2) + b(x_1, x_2)u \end{cases} \tag{6.21}$$

with:

$$f(x_1, x_2) = \frac{(M + m)\, g \sin x_1 - mlx_2^2 \sin x_1 \cos x_1}{\frac{4}{3}\,(M + m)\, l - ml \cos^2 x_1}$$

$$b(x_1, x_2) = \frac{\cos x_1}{\frac{4}{3}\,(M + m)\, l - ml \cos^2 x_1}$$

such that $M = 1$ kg is the cart mass, $m = 0.1$ kg is the pendulum mass, $l = 0.5$ m is the half length of the pendulum, $g = 9.8\ \mathrm{m/s^2}$ is the acceleration of gravity (Wang 1994).

6.4.1 Application of First Order Sliding Mode Control

6.4.1.1 Classic Sliding Mode Control

By referring to Slotine and Li (1991), the sliding function is chosen as:

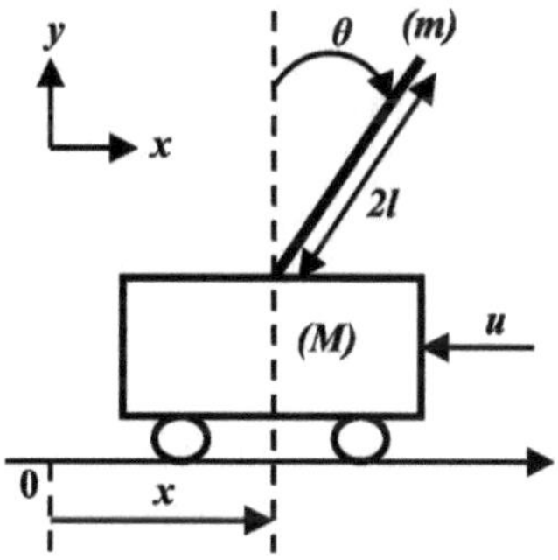

Fig. 6.4 Schematic of the inverted pendulum system

$$s = \lambda e + \dot{e} \tag{6.22}$$

where $e = x_1 - x_{d1}$ is the tracking error and λ is a positive constant. The first time derivative of s is:

$$\begin{aligned} \dot{s} &= \lambda (x_2 - \dot{x}_{d1}) + (f - \ddot{x}_{d1}) + b\,u \\ &= h + b\,u \end{aligned} \tag{6.23}$$

Using (6.8) and (6.23), the control law of the classic sliding mode controller is expressed as:

$$u = -b^{-1}\,(h + k\,\text{sign}\,(s)) \tag{6.24}$$

For simulation results, the design parameters are chosen as follows: $\lambda = 5$ and $k = 18$. The reference signal and the initial conditions are respectively $x_{d1} = \frac{\pi}{30}\sin t$ and $x_0 = [0.2\ 0]^\top$. Thus, by applying the classic sliding mode to the inverted pendulum, we obtain results shown in Fig. 6.5.

Figure 6.5 shows clearly the chattering phenomenon present in the control input. Thus, in the next step, the signum function will be replaced by the saturation function in order to attenuate the chatter effect.

6.4.1.2 Sliding Mode Control with Saturation Function

Substituting the signum function by the saturation function defined in (6.11) in the control law (6.24) yields to:

$$u = -b^{-1}\,(h + k\,\text{sat}\,(s, \varphi)) \tag{6.25}$$

where φ is chosen to be equal to 0.5. Simulation results are shown in Fig. 6.6.

These results show the efficiency of this approach in reducing the chatter effect. However, the existence of the boundary layer neighboring the sliding surface may affect the main characteristics of the sliding mode control such as robustness.

6.4.2 Application of Second Order Sliding Mode Control

6.4.2.1 Twisting Algorithm

Using (6.23), the second time derivative of the sliding function is given by:

$$\ddot{s} = \lambda\,(f + b\,u - \ddot{x}_{d1}) + \left(\dot{f} + \dot{b}\,u - \dddot{x}_{d1}\right) + b\,\dot{u} = d + b\,\dot{u} \tag{6.26}$$

where d and b must verify relations given in (6.16).

Then, using (6.17) the control law of twisting algorithm is expressed as:

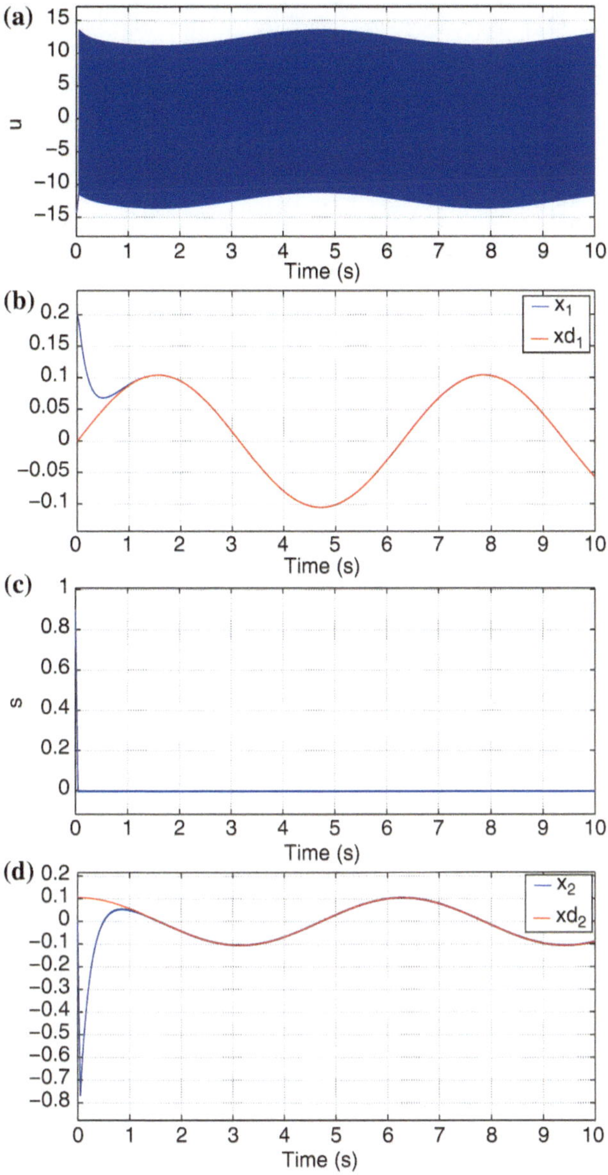

Fig. 6.5 **a** Control input, **b** angular displacement, **c** sliding function, **d** angular velocity for classic SMC

$$\dot{u} = b^{-1}\left(-d + \dot{u}_{Tw}\right) \tag{6.27}$$

By fulfilling the necessary conditions (6.16) and the sufficient conditions (6.18), the optimum value of k_m and k_M, chosen after some trials, are respectively 7 and 35.

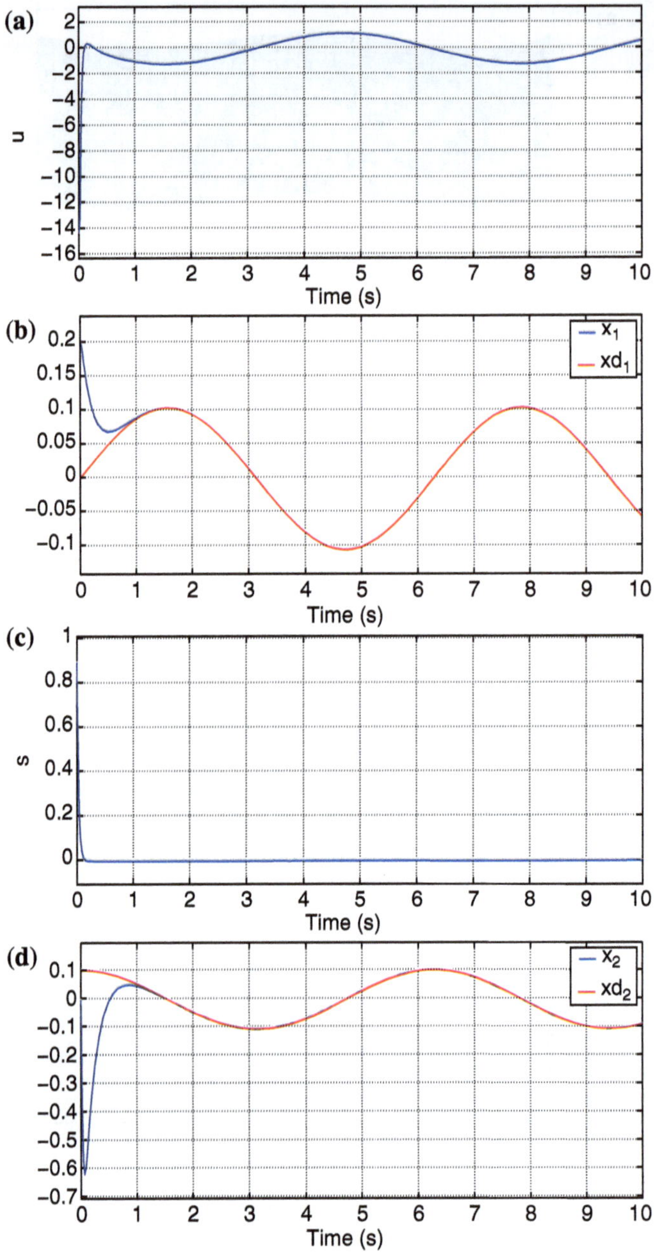

Fig. 6.6 **a** Control input, **b** angular displacement, **c** sliding function, **d** angular velocity for SMC with saturation function

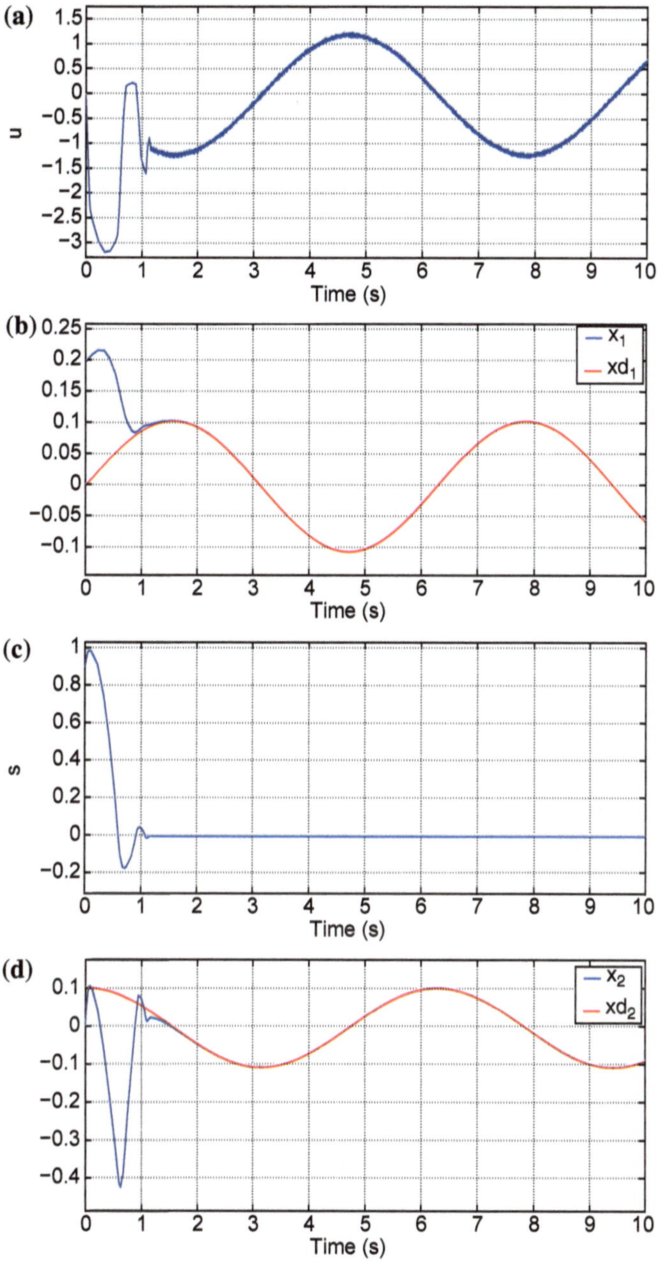

Fig. 6.7 **a** Control input, **b** angular displacement, **c** sliding function, **d** angular velocity for twisting algorithm

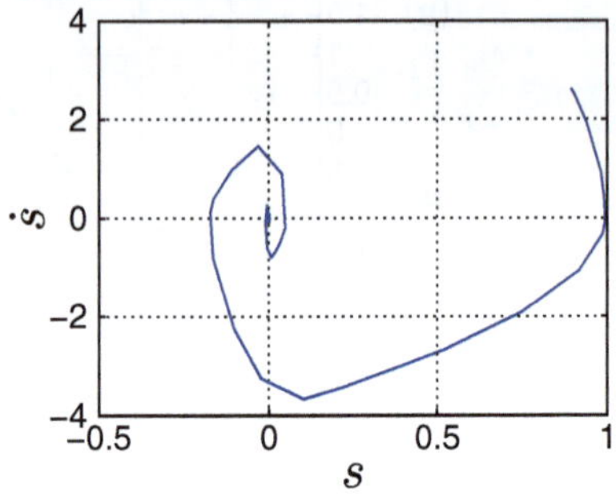

Fig. 6.8 Twisting algorithm phase trajectory

Figure 6.7 shows that the twisting Algorithm reduces the chattering phenomenon while Fig. 6.8 shows that the system state trajectory converges to the origin of the phase plane $(s, \dot{s})$.

6.4.2.2 Super-Twisting Algorithm

Using (6.19) and (6.23), the control law for the super-twisting algorithm is given by:

$$u = b^{-1}\left(-h + u_{ST}\right) \tag{6.28}$$

In accordance with the conditions given in (6.16) and (6.20), and after performing some simulations, we choose $w = 8$, $\alpha = 2$ and $\rho = 0.5$.

Both Figs. 6.9 and 6.10 show that the super-twisting algorithm ensures a finite time convergence of the system trajectory to the origin of the phase plane. In addition, by comparing to Figs. 6.7a and 6.9a, we can conclude that super-twisting algorithm is better than twisting algorithm with regards to the reduction of chattering effect in the control input.

6.5 Conclusion

This paper has proposed some solutions to attenuate chattering phenomenon present in classic sliding mode control. At first, a continuous approximation has been made in the vicinity of the sliding surface by using the saturation instead of the signum function. However, this introduces a boundary layer that may affect the main features of sliding mode such as robustness. Then, a second order sliding mode control algorithms: twisting and super-twisting algorithms, have been applied to the inverted pendulum system. The simulations results have shown their efficiency in attenuating chatter effect while ensuring a finite time convergence and high accuracy.

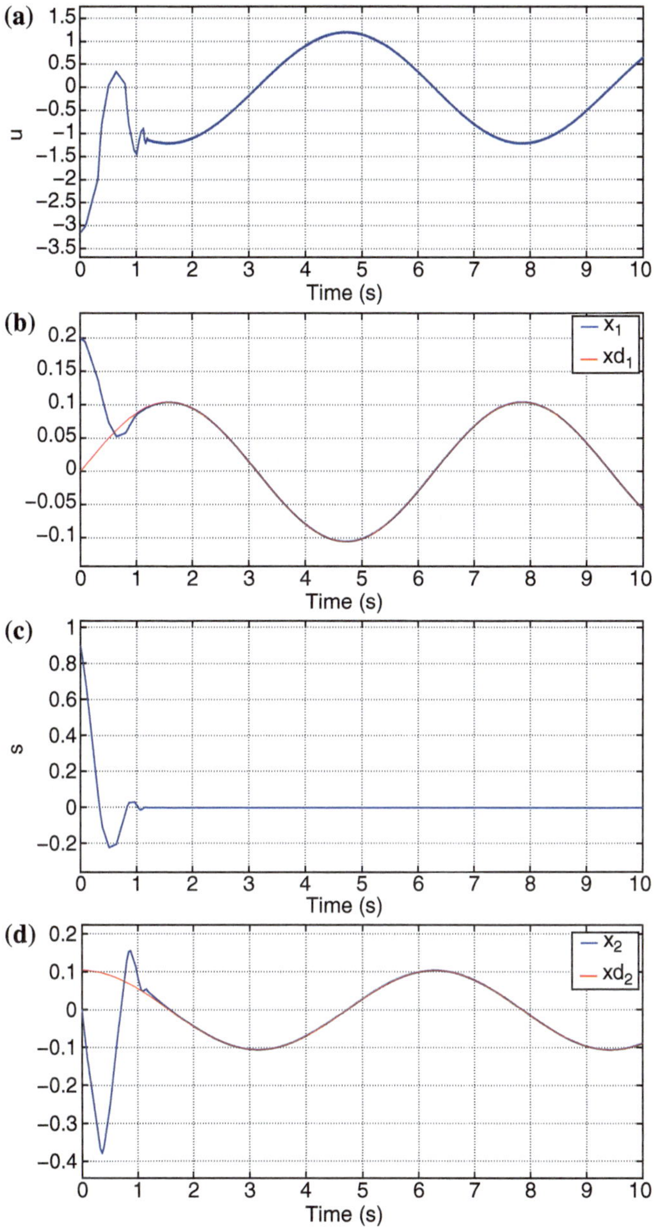

Fig. 6.9 **a** control input, **b** angular displacement, **c** sliding function, **d** angular velocity for super-twisting algorithm

Fig. 6.10 Super-twisting algorithm phase trajectory

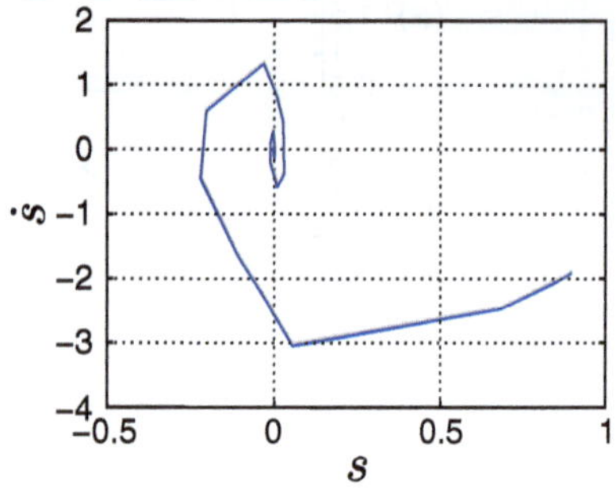

References

Bandyopadhyay, B., Deepak, F., & Kim, K. (2009). *Sliding mode control using novel sliding surfaces*. Berlin: Springer.

Bregeaut, V. (2010). Quelques contributions a la theorie de la commande par modes glissants. Ph.D. thesis, Central School of Nantes, Nantes, France.

DeCarlo, R. A., Zak, S. H., & Matthews G. P. (1988). Variable structure control of nonlinear multivariable systems: A tutorial. *Proceedings of the IEEE*, *76*, 212–232.

Edwards, C., Spurgeon, S. K. (1998). *Sliding mode control: Theory and applications*. London: Taylor & Francis.

Emelyanov, S., Korovin, S., & Levant, A. (1996). High-order sliding modes in control systems. *Computational Mathematics and Modeling*, *7*, 294–318.

Gao, W., & Hung, J. (1992). Variable structure control of nonlinear systems: A new approach. *IEEE Transactions on Industrial Electronics*, *40*, 45–55.

Hung, J., Gao, W., & Hung, J. (1993). Variable structure control: A survey. *IEEE Transactions on Industrial Electronics*, *40*, 2–22.

Levant, A. (1993). Sliding order and sliding accuracy in sliding mode control. *International Journal of Control*, *58*, 1247–1263.

Levant, A. (2003). High-order sliding modes, differentiation and output-feedback control. *International Journal of Control*, *76*, 924–941.

Perruquetti, W., & Barbot, J. (2002). *Sliding mode control in engineering*. New York: Marcel Dekker.

Riachy, S. (2008). Contribution a l'estimation et a la commande de systemes mecaniques sous-actionnes. Ph.D. thesis, Central School of Lille, Lille, France.

Slotine, J. (1984). Sliding controller design for non-linear systems. *International Journal of Control*, *40*, 421–434.

Slotine, J., & Sastry, S. (1983). Tracking control of non-linear systems using sliding surfaces, with application to robot manipulators. *International Journal of Control*, *83*, 465–492.

Slotine, J., & Li, W. (1991). *Applied nonlinear control*. New Jersey: Prentice Hall.

Utkin, V. (1977). Variable structure systems with sliding modes. *IEEE Transactions on Automatic Control*, *22*, 212–222.

Utkin, V. (1992). *Sliding modes in control and optimization*. Berlin: Springer.

Wang, L. (1994). A supervisory controller for fuzzy control systems that guarantees stability. *IEEE Transactions on Automatic Control*, *39*, 1845–1847.

Young, K., Utkin, V., & Ozguner, U. (1999). A control engineers guide to sliding mode control. *IEEE Transactions on Control Systems Technology*, *7*, 328–342.

Shtessel, Y., Edwards, C., Fridman, L., & Levant, A. (2014). *Sliding mode control and observation*. New York: Springer.

Chapter 7
Robust Adaptive Manoeuvering Control of an Autonomous Surface Vessel in the Presence of Ocean Currents and Parametric Model Uncertainty

Jawhar Ghommam and Faiçal Mnif

Abstract This chapter considers the problem of path-following control of autonomous surface vessels in the presence of ocean currents and parametric model uncertainty. The problem at hand consists of steering a vehicle surface ship along a geometric path with a desired speed profile. The Lyapunov technique is used to derive a robust architecture. To ensure path-following of the surface vessel, robust controller is designed based on adaptive sliding mode control in combination with the radial basis function neural network (RBFNN) to suppress the effect of parameter variations and external disturbances. Closed-loop tracking errors are shown to be asymptotically stable. Simulation results show that the proposed control algorithm attains a satisfied performance and is robust against parameter variations and external disturbances.

Keywords Distributed sliding motion control · Path following · Surface vessels · Backstepping design · Radial basis function neural network (RBFNN)

7.1 Introduction

Over the last decade, trajectory tracking and path following issues of autonomous marine vehicles have received a lot of attention from the control community. Trajectory tracking refers to the case where the autonomous vehicle must track a reference trajectory generated by a suitable virtual vehicle. The problem of path following is concerned with forcing the vehicle to follow a given path, which is not necessarily generated by the suitable virtual model. In this chapter, the latter control problem is considered.

J. Ghommam (✉)
Control and Energy Management-Lab, National School of Engineering of Sfax, Sfax, Tunisia
e-mail: jawhar.ghommam@gmail.com

F. Mnif
College of Engineering, Sultan Qaboos University, Muscat, Oman
e-mail: mnif@squ.edu.om

N. Derbel et al. (eds.), *Applications of Sliding Mode Control*,
Studies in Systems, Decision and Control 79, DOI 10.1007/978-981-10-2374-3_7

Indeed, the path-following control for marine autonomous vehicles is functionally divided into three subsystems in marine engineering: navigation, guidance and control (Fossen 1994). Navigation consists in operating the vehicle so as it points out to a given direction. This operation is usually executed through a simple program implemented using the nautical instrument onboard. Guidance on the other hand is a system that provides the desired reference used in the control derivation. Finally, control is the action of determining the necessary control orders to be provided by actuators onboard.

Considerable research has been conducted to address the path following control problem for autonomous marine vehicles, and various robust control strategies have been proposed in the literature. An outstanding research was conducted by (Skjetne et al. 2004), the authors proposed a framework for output maneuvering control for a class of strict feedback nonlinear systems and applied to maneuver fully actuated ships. The underlying assumption in this framework for path-following is that the vehicle's forward speed is bounded to track a specified desired speed profile, while the controller acts on the vehicle's orientation to drive it to the path. The authors in (Aguiar and Hespanha 2007) addressed the control problem of the path-following for more general class of autonomous vehicles in the presence of possibility large modeling parametric uncertainty. The authors in this work proposed an adaptive switching supervisory control to solve this kind of problem. In (Ihle et al. 2007) a passivity based controller is developed to make fully actuated ships follow parameterized paths while path following of fully actuated ships in presence of parameter uncertainties is addressed in (Kaminer et al. 2005). Also the authors in (Zereik et al. 2013) introduced the Jacobian task priority-based approach to solve the path following control problem. To account for uncertainties such as the disturbance forces and modeling errors, the authors in (Zhang et al. 2000) proposed a path following controller for an autonomous ship in restricted waters using sliding mode techniques to ensure the system's robustness and better performance. Constructive maneuvering control design for uncertain autonomous vehicles was presented in (Skjetne and Teel 2004) where sliding mode technique was shown to be effective with respect to additive and multiplicative structural uncertainties that affect the autonomous marine vehicle. However, in practical control implementation, the standard sliding mode controller may suffer from high frequency oscillations due to the discontinuous switching control action. In order to decrease this high frequency oscillations (most often called chattering phenomena), some control structures are introduced in (Slotine 1984) and (Utkin and Shi 1996). None of these alleviating algorithms have been applied to maneuvering of autonomous marine vehicles.

Among the aforementioned controllers the radial basis function neural network has attracted attention due to its superior ability to approximate nonlinear continuous function with certain precision. Moreover, it is characterized by less computational burden compared to multilayer perceptron since only the connective weights between the hidden layer and the output layer of the network are adjusted during training. The use of the approximation-based control for fully actuated ocean vessels in handling uncertainties had gained recognition in the work of (Tee and Ge 2006) and (Zhao et al. 2014). Motivated by the use of the RBFNN and sliding mode techniques, this chapter proposes a robust maneuvering control technique for a fully actuated autonomous marine vehicle that combines the advantage of the integral sliding mode control method, RBFNN and the backstepping procedure. In the first stage, the kinematic model of the autonomous vehicle is considered, an adaptive backstepping control is proposed to suppress the effect of the unknown ocean current. In the second stage based on the virtual control designed previously, an integral sliding surface is defined and integrated along with the RBFNN in the backstepping procedure to ensure robust estimation of the hydrodynamics uncertainties. The stability of the closed loop system is proved based on Lyapunov stability theory.

The reminder of the chapter is as follows: Sect. 7.2 presents some mathematical notions related to RBFNN. Section 7.3 briefly describes the fully autonomous marine vehicle by its kinematics and dynamics. Section 7.4 formulates the control objective of the paper. Section 7.5 gives details about the robust backstepping procedure that solves the maneuvering control problem for the autonomous fully actuated marine vehicle. Section 7.6 shows some numerical simulations to validate the proposed approach. Finally Sect. 7.7, concludes the paper.

7.2 Preliminaries

7.2.1 Notations

Throughout this chapter, $|.|$ denotes the absolute value of a scalar and $\|.\|$ denotes the Euclidean norm of a vector. For a matrix $\boldsymbol{X} \in \mathbb{R}^{n\times n}$, tr denotes its trace with the property $\mathrm{tr}(X^\top X) = \|X\|^2$.

7.2.2 RBFNN Approximation

Consider a function $f(\boldsymbol{x}) : \mathbb{R}^m \to \mathbb{R}$. Suppose that $f(\boldsymbol{x})$ is unknown smooth nonlinear function and it can be approximated over a compact set $\Omega \subseteq \mathbb{R}^m$ with the following RBFNN:

$$f(\boldsymbol{x}) = W^{*^\top} \phi(\boldsymbol{x}, \boldsymbol{\theta}^*) + \delta_f(x) \tag{7.1}$$

where the node number of the NN is l. More nodes mean more accurate approximation. $W^* \in \mathbb{R}^l$ represents the optimal weight vector, which is defined by

$$W^* = \arg\min_{\widehat{W}} \left\{ \sup_{x\in\Omega} |f(\boldsymbol{x}) - \widehat{W}^\top \phi(\boldsymbol{x}, \boldsymbol{\theta})| \right\} \tag{7.2}$$

where $\widehat{W}$ represents the estimate of W^*, $\phi(\boldsymbol{x}, \boldsymbol{\theta}) = [\phi_1, \phi_2, \ldots, \phi_l]^\top : \Omega \to \mathbb{R}^l$ represents the radial basis function vector, the element of which is chosen as the Gaussian function:

$$\phi(\boldsymbol{x}, \boldsymbol{\theta}) = \exp\left[\frac{-(\boldsymbol{x}-\theta_i)^\top(\boldsymbol{x}-\theta_i)}{\sigma^2}\right] \tag{7.3}$$

where $\boldsymbol{\theta} = [\theta_1, \theta_2, \ldots, \theta_l]^\top$ is the center vector of the Gaussian basis function, and σ is the spread of the Gaussian basis function. $\delta_f(\boldsymbol{x})$ is the approximation error that is bounded over Ω, such that $|\delta_f(\boldsymbol{x})| \leq \bar{\delta}_f$, where $\bar{\delta}_f$ is an unknown constant.

7.3 Vehicle Model

The autonomous surface vessel model is shown in Fig. 7.1, which is discussed in two reference frames (1) the inertial coordinate frame denoted by $\{\mathcal{I}\}$ and (2) the body-fixed coordinate frame with its origin at the center of mass of the surface vessel and is denoted by $\{\mathcal{B}\}$. The generalized position of the vehicle is $\boldsymbol{\eta} = [x, y, \psi]^\top \in \mathbb{R}^3$ where (x, y) are the coordinates of the origin of $\{\mathcal{B}\}$ in $\{\mathcal{I}\}$ and ψ is the orientation of the vehicle that defines the rotation matrix $R(\psi)$ which transforms the body coordinates into inertial coordinates.

$$\boldsymbol{R}(\psi) = \begin{bmatrix} \cos(\psi) & -\sin(\psi) & 0 \\ \sin(\psi) & \cos(\psi) & 0 \\ 0 & 0 & 1 \end{bmatrix} \tag{7.4}$$

Denote by $\boldsymbol{\nu} = [u, v, r]^\top \in \mathbb{R}^3$ the generalized velocity of the vehicle relative to $\{\mathcal{I}\}$ and expressed in $\{\mathcal{B}\}$, where u is the surge velocity, v is the sway velocity and r is the yaw rate. The ocean current velocity in the inertial frame $\{\mathcal{I}\}$ is denoted by $\boldsymbol{V}_c = [V_{cx}, V_{cy}]^\top \in \mathbb{R}^2$ and assumed to be constant and irrotational. The ocean current

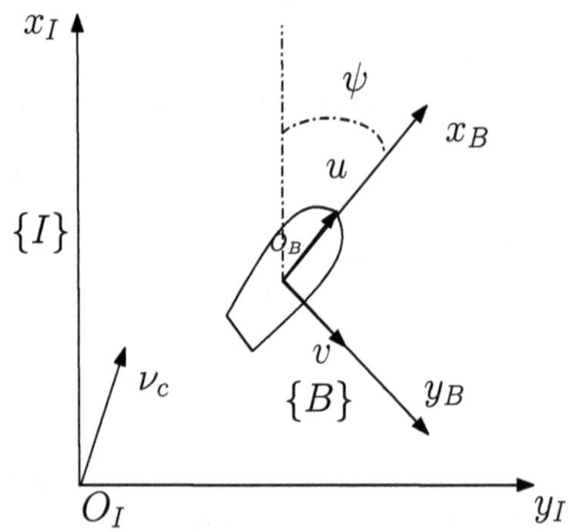

Fig. 7.1 Inertial and body-fixed coordinate frames

velocity in the body frame $\{\mathcal{B}\}$ denoted by $\boldsymbol{\nu}_c = [u_c, v_c, 0]^\top \in \mathbb{R}^3$ is obtained from $\boldsymbol{\nu}_c = \boldsymbol{R}(\psi)^\top[\boldsymbol{V}_c^\top, 0]^\top$. Since the ocean current is assumed irrotational and constant then $\dot{\boldsymbol{V}}_c = 0$ which also implies that $\dot{\boldsymbol{\nu}}_c = [rv_c, -ru_c, 0]^\top$. In navigation problems of autonomous surface vessels involving ocean currents, most often the kinematic model is expressed in terms of relative velocity between the vessel and the ocean current. Let the vector $\boldsymbol{\nu}_r = \boldsymbol{\nu} - \boldsymbol{\nu}_c = [u_r, v_r, r]^\top$ denotes is the relative velocity of the vehicle defined in the body fixed frame $\{\mathcal{B}\}$. Thus the following kinematic for the surface vessel model applies

$$\dot{\eta} = \boldsymbol{R}(\psi)\boldsymbol{\nu}_r + [\boldsymbol{V}_c^\top, 0]^\top \tag{7.5}$$

$$\dot{\boldsymbol{R}}(\psi) = r\boldsymbol{R}(\psi)S \tag{7.6}$$

where S is the skew symmetric matrix such that

$$S = \begin{bmatrix} 0 & -1 & 0 \\ 1 & 0 & 0 \\ 0 & 0 & 0 \end{bmatrix}, S^\top = -S$$

In the following a 3-DOF manoeuvering dynamic equations of motion for a fully actuated autonomous surface vessel is considered (Fossen 1994)

$$\boldsymbol{M}\dot{\boldsymbol{\nu}}_r = \boldsymbol{\tau} - \boldsymbol{C}(\boldsymbol{\nu}_r)\boldsymbol{\nu}_r + \boldsymbol{f}(\boldsymbol{\nu}_r) \tag{7.7}$$

where $\boldsymbol{M} = \boldsymbol{M}^\top \in R^{3\times 3}$ is the mass and inertia matrix which also includes hydrodynamic added mass. The matrix $\boldsymbol{C}(\boldsymbol{\nu}_r) \in \mathbb{R}^{3\times 3}$ is the Coriolis and centripetal matrix, $\boldsymbol{\tau} = [\tau_u, \tau_v, \tau_r]^\top \in \mathbb{R}^3$ is the generalized control input consisting of forces τ_u, τ_v and torque input τ_r. The vector $\boldsymbol{f}(\boldsymbol{\nu}_r)$ denotes the hydrodynamic damping forces and torques acting on the body. Using semi-empirical methods or hydrodynamic computation programs, the coefficients in $\boldsymbol{M}$ and $\boldsymbol{C}(\boldsymbol{\nu}_r)$ are determined quite accurately (Sun and Ge 2014). There exists difficulty however in finding the coefficients in $\boldsymbol{f}(\boldsymbol{\nu}_r)$ therefore they should be considered as unknown or uncertain and have to be estimated by an RBFNN adaptive controller to be designed.

The overall equation of motion of the autonomous surface vessel subject to a constant ocean current can be written as

$$\dot{\eta} = \boldsymbol{R}(\psi)\boldsymbol{\nu}_r + \boldsymbol{\Phi}_1 \tag{7.8}$$

$$\boldsymbol{M}\dot{\boldsymbol{\nu}}_r = \boldsymbol{\tau} - \boldsymbol{C}(\boldsymbol{\nu}_r)\boldsymbol{\nu}_r + \boldsymbol{\Phi}_2 \tag{7.9}$$

where $\boldsymbol{\Phi}_1 = [\boldsymbol{V}_c^\top, 0]^\top$ and $\boldsymbol{\Phi}_2 = \boldsymbol{f}(\boldsymbol{\nu}_r)$ are compound uncertainties, which contain parameter variations and unknown ocean currents, that will be estimated in the control design. The following Assumption will be useful for the stability analysis of the chapter.

Assumption 1 The uncertainties in the autonomous surface vessel dynamic model (7.8)–(7.9) satisfy $\|\boldsymbol{\Phi}_k\| \leq \varphi_k$, where φ_k is an unknown constant, $k = 1, 2$.

7.4 Problem Statement

The path manoeuvering control problem is primarily concerned with the design of control law that is able to steer a marine vehicle to reach then keep following a geometric path (see Fig. 7.2) without a temporal law as opposed to trajectory tracking control problem (Aguiar and Hespanha 2007). Once in the path, the vehicle should follow it with a desired speed profile. By using the convenient task classification scheme of (Skjetne et al. 2004), the path following problem can be split into two distinct task objectives: The *geometric task* where the position of the surface vessel is required to converge to and follow a desired geometric path and the *dynamic Task* where the speed of the vehicle is required to converge to and track a desired speed assignment.

The problem at hand can be formally formulated to tackle as follows:

Path-Following Control Problem

Let $\boldsymbol{\eta}_d(\gamma) \in \mathbb{R}^3$ be a desired path parameterized by a continuous variable $\gamma \in R$ and $v_d(t) \in \mathbb{R}$ is the desired speed profile for the autonomous surface vessel. Suppose also that $\boldsymbol{\eta}_d(\gamma)$ is sufficiently smooth and its derivatives with respect to γ are bounded. Design a path following controller $\boldsymbol{\tau}$ such that all closed loop signals are bounded, the geometric as well as the dynamic tasks are satisfied, that is:

$$\lim_{t\to\infty} \|\boldsymbol{\eta}(t) - \boldsymbol{\eta}_d(\gamma(t))\| = 0 \tag{7.10}$$

$$\lim_{t\to\infty} |\dot{\gamma}(t) - v_d(t)| = 0 \tag{7.11}$$

Note that the parameterizing variable γ in this case of application is chosen to be the arc length of the path. Some other parametrization can also be selected.

Remark 1 The path following control problem 1 has been solved by many authors (see for instance (Aguiar and Hespanha 2007; Almeida et al. 2010; Ghabcheloo et al. 2007)) to name a few. Most of the outstanding work in the aforementioned references either estimated the ocean current using simple kinematic observer or resorting to an adaptive scheme. In this chapter however we propose instead a compensation term that is able to suppress the effect of this ocean current.

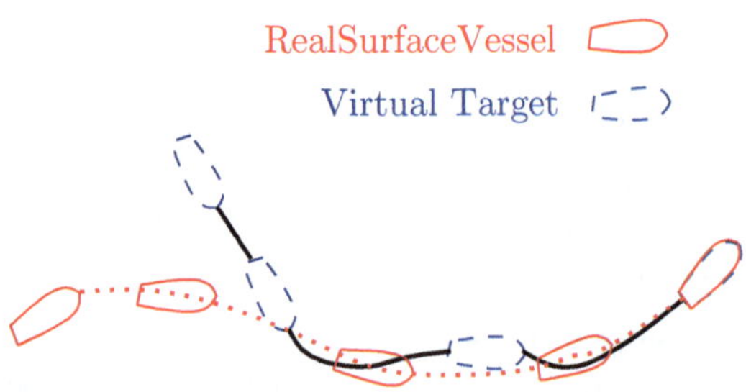

Fig. 7.2 Path-following concept for the autonomous surface vessel

7.5 Control Design

7.5.1 Path Following Control Design

The path following controller proposed in this section is mainly inspired by the work of (Almeida et al. 2010). However, it differs in the control scheme. In our approach, the RBFNN and adaptive sliding mode control are introduced to increase robustness of the backstepping procedure against parameter variations and external disturbances. In contrast to (Almeida et al. 2010), our control approach allows to efficiently reduce the control efforts.

The backstepping procedure for the path following of the autonomous surface vessel is designed following three steps:

Step 1: Consider the first equation of the dynamic model (7.8), then define the position error in the body-fixed frame as $z_1 = \boldsymbol{R}(\psi)^\top(\eta - \eta_d)$, the dynamic equation of z_1 is given as

$$\begin{aligned}\dot{z}_1 &= \dot{\boldsymbol{R}}(\psi)^\top(\eta - \eta_d) + \boldsymbol{R}(\psi)^\top(\dot{\eta} - \eta_d^\gamma\dot{\gamma})\\ &= -rS\dot{\boldsymbol{R}}^\top(\psi)(\eta - \eta_d) + \boldsymbol{R}(\psi)^\top(\boldsymbol{R}(\psi)\boldsymbol{\nu}_r + \boldsymbol{\Phi}_1 - \eta_d^\gamma\dot{\gamma})\\ &= -rSz_1 + \boldsymbol{\nu}_r - \boldsymbol{R}(\psi)^\top\eta_d^\gamma\dot{\gamma} + \boldsymbol{R}(\psi)^\top\boldsymbol{\Phi}_1\end{aligned} \tag{7.12}$$

where we have defined $\eta_d^\gamma = \frac{\partial\eta_d}{\partial\gamma}$. Furthermore, let the along-path speed tracking error be represented as $\varpi = \dot{\gamma} - v_d(\gamma)$, with $v_d(\gamma)$ being the desired speed profile on the path. The dynamic of z_1 rewrites

$$\dot{z}_1 = -rSz_1 + \boldsymbol{\nu}_r - \boldsymbol{R}(\psi)^\top\eta_d^\gamma v_d(\gamma) + \boldsymbol{R}(\psi)^\top\boldsymbol{\Phi}_1 - \boldsymbol{R}(\psi)^\top\eta_d^\gamma\varpi \tag{7.13}$$

At this level we regard $\boldsymbol{\nu}_r$ as a control input to stabilize z_1. Define the velocity error $z_2 = \boldsymbol{\nu}_r - \boldsymbol{\alpha}_r$, where $\boldsymbol{\alpha}_r$ is the stabilizing function chosen as follows:

$$\boldsymbol{\alpha}_r = -K_1z_1 + \boldsymbol{R}(\psi)^\top\eta_d^\gamma v_d(\gamma) - u_{es} \tag{7.14}$$

where K_1 is a diagonal positive gain matrix and u_{es} is the estimation of uncertainty $\boldsymbol{R}(\psi)^\top\boldsymbol{\Phi}_1$ and is composed of an RBFNN estimator u_{es1} and a robust term u_{es2} to confront the approximation error $\epsilon_1 = \boldsymbol{R}(\psi)^\top\boldsymbol{\Phi}_1 - u_{es1}$. The auxiliary control input u_{es} is therefore given as follows:

$$u_{es} = \underbrace{W_1^\top\phi(z_1, \boldsymbol{\theta}_1)}_{u_{es1}} + \underbrace{\frac{\hat{\epsilon}_1^2 z_1}{\hat{\epsilon}_1\|z_1\| + a_1 exp(-a_2 t)}}_{u_{es2}} \tag{7.15}$$

where W_1, $\boldsymbol{\theta}_1$ and the estimated value $\hat{\epsilon}_1$ of ϵ_1 are calculated as follows:

$$\begin{aligned}\dot{W}_1 &= \lambda_1 \phi(z_1, \boldsymbol{\theta}_1) z_1^\top \\ \dot{\boldsymbol{\theta}}_1 &= \lambda_2 (z_1^\top W_1^\top \phi'(z_1, \boldsymbol{\theta}_1))^\top \\ \dot{\hat{\epsilon}}_1 &= \kappa_1 \|z_1\|, \qquad \kappa_1 > 0\end{aligned} \tag{7.16}$$

where $\phi'(z_1, \boldsymbol{\theta}_1)$ denotes the partial derivative of $\phi(z_1, \boldsymbol{\theta}_1)$ with respect to $\boldsymbol{\theta}_1$ and λ_1, λ_2 are learning rates. For convenience of the development, we will drop the argument of the partial derivative of $\phi(z_1, \boldsymbol{\theta}_1)$. The uncertainty function $\boldsymbol{R}(\psi)^\top \boldsymbol{\Phi}_1$ can be written as

$$\boldsymbol{R}(\psi)^\top \boldsymbol{\Phi}_1 = W_1^{*\top} \phi(z_1, \boldsymbol{\theta}_1^*) + \delta_{\Phi_1}(z_1) \tag{7.17}$$

Define the estimate error $\widetilde{W}_1 = W_1^* - W_1$, the error of the center values as $\widetilde{\boldsymbol{\theta}}_1 = \boldsymbol{\theta}_1^* - \boldsymbol{\theta}_1$ and $\tilde{\epsilon}_1 = \hat{\epsilon}_1 - \epsilon_1$. The RBFNN approximation error can be calculated as follows:

$$\begin{aligned}\boldsymbol{R}(\psi)^\top \boldsymbol{\Phi}_1 - u_{es1} &= W_1^{*\top} \phi(z_1, \boldsymbol{\theta}_1^*) + \delta_{\Phi}(z_1) - W_1^\top \phi(z_1, \boldsymbol{\theta}_1) \\ &= \widetilde{W}_1^\top \phi(z_1, \boldsymbol{\theta}_1) + W_1^\top \phi' \widetilde{\boldsymbol{\theta}}_1 + \varrho(z_1, \widetilde{\boldsymbol{\theta}}_1)\end{aligned} \tag{7.18}$$

where $\widetilde{W}_1 = W_1^* - W_1$, $\widetilde{\boldsymbol{\theta}} = \boldsymbol{\theta}_1^* - \boldsymbol{\theta}$ and $\varrho(z_1, \widetilde{\boldsymbol{\theta}}_1)$ encompasses some bounded terms and is given as $\varrho(z_1, \widetilde{\boldsymbol{\theta}}_1) = \widetilde{W}_1^\top \phi' \widetilde{\boldsymbol{\theta}}_1 + W_1^{*\top} \boldsymbol{o}(z_1, \widetilde{\boldsymbol{\theta}}) + \delta_{\Phi_1}(z_1)$. With $\boldsymbol{o}(z_1, \widetilde{\boldsymbol{\theta}})$ being a higher order term of the Taylor expansion of $\phi(z_1, \boldsymbol{\theta}^*)$. It is further shown that $\varrho(z_1, \widetilde{\boldsymbol{\theta}}_1)$ is bounded, with $\|\varrho(z_1, \widetilde{\boldsymbol{\theta}}_1)\| \leq \epsilon_1$.

To study the stability of z_1-dynamics (7.12) with (7.14) being the stabilizing function, we consider the following Lyapunov function candidate:

$$V_1 = \frac{1}{2} z_1^\top z_1 + \frac{1}{2\lambda_1} \mathrm{tr}\big(\widetilde{W}_1^\top \widetilde{W}_1\big) + \frac{1}{2\lambda_2} \widetilde{\boldsymbol{\theta}}_1^\top \widetilde{\boldsymbol{\theta}}_1 + \frac{1}{2\kappa_1} \tilde{\epsilon}^2 + \frac{a_1}{a_2} \exp^{-a_2 t} \tag{7.19}$$

where $\tilde{\epsilon}_1 = \hat{\epsilon}_1 - \epsilon_1$. Taking the time derivative of (7.19) along the solutions of (7.13) and (7.16) and using the expression (7.18) results in:

$$\begin{aligned}\dot{V}_1 &= z_1^\top \dot{z}_1 - \frac{1}{\lambda_1} \mathrm{tr}\big(\widetilde{W}_1^\top \dot{W}_1\big) - \frac{1}{\lambda_2} \widetilde{\boldsymbol{\theta}}_1^\top \dot{\boldsymbol{\theta}}_1 + \frac{1}{\kappa_1} \tilde{\epsilon}_1 \dot{\hat{\epsilon}}_1 - a_1 \exp(-a_2 t) \\ &= z_1^\top \big(- rSz_1 - K_1 z_1 - u_{es1} - u_{es2} + z_2 + \boldsymbol{R}(\psi)^\top \boldsymbol{\Phi}_1 \big) \\ &\quad - \frac{1}{\lambda_1} \mathrm{tr}\big(\widetilde{W}_1^\top \dot{W}_1\big) - \frac{1}{\lambda_2} \widetilde{\boldsymbol{\theta}}_1^\top \dot{\boldsymbol{\theta}}_1 + \frac{1}{\kappa_1} \tilde{\epsilon}_1 \dot{\hat{\epsilon}}_1 - a_1 \exp(-a_2 t) \\ &\quad - z_1^\top \boldsymbol{R}(\psi)^\top \boldsymbol{\eta}_d^\gamma \varpi \\ &= z_1^\top \big(- rSz_1 - K_1 z_1 - u_{es2} + z_2 + \widetilde{W}_1^\top \phi(z_1, \boldsymbol{\theta}_1) + W_1^\top \phi' \widetilde{\boldsymbol{\theta}}_1 \\ &\quad + \varrho(z_1, \boldsymbol{\theta}_1)\big) - \mathrm{tr}\big(\widetilde{W}_1^\top \phi(z_1, \boldsymbol{\theta}_1) z_1^\top\big) - \widetilde{\boldsymbol{\theta}}_1^\top z_1^\top W_1^\top \psi'^\top \\ &\quad + \hat{\epsilon}_1 \|z_1\| - \epsilon_1 \|z_1\| - a_1 \exp(-a_2 t) - z_1^\top \boldsymbol{R}(\psi)^\top \boldsymbol{\eta}_d^\gamma \varpi \\ &= z_1^\top \big(- K_1 z_1 - u_{es2} + z_2 + \varrho(z_1, \widetilde{\boldsymbol{\theta}}_1)\big) + \hat{\epsilon}_1 \|z_1\| \\ &\quad - \epsilon_1 \|z_1\| - a_1 \exp(-a_2 t) - z_1^\top \boldsymbol{R}(\psi)^\top \boldsymbol{\eta}_d^\gamma \varpi\end{aligned}$$

where we made use of the fact that $z_1^\top S z_1 = 0$ for all z_1 and the fact that $z_1^\top \widetilde{W}_1^\top \phi(z_1, \theta_1) = \mathrm{tr}\big(\widetilde{W}_1^\top \phi(z_1, \theta_1) z_1\big)$. Further manipulation of the above equation by noticing the following

$$\begin{aligned} z_1{}^\top \varrho(z_1, \widetilde{\theta}_1) - u_{es2} + \hat{\epsilon}_1 \|z_1\| - \epsilon_1 \|z_1\| - a_1 \exp(-a_2 t) \leq \\ - \frac{\hat{\epsilon}_1^2 z_1}{\hat{\epsilon}_1 \|z_1\| + a_1 exp(-a_2 t)} - \hat{\epsilon}_1 \|z_1\| - a_1 \exp(-a_2 t) \leq 0 \end{aligned}$$

implies the following

$$\dot{V}_1 \leq -z_1^\top K_1 z_1 + z_1^\top z_2 - z_1^\top \boldsymbol{R}(\psi)^\top \boldsymbol{\eta}_d^\gamma \varpi \tag{7.20}$$

It becomes apparent from (7.20) that if z_2 converges to zero and the last term of (7.20) vanishes then the tracking error z_1 asymptotically converges to zero despite the presence of unknown ocean current.

Step 2: To ensure the last term of (7.20) vanishes we need to define a feedback law for $\ddot{\gamma}$ which implicitly sets up the dynamic task required for the autonomous surface while traversing its path. Augmenting the Lyapunov function V_1 by the following:

$$V_2 = V_1 + \frac{1}{2k_1\beta}\varpi^2 \tag{7.21}$$

where k_1 and β are positive constants. Taking its time derivative along the solutions of (7.20) gives

$$\begin{aligned} \dot{V}_2 &= \dot{V}_1 + \frac{1}{k_1\beta}\varpi\dot{\varpi} \\ &\leq -z_1^\top K_1 z_1 + z_1^\top z_2 - z_1^\top \boldsymbol{R}(\psi)^\top \boldsymbol{\eta}_d^\gamma \varpi + \frac{1}{k_1\beta}\varpi\dot{\varpi} \\ &\leq -z_1^\top K_1 z_1 + z_1^\top z_2 + \varpi\left(\frac{1}{k_1\beta}\dot{\varpi} - z_1^\top \boldsymbol{R}(\psi)^\top \boldsymbol{\eta}_d^\gamma\right) \end{aligned} \tag{7.22}$$

By imposing given dynamic profile for the along-path speed tracking error ϖ, we implicitly set a desired motion dynamic along the path for the autonomous surface vessel following (Skjetne et al. 2004) a filtered-gradient law can be designed as follows

$$\dot{\varpi} = -k_1\varpi + k_1\beta z_1^\top \boldsymbol{R}(\psi)^\top \boldsymbol{\eta}_d^\gamma \tag{7.23}$$

Since by definition $\dot{\gamma} = \varpi + v_d(\gamma)$, then the feedback control law for $\ddot{\gamma}$ is given as

$$\ddot{\gamma} = -k_1\varpi + k_1\beta z_1^\top \boldsymbol{R}(\psi)^\top \boldsymbol{\eta}_d^\gamma + v_d(\gamma)^\gamma \dot{\gamma} \tag{7.24}$$

substituting (7.23) into (7.22) results in

$$\dot{V}_2 \leq -z_1^\top K_1 z_1 - \frac{1}{\beta}\varpi^2 + z_1^\top z_2 \tag{7.25}$$

Clearly from (7.25), if z_2 tends to zero asymptotically then z_1 and ϖ converge asymptotically to zero.

Step 3: In aid of the sequel development, we need an analytic differentiation of the virtual control $\boldsymbol{\alpha}_r$. Its much difficult however to obtain such analytic differentiation. To facilitate the subsequent development and as to avoid analytic differentiation, a robust differentiator is then introduced to overcome this problem:

$$\mathcal{D} \begin{cases} \dot{\zeta}_{1\ell} = -\mu_{1\ell}|\zeta_{1\ell} - \alpha_{r\ell}|^{0.5}\text{sign}(\zeta_{1\ell} - \alpha_{r\ell}) + \zeta_{2\ell} \\ \dot{\zeta}_{2\ell} = -\mu_{2\ell}\text{sign}(\zeta_{2\ell} - \alpha_{r\ell}) \end{cases}$$

where $\boldsymbol{\alpha}_r = [\alpha_{r\ell}]_{\ell=1,\dots,3}$ and $\mu_{1\ell}, \mu_{2\ell}$ are positive gains. It has been proven that the differentiator $\mathcal{D}$ can exactly approximate the first derivative of $\alpha_{r\ell}, \ell = 1, \dots, 3$ within a finite time (Levant 2003). Therefore in the subsequent development, the first derivative of $\boldsymbol{\alpha}_r$ will be substituted by $\boldsymbol{\zeta} = [\zeta_{21}, \zeta_{22}, \zeta_{23}]^\top$. The dynamic equation of z_2 is then

$$\begin{aligned} \boldsymbol{M}\dot{z}_2 &= \boldsymbol{M}\boldsymbol{\nu}_r - \boldsymbol{M}\dot{\boldsymbol{\alpha}}_r \\ &= \boldsymbol{M}\boldsymbol{\nu}_r - \boldsymbol{M}\boldsymbol{\zeta} \\ &= \boldsymbol{\tau} - \boldsymbol{C}(\boldsymbol{\nu}_r)\boldsymbol{\nu}_r + \boldsymbol{\Phi}_2 - \boldsymbol{M}\boldsymbol{\zeta} \end{aligned} \tag{7.26}$$

In this step, the actual control will be designed to make the tracking error $\boldsymbol{z}_2$ converges asymptotically to zero. To chase away the parameters variation in the model as well as the external disturbances, A sliding surface will be employed in the design.

Define the integral sliding surface as follows:

$$\boldsymbol{s} = \boldsymbol{M}z_2 + \lambda_3 \int z_2 d\sigma \tag{7.27}$$

where $\lambda_3 > 0$ is a gain controller that adjusts the bandwidth of the error state s. Based on (7.27) the actual control $\boldsymbol{\tau}$ can be designed as follows:

$$\boldsymbol{\tau} = \boldsymbol{C}(\boldsymbol{\nu}_r)\boldsymbol{\nu}_r + \boldsymbol{M}\boldsymbol{\zeta} - \lambda_3 z_2 - \rho\mathbf{sgn}(\boldsymbol{s}) + \boldsymbol{u}_{des} \tag{7.28}$$

where ρ is a positive gain, $\mathbf{sgn}(\boldsymbol{s}) = [\text{sgn}(s_1), \text{sgn}(s_2), \text{sgn}(s_3)]^\top$ and is defined as

$$\text{sgn}(s_i) = \begin{cases} -1, & s_i < 0 \\ 0, & s_i = 0 \\ 1, & s_i > 0 \end{cases} \tag{7.29}$$

the auxiliary input $\boldsymbol{u}_{des}$ is designed to estimate the uncertainty $\boldsymbol{\Phi}_2$ and has the following expression

$$\boldsymbol{u}_{des} = W_2^\top \phi(z_2, \boldsymbol{\theta}_2) \tag{7.30}$$

where $W_2, \boldsymbol{\theta}_2$ are computed according to the the following update laws

$$\begin{aligned} \dot{W}_2 &= \lambda_4 \phi(z_2, \boldsymbol{\theta}_2)\boldsymbol{s}^\top \\ \dot{\boldsymbol{\theta}}_2 &= \lambda_5 \big(\boldsymbol{s}^\top W_2^\top \phi^{'}(z_2, \boldsymbol{\theta}_2)\big)^\top \end{aligned} \tag{7.31}$$

and where $\lambda_4 > 0$ and $\lambda_5 > 0$ are learning coefficients. Adopting the same notation as in step 1, the uncertainty $\boldsymbol{\Phi}_2$ can be rewritten as

$$\boldsymbol{\Phi}_2 = W_2^{*^\top} \phi(z_2, \boldsymbol{\theta}_2^*) + \delta_{\Phi_2} \tag{7.32}$$

The residue $\bar{\varrho}(z_2, \boldsymbol{\theta}_2)$ due to the difference between the uncertainty and its estimator is defined as $\bar{\varrho}(z_2, \boldsymbol{\theta}_2) = \widetilde{W}_2^\top \phi^{'} \widetilde{\boldsymbol{\theta}}_2 + W_2^{*^\top} \boldsymbol{o}(z_2, \widetilde{\boldsymbol{\theta}}_2) + \delta_{\Phi_2}$ and verifies $\|\bar{\varrho}(z_2, \boldsymbol{\theta}_2)\| \leq \epsilon_2$.

Based on this design, we can now state our main theorem of the control design for the path following problem.

Theorem 1 *Consider an individual smooth path $\boldsymbol{\eta}_d$ parameterized by a variable γ, let the autonomous surface vessel be described by the equation of motion (7.9). The control law (7.28) together with the update laws (7.16), (7.31) and (7.24) solve the path following control problem defined in (7.10)–(7.11). In particular, the trajectories of the system closed-loop are exponentially convergent and the sliding surface $\boldsymbol{s} = 0$ is reachable. Furthermore the the position of the surface vessel $\boldsymbol{\eta}$ asymptotically follows the desired parameterized path $\boldsymbol{\eta}_d$ with speed assignment $\dot{\gamma} = v_d(\gamma)$.*

Proof To show Theorem 1, let first rewrite the closed loop system dynamics in terms of the variable states z_1, z_2 and ϖ:

$$\begin{aligned} \dot{z}_1 &= -K_1 z_1 + z_2 - rSz_1 + \boldsymbol{R}(\psi)^\top \boldsymbol{\Phi}_1 - u_{es} - \boldsymbol{R}(\psi)^\top \boldsymbol{\eta}_d^\gamma \varpi \\ \dot{\varpi} &= -k_1 \varpi + k_1 \beta z_1^\top \boldsymbol{R}(\psi)^\top \boldsymbol{\eta}_d^\gamma \\ \boldsymbol{M}\dot{z}_2 &= \boldsymbol{\Phi}_2 - \lambda_3 z_2 - \rho\,\mathbf{sgn}(\boldsymbol{s}) + \boldsymbol{u}_{des} \end{aligned} \tag{7.33}$$

It is clear from (7.33), that the closed loop system has a cascaded form. Consider the Lyapunov function V_2 defined in (7.21) for the first two equations of (7.33), Its time derivative along the solutions of the above two first equations yields:

$$\begin{aligned} \dot{V}_2 &\leq -z_1^\top K_1 z_1 - \frac{1}{\beta}\varpi^2 + z_1^\top z_2 \\ &\leq -z_1^\top (K_1 - \varsigma \mathbf{I}_{3\times 3}) z_1 - \frac{1}{\beta}\varpi^2 + \frac{1}{4\varsigma} z_2^\top z_2, \qquad \varsigma > 0 \\ &= -q_{\min}\|q\|^2 + \frac{1}{4\varsigma}\|z_2\|^2, \qquad \forall \|q\| > \frac{1}{2}\sqrt{\frac{1}{\varsigma q_{\min}}}\|z_2\| \end{aligned} \tag{7.34}$$

where $\mathbf{I}_{3\times 3}$ is a 3 by 3 identity matrix, $q_{\min} = \min\left(\lambda_{\min}(K_1 - \varsigma \mathbf{I}_{3\times 3}), \frac{1}{\beta}\right)$ and $q = [z_1^\top, \varpi]^\top$. Then, the subsystem formed by the two first equations of (7.33) is (Input to State Stable) w.r.t. z_2 (Khalil 2002), which essentially expresses that for any bounded input z_2, the states z_1 and ϖ will be ultimately bounded by a class $\mathcal{K}$ function of $\sup_{t>0} \|z_2\|$. If furthermore z_2 converges asymptotically to zero, then z_1 and ϖ converge to zero as well (Khalil 2002). In the sequel, we will prove effectively that z_2 converges asymptotically to zero.

Consider the last equation of (7.33). To show that z_2 asymptotically converges to zero, we will consider two cases: the first case is when $\boldsymbol{s}^\top \mathbf{sgn}(\boldsymbol{s}) \neq 0$ and the second case is when $\boldsymbol{s}^\top \mathbf{sgn}(\boldsymbol{s}) = 0$. For the first case, let the following Lyapunov function candidate:

$$V_3 = \boldsymbol{s}^\top \boldsymbol{s} + \frac{1}{2\lambda_4} \mathrm{tr}\big(\widetilde{W}_2^\top \widetilde{W}_2\big) + \frac{1}{2\lambda_5} \widetilde{\boldsymbol{\theta}}_2^\top \widetilde{\boldsymbol{\theta}}_2 \tag{7.35}$$

Taking the time derivative of (7.35) along the solutions of (7.33) and (7.31) results in:

$$\begin{aligned}
\dot{V}_3 &= \boldsymbol{s}^\top \dot{\boldsymbol{s}} - \frac{1}{\lambda_4} \mathrm{tr}(\widetilde{W}_2^\top \dot{W}_2) - \frac{1}{\lambda_5} \widetilde{\boldsymbol{\theta}}_2 \top \dot{\boldsymbol{\theta}}_2 \\
&= \boldsymbol{s}^\top \left(\boldsymbol{M}\dot{z}_2 + \lambda_3 z_2\right) - \mathrm{tr}(\widetilde{W}_2^\top \phi(z_2, \boldsymbol{\theta}_2)\boldsymbol{s}^\top) - \widetilde{\boldsymbol{\theta}}_2^\top \left(\boldsymbol{s}^\top W_2^\top \phi'(z_2, \boldsymbol{\theta}_2)\right)^\top \\
&= \boldsymbol{s}^\top \left(\boldsymbol{\Phi}_2 - \rho \mathbf{sgn}(\boldsymbol{s}) + \boldsymbol{u}_{des}\right) - \mathrm{tr}(\widetilde{W}_2^\top \phi(z_2, \boldsymbol{\theta}_2)\boldsymbol{s}^\top) \\
&\quad - \widetilde{\boldsymbol{\theta}}_2^\top \left(\boldsymbol{s}^\top W_2^\top \phi'(z_2, \boldsymbol{\theta}_2)\right)^\top \\
&= \boldsymbol{s}^\top \left(\widetilde{W}_2^\top \phi(z_2, \boldsymbol{\theta}_2) + \bar{\varrho}(z_2, \boldsymbol{\theta}_2) - \rho \mathbf{sgn}(\boldsymbol{s})\right) - \mathrm{tr}(\widetilde{W}_2^\top \phi(z_2, \boldsymbol{\theta}_2)\boldsymbol{s}^\top)
\end{aligned} \tag{7.36}$$

Noticing that $\boldsymbol{s}^\top \widetilde{W}_2^\top \phi(z_2, \boldsymbol{\theta}_2) = \mathrm{tr}\big(\widetilde{W}_2^\top \phi(z_2, \boldsymbol{\theta}_2)\boldsymbol{s}^\top\big)$, it follows from (7.36) that

$$\begin{aligned}
\dot{V}_3 &= \boldsymbol{s}^\top \left(\bar{\varrho}(z_2, \boldsymbol{\theta}_2) - \rho \mathbf{sgn}(\boldsymbol{s})\right) \\
&\leq \epsilon_2 \boldsymbol{s}^\top \mathbf{sgn}(\boldsymbol{s}) - \boldsymbol{s}^\top \rho \mathbf{sgn}(\boldsymbol{s}) \\
&\leq -(\rho - \epsilon_2)\boldsymbol{s}^\top \mathbf{sgn}(\boldsymbol{s})
\end{aligned} \tag{7.37}$$

where we have used the fact that $\boldsymbol{s}^\top \bar{\varrho}(z_2, \boldsymbol{\theta}_2) \leq \epsilon_2 \|\boldsymbol{s}\|$. If ρ is selected such that $\rho > \epsilon_2$, then there exists $c_0 > 0$ such that:

$$\dot{V}_3 \leq -(\rho - \epsilon_2)(\boldsymbol{s}^\top \boldsymbol{s})^{\frac{1}{2}} \leq -c_0 V_3^{\frac{1}{2}} \leq 0 \tag{7.38}$$

From (7.38), it can be concluded that the sliding surface reaches zero within a finite time $t_f = \frac{2V_3(0)^{\frac{1}{2}}}{c_0}$ and thus the path following control design satisfies the sliding condition.

In the second case, if $\boldsymbol{s}^\top \mathbf{sgn}(\boldsymbol{s}) = 0$, it can be inferred from Theorem 1 in (Plestan et al. 2010) that the control law (7.28) can guarantee the states of the system (7.33) stay on the sliding surface $\boldsymbol{s} = 0$.

From the two cases studied above, it can be concluded that the trajectories of the closed loop system (7.28) are asymptotically convergent and the sliding surface $s = 0$ is reachable. This concludes the proof.

7.6 Numerical Simulations

The main purpose of this section is to illustrate the effectiveness of the stabilisation and tracking control laws presented above via simulation example. The model ship used in simulation is CybershipII (Skjetne et al. 2004).

The matrices $\boldsymbol{M}$ and $\boldsymbol{C}(\boldsymbol{\nu}_r)$ are given by:

$$\boldsymbol{M} = \begin{bmatrix} m - X_{\dot{u}} & 0 & 0 \\ 0 & m - Y_{\dot{v}} & mx_g - Y_{\dot{r}} \\ 0 & mx_g - N_{\dot{v}} & I_z - N_{\dot{r}} \end{bmatrix}, \quad \boldsymbol{C}(\boldsymbol{\nu}_r) = \begin{bmatrix} 0 & 0 & c_{13}(\boldsymbol{\nu}) \\ 0 & 0 & c_{23}(\boldsymbol{\nu}) \\ -c_{13}(\boldsymbol{\nu}) & -c_{23}(\boldsymbol{\nu}) & 0 \end{bmatrix}$$

where $c_{13}(\boldsymbol{\nu}) = -(m - Y_{\dot{v}})v - (mx_g - Y_{\dot{r}})r$ and; $c_{23}(\boldsymbol{\nu}) = (m - X_{\dot{u}})u$ and the unknown hydrodynamic damping forces $\boldsymbol{f}(\boldsymbol{\nu}_r)$ expression is given by

$$\boldsymbol{f}(\boldsymbol{\nu}_r) = \begin{bmatrix} d_{11}(\boldsymbol{\nu}) & 0 & 0 \\ 0 & d_{22}(\boldsymbol{\nu}) & d_{23}(\boldsymbol{\nu}) \\ 0 & d_{32}(\boldsymbol{\nu}) & d_{33}(\boldsymbol{\nu}) \end{bmatrix}$$

where $d_{11}(\boldsymbol{\nu}) = -X_u - X_{|u|u}|u| - X_{uuu}u^2$, $d_{22} = -Y_v - Y_{|v|v}|v| - Y_{|r|v}|r|$, $d_{23}(\boldsymbol{\nu}) = -Y_r - Y_{|v|r}|v| - Y_{|r|r}|r|$, $d_{32}(\boldsymbol{\nu}) = -N_v - N_{|v|v}|v| - N_{|r|v}|r|$ and $d_{33}(\boldsymbol{\nu}) = -N_r - N_{|v|r}|v - N_{|r|r}|r|$. The coefficients $X_{\{.\}}, Y_{\{.\}}, N_{\{.\}}$ are the hydrodynamic parameters. Only the coefficients in $\boldsymbol{M}$ and $\boldsymbol{C}(\boldsymbol{\nu})$ are determined quite accurately using semi-empirical methods, or system identification, whereas $\boldsymbol{f}(\boldsymbol{\nu}_r)$ contains the uncertain (non-identified) constant parameters. The physical parameters of the vessel are given (Fossen 1994).

In the simulation, the desired path is generated based on polynomial interpolation. We assume that the surface vessel starts at the following initial position $x(0) = -40$ m, $y(0) = 20$ m and $\psi(0) = 0$ rad, it's initial velocity is $u(0) = 0$ m/s, $v(0) = 0$ m/s and $r(0) = 0$ rad/s. The initial value for γ is $\gamma(0) = 0$, the desired speed assignment is set to $v_d(\gamma) = v_0 * (1 - \ell_1 \exp^{-\ell_2(t-t_0)})\exp(-\ell_3\|z_1\|)$, where $v_0 \neq 0$, $\ell_i > 0$, $i = 1, 2, 3$ and $\ell_1 < 1$. This choice for the speed assignment has the feature that when the path following error z_1 is large, the virtual target will wait for the real ship and when the z_1 is small the virtual target will move along the path at the speed close to v_0. The ocean current is set to $\boldsymbol{V}_c = [1, -1, 0]^\top$ m/s. The controller gains are $K_1 = \text{diag}([10, 10, 10])$, $\lambda_1 = 10$, $\lambda_2 = 10$, $\lambda_3 = 10$, $\lambda_4 = 10$, $\lambda_5 = 10\kappa_1 = 20$, $k_1 = 15$, $\beta = 5$, $\mu_{1\ell} = 30$, $\mu_{2\ell} = 30$, $\rho = 10$. Simulations are carried out in the presence of noise, all signals measured by the controller are affected by disturbed additive gaussian white noise. Figure 7.3 illustrates the convergence of the vehicle to its desired path despite the presence of uncertainties and external per-

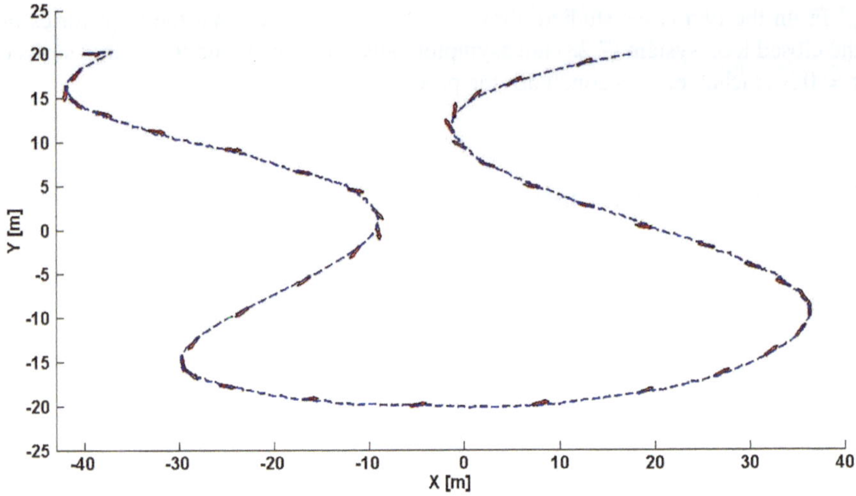

Fig. 7.3 Desired and actual vehicle paths

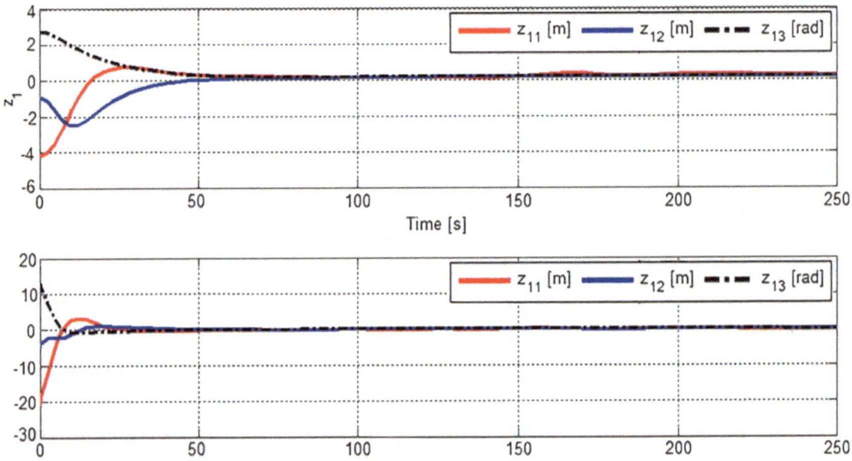

Fig. 7.4 Backstepping error variables

turbation. This can also be seen in Figs. 7.4 and 7.5 where the backstepping errors state variables as well as the position and orientation errors given by $\|\eta - \eta_d\|$ and $\psi - \psi_e$ converge robustly to the origine respectively as shown formally in Theorem 1. The control inputs applied to the vehicle are represented in Fig. 7.6.

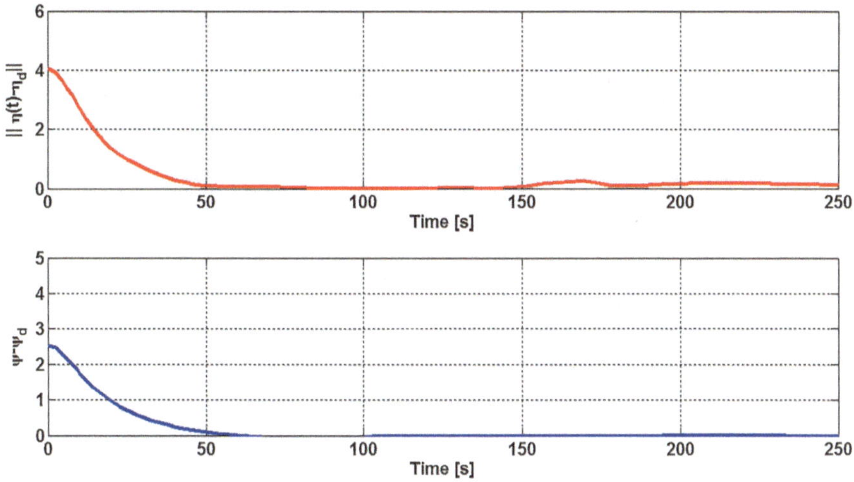

Fig. 7.5 Position and orientation error

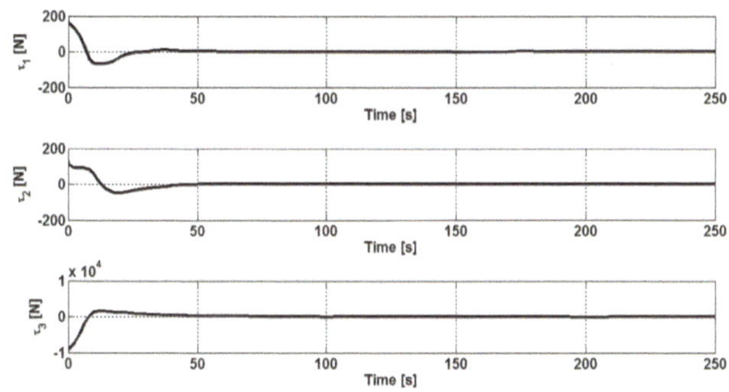

Fig. 7.6 Temporal evolution of the control inputs

7.7 Conclusion

In this paper a robust adaptive sliding mode control system is proposed for the maneuvering control of fully actuated surface vessel subjected to unknown but constant ocean currents and parametric model uncertainty. The path-following controller was designed based on a compensation term plus an RBFNN estimator that suppresses the effects of unknown constant ocean currents and parametric model uncertainty. Stability and asymptotic convergence of the closed-loop system were formally proved and are guaranteed for any set of initial conditions based on Lyapunov-based techniques. Illustrative example for maneuvering an autonomous surface vessel along a complicated path has been shown and discussed. Further research directions is to extend the results obtained to under-actuated vehicles under the influence of time-varying ocean currents and model uncertainties.

References

Aguiar, P., & Hespanha, J. P. (2007). Trajectory-tracking and path-following of underactuated autonomous vehicles with parametric modeling uncertainty. *IEEE Transactions on Automatic Control, 52*, 1362–1379.

Almeida, J., Silvestre, C., & Pascoal, A. (2010). Cooperative control of multiple surface vessels in the presence of ocean currents and parametric model uncertainty. *International Journal of Robust and Nonlinear Control, 20*, 1549–1565.

Fossen, T. I. (1994). *Guidance and control of ocean vehicles*. New York: Wiley.

Ghabcheloo, R., Pascoal, A., Silvestre, C., & Kaminer, I. (2007). Nonlinear coordinated path following control of multiple wheeled robots with bidirectional communication constraints. *International Journal of Adaptive Control and Signal Processing, 21*, 133–157.

Ihle, I.-A. F., Arcak, M., & Fossen, T. I. (2007). Passivity-based designs for synchronized path-following. *Automatica, 43*, 1508–1518.

Kaminer, I., Pascoal, A. M. & Yakimenko, O. (2005). Nonlinear path following control of fully actuated marine vehicles with parameter uncertainty. In *Proceedings of the 16th IFAC World Congress*.

Khalil, H. K. (2002). *Nonlinear Systems*. Englewood Cliffs: Prentice-Hall.

Levant, A. (2003). Higher-order sliding modes, differentiation and output-feedback control. *International Journal of Control, 76*, 924–941.

Plestan, F., Shtessel, Y., Bregeault, V., & Poznyak, A. (2010). New methodologies for adaptive sliding mode control. *International Journal of Control, 83*, 1907–1919.

Skjetne, R., Fossen, T. I., & Kokotovic, P. V. (2004). Robust output maneuvering for a class of nonlinear systems. *Automatica, 40*, 373–383.

Skjetne, R. & Teel, A. R. (2004). Maneuvering dynamical systems by sliding-mode control. In *Proceedings of the American Conference on Control*.

Slotine, J. (1984). Sliding controller design for non-linear systems. *International Journal of Control, 40*, 421–434.

Sun, X., & Ge, S. S. (2014). Adaptive neural region tracking control of multi-fully actuated ocean surface vessels. *IEEE/CAA Journal of Automatica Sinca, 1*, 1536–1543.

Tee, K., & Ge, S. (2006). Control of fully-actuated ocean surface vessels using a class of feedforward approximators. *IEEE Transactions on Control Systems Technology, 14*, 750–756.

Utkin, V. & Shi, J. (1996). Integral sliding mode in systems operating under uncertainty conditions. In *Proceedings of the 35th IEEE on Decision and Control*.

Zereik, I., Bibuli, M., Bruzzone, G., & Caccia, M. (2013). Jacobian task priority-based approach for path following of unmanned surface vehicles. In *Proceedings of the 9th IFAC Conference on Control Applications in Marine Systems*.

Zhang, R., Chen, Y., Sun, Z., Sun, F., & Xu, H. (2000). Path control of a surface ship in restricted waters using sliding mode. *IEEE Transactions on Control Systems Technology, 8*, 722–732.

Zhao, Z., He, W., & Ge, S. S. (2014). Adaptive neural network control of a fully actuated marine surface vessel with multiple output constraints. *IEEE Transactions on Control Systems Technology, 22*, 1536–1543.

Chapter 8
Sliding Mode with Time Delay Control for Robot Manipulators

Yassine Kali, Maarouf Saad, Khalid Benjelloun and Mohammed Benbrahim

Abstract This chapter introduces two controllers design for the trajectory tracking of robot manipulators with unknown dynamics and external disturbances, including:

- First Order Sliding Mode with Time Delay Control (FOSMTDC),
- Second Order Sliding Mode with Time Delay Control (SOSMTDC).

For both proposed controllers, the Lyapunov function are invoked to establish the stability of the closed loop system and finite time convergence. Simulation results are presented to show the effectiveness of the proposed controllers, regarding particularly the unknown dynamics and external disturbances and the chattering reduction on control input.

Keywords Sliding mode · Second order sliding mode · Time delay · Robot manipulators · Unknown dynamics · External disturbances

Y. Kali (✉) · K. Benjelloun
AII-Lab, Mohammadia School of Engineers, Mohammed V University, Rabat, Morocco
e-mail: y.kali88@gmail.com

K. Benjelloun
e-mail: bkhalid@emi.ac.ma

M. Saad
GREPCI-Lab, Ecole de Technologie Supérieure, Quebec University, Montreal, Canada
e-mail: maarouf.saad@etsmtl.ca

M. Benbrahim
LISTA-Lab, Sciences Faculty, Sidi Mohammed Ben Abdallah University, Fes, Morocco
e-mail: mohammed.benbrahim@usmba.ac.ma

N. Derbel et al. (eds.), *Applications of Sliding Mode Control*,
Studies in Systems, Decision and Control 79, DOI 10.1007/978-981-10-2374-3_8

8.1 Introduction

Robot manipulator dynamics are governed by a set of highly nonlinear and strongly coupled second order differential equations which exhibit numerous problems, such as a variety of external disturbances and/or nonlinear unknown dynamics. Thus it is well known that accurate robot motion control requires complex nonlinear controllers based on the mathematical model which is always an approximation of reality. In literature, we can find many nonlinear control techniques such as fuzzy feedback linearization (Park and Cho 2007), backstepping (Zhou and Wen 2008), Sliding Mode Control (SMC) (Liu and Wang 2012; Utkin 1992; Utkin et al. 1999), etc.

The SMC is a robust technique used for uncertain robot manipulators. This technique is based on the design of a high-speed switching control law that drives the system's trajectory onto a user-chosen hyperplane in the state space, also known as sliding surface. However, its design requires that the switching gain should be larger than the superior bound of unknown dynamics and disturbances, but the problem is that this bound is always overestimated, which makes the choice of the switching gain excessive. Then, the major drawback of SMC, the well-known chattering phenomenon (Boiko and Fridman 2005; Fridman 2001) will be important and couldn't be accepted by the actuators and may deteriorate the controlled systems if the control has any physical sense. In order to avoid this phenomenon, some works have considered continuous functions (Liu and Wang 2012) instead of the signum one, but the provided results lead to a large steady state error while others proposed the use of higher order sliding mode (HOSM) controller (Fridman and Levant 2002). Even for these approaches, the knowledge of the superior bound of the unknown dynamics and disturbances is always required.

Another solution have been presented, it consists on using adaptive sliding mode, which provide an adaptation of the control gain to be as small as possible and sufficient to eliminate the effect of unknown dynamics and external disturbances, but this solution didn't take into account the chattering phenomenon (Sastry and Bodson 1994). Furthermore, the necessity to cancel unknown dynamics and unexpected disturbances has led to the introduction of sliding mode observers and sliding mode disturbance observers (Shtessel et al. 2014), therefore the unknown dynamics and unexpected disturbances are supposed to be bounded and a prior of these bound is also required here. Otherwise, a combination of SMC with intelligent controller as neural-network and fuzzy logic have been considered (Jezernik et al. 1997; Palm 1992). These techniques can approximate unknown dynamics and disturbances; however, they introduce a number of parameters or fuzzy rules which are difficult to implement and they don't guarantee a good tracking performance or overestimate the switching gains.

The objective of this chapter is to discuss two techniques for independent joint control of robot manipulators with unknown dynamics and external disturbances. They consist on controlling each joint as a Single-Input-Single-Output (SISO) system without a prior knowledge of the model. Each SISO subsystem is treated as a linear system and the coupling effects are treated as disturbances. The first controller is a combination of First Order Sliding Mode (FOSM) and Time Delay Control (TDC) (Youcef-Toumi and Ito 1990) while the second is a combination of Second Order Sliding Mode (SOSM) (Cherrid et al. 2000) and TDC. The choice of these combinations is related to the fact that the Time Delay Estimation (TDE) can estimate unknown dynamics and disturbances simply under the assumption that these unknown dynamics do not vary largely for a sufficient small time. The TDE uses time delayed signals of control inputs and states. Furthermore, the TDE is very fast but introduces a time delay estimation error which will be removed by SMC and SOSM.

The rest of the chapter is structured as follows. In Sect. 8.2, the robot model is presented with some assumptions. The first proposed technique is proposed in Sect. 8.3.1 and a stability analysis based on Lyapunov method is provided in Sect. 8.3.2. Section 8.4 introduces the second method with stability analysis in Sect. 8.4.2. The Sect. 8.5 contains four simulations on ANAT robot 3-DOF, the Sect. 8.5.4 shows the simulation results and a comparative study between the conventional SMC and the proposed controllers, respectively. Finally, Sect. 8.6 concludes the chapter.

8.2 System Description

Consider the dynamics of robot manipulators given by the well-known equation for rigid manipulators:

$$M(q)\ddot{q} + C(q, \dot{q})\dot{q} + G(q) + F(\dot{q}) = \tau + \tau_d \tag{8.1}$$

where $q, \dot{q}, \ddot{q} \in R^n$ are the joint position, velocity and acceleration vectors, respectively, $M(q) \in R^{n \times n}$ is symmetric positive definite inertia matrix such as $M(q)^{-1}$ always exists, $C(q, \dot{q})\dot{q} \in R^n$ is Coriolis and centrifugal forces vector, $G(q) \in R^n$ is the gravitational forces vector, $F(\dot{q}) \in R^n$ is the frictional forces, $\tau_d \in R^n$ is the disturbance vector and $\tau \in R^n$ is the torque input vector.

For ease of control design, denote $x_1 = q$, $x_2 = \dot{q}$, we rewrite the Eq. (8.1) into the following form:

$$\begin{cases} \dot{x}_1 = x_2 \\ \dot{x}_2 = f(x, t) + g(x, t)u(t) \end{cases} \tag{8.2}$$

where $x = \left[x_1^T \, x_2^T\right]^T$ is the state variable, $g(x, t) = M(q)^{-1}$ is the control matrix, $f(x, t) = M(q)^{-1}[\tau_d - C(q, \dot{q})\dot{q} - G(q) - F(\dot{q})]$ is the nonlinear dynamics which may be uncertain and $u(t) = \tau$ is the control input. We can rewrite the model in Eq. (8.2) as:

$$\begin{cases} \dot{x}_1 = x_2 \\ \dot{x}_2 = h(t) + \bar{g}u(t) \end{cases} \tag{8.3}$$

where $h(t) = f(x, t) + (g(x, t) - \bar{g})u(t)$ and $\bar{g}$ is a constant diagonal $(n \times n)$ matrix (Determination of $\bar{g}$ is discussed in (Youcef-Toumi and Ito 1990; Hsia and Gao 1991)).

The control objective is to ensure that the joint position x_1 tracks a desired trajectory x_{1d} even in the presence of unknown dynamics and external disturbances. In what follows, we will design the controller and carry out its stability analysis based on the following assumptions:

- **Assumption 1.** The system states and their derivatives are measurable.
- **Assumption 2.** $h(t)$ and its time derivative $\frac{d}{dt}[h(t)]$ are globally lipschitz functions.

8.3 First Order Sliding Mode with Time Delay

8.3.1 Design of Controller

For the design of FOSMTDC, we consider the system in (8.3). The first step in this method is to choose the switching function S in terms of the tracking error. The choice of S in the case of second-order nonlinear system (Kali et al. 2015) is:

$$S = \dot{e} + \lambda e \tag{8.4}$$

where $e = x_1 - x_{1d}$ is the tracking position error with x_{1d} is the desired trajectory, $\dot{e}$ is the tracking velocity error and $\lambda = diag(\lambda_1, \lambda_2, \ldots, \lambda_n)$ with λ_i for $i = 1, \ldots, n$ are positif constants.

The next step would be to find the control law $u(t)$ that will allow error vector $(e, \dot{e})$ to reach the sliding surface. To achieve this objective, the control law should be designed such as the reaching condition is fulfilled (Utkin et al. 1999):

$$S^T \dot{S} < 0 \tag{8.5}$$

In order to satisfy condition (8.5), $\dot{S}$ is chosen as:

$$\dot{S} = -K\, sign(S) \tag{8.6}$$

where $K = diag(k_1, k_2, \ldots, k_n)$ with k_i for $i = 1, \ldots, n$ are positif constants (determination of K will be discussed in the Sect. 8.3.2) and the signum function $sign(S) = [sign(S_1), \ldots, sign(S_n)]^T$ is defined as:

$$sign(S_i) = \begin{cases} 1, & if\ S_i > 0 \\ 0, & if\ S_i = 0 \quad for\ i = 1, \cdots, n. \\ -1, & if\ S_i < 0 \end{cases} \tag{8.7}$$

Therefore, the time derivative of S is:

$$\begin{aligned}\dot{S} &= \ddot{e} + \lambda\dot{e} \\ &= \dot{x}_2 - \dot{x}_{2d} + \lambda\dot{e} \\ &= h(t) + \bar{g}u(t) - \dot{x}_{2d} + \lambda\dot{e}\end{aligned} \tag{8.8}$$

From Eqs. (8.6) and (8.8), we get the controller as:

$$u = \bar{g}^{-1}\left(-h(t) + \dot{x}_{2d} - \lambda\dot{e} - K\, sign(S)\right) \tag{8.9}$$

As $h(t)$ has unknown parts, the control law given by Eq. (8.9) can't be used for systems given by Eq. (8.3). Then, if **Assumption 2** given in Sect. 8.2 is verified, we estimate $h(t)$ by using a TDE (Youcef-Toumi and Ito 1990):

$$\begin{aligned}\hat{h}(t) &\cong h(t - L) \\ &= \ddot{y}(t - L) - \bar{g}u(t - L)\end{aligned} \tag{8.10}$$

where L is the estimation time delay. Clearly the accuracy of $\hat{h}(t)$ improves as L decreases. In practice, the smallest achievable L is the sampling period.

Finally, replacing $h(t)$ in Eq. (8.9) by its estimate $\hat{h}(t)$ given in Eq. (8.10). Then, with sufficient condition given in the next Subsection, the FOSMTDC is obtained as:

$$u(t) = u(t - L) + \bar{g}^{-1}\left(-\dot{x}_2(t - L) + \dot{x}_{2d} - \lambda\dot{e} - K\, sign(S)\right) \tag{8.11}$$

8.3.2 *Stability Analysis*

For the stability analysis of the overall system, the method of Lyapunov is used. We select the Lyapunov function as:

$$V = \frac{1}{2}S^T S \tag{8.12}$$

Therefore, its time derivative is:

$$\begin{aligned}\dot{V} &= S^T\dot{S} \\ &= S^T\left(h(t) + \bar{g}u(t) - \dot{x}_{2d} + \lambda\dot{e}\right)\end{aligned} \tag{8.13}$$

Replacing $u(t)$ by its expression given in Eqs. (8.11) and (8.13), we obtain:

$$\dot{V} = S^T\left(h(t) - h(t - L) - K\, sign(S)\right)$$

$$= \sum_{i=1}^{n} [S_i \Delta h_i - k_i S_i sign(S_i)]$$
$$\leq \sum_{i=1}^{n} |S_i|(|\Delta h_i| - k_i) \tag{8.14}$$

To ensure $\dot{V}$ is a negative-definite function, the following condition is needed:

$$k_i > |\Delta h_i|, \; for \; i = 1, \dots, n \tag{8.15}$$

where $\Delta h_i = h_i(t) - h_i(t - L)$ is the term due to TDE error. Otherwise, as $h(t)$ is a Lipschitz function, then:

$$\begin{aligned} |\Delta h_i| &= |h_i(t) - h_i(t - L)| \\ &\leq c_i|(t) - (t - L)| \\ &\leq c_i L \end{aligned} \tag{8.16}$$

where $c_i > 0$ is the Lipschitz constant. Hence, for $k_i > c_i L$, the stability of the closed loop system is proven. The finite-time convergence of the sliding surfaces can be shown by recalling that:

$$\begin{aligned} \sum_{i=1}^{n} S_i \dot{S}_i &\leq \sum_{i=1}^{n} |S_i|(|\Delta h_i| - k_i) \\ &\leq \sum_{i=1}^{n} |S_i|(c_i L - k_i) \end{aligned} \tag{8.17}$$

Then, dividing by $|S_i|$ and integrating both sides between 0 and t gives:

$$\begin{aligned} \int_0^t \frac{S_i}{|S_i|} \dot{S}_i dt &\leq \int_0^t (c_i L - k_i) dt \\ |S_i(t)| - |S_i(0)| &\leq (c_i L - k_i)t \end{aligned} \tag{8.18}$$

Hence, by considering t_r the time required to hit S_i and noting that $|S_i(t_r)| = 0$, one has:

$$t_r \leq \frac{|S_i(0)|}{k_i - c_i L}, \; for \; i = 1, \dots, n \tag{8.19}$$

which guarantees the convergence of the sliding surfaces in a time smaller than $|S_i(0)|/(k_i - c_i L)$.

8.3.3 Effect of Switching Action

Even if the switching gains are small, the proposed FOSMTDC still have chattering in the control input. This is why instead of the $sign(S)$ function, we use saturation function $sat(S) = [sat(S_1), \ldots, sat(S_n)]^T$ defined as:

$$sat(S_i) = \begin{cases} sign(S_i) \ if \ |S_i| > \varphi_i \\ \frac{S_i}{\varphi_i} \ if \ |S_i| \leq \varphi_i \end{cases} \quad for \ i = 1, \ldots, n. \tag{8.20}$$

where φ_i is a positive constant that defines the thickness of the boundary layer. Then, the FOSMTDC with saturation function is as follows:

$$u = u(t-L) + \bar{g}^{-1}\left(-\dot{x}_2(t-L) + \dot{x}_{2d} - \lambda\dot{e} - Ksat(S)\right) \tag{8.21}$$

By using saturation function, we can expect that the chattering is reduced but the tracking error increases.

8.4 Second Order Sliding Mode with Time Delay

8.4.1 Design of Controller

The sliding set of order $r - th$ associated to manifold is defined in (Shtessel et al. 2014) by:

$$S = \dot{S} = \ddot{S} = \cdots = S^{(r-1)} = 0 \tag{8.22}$$

Note that Eq. (8.22) represents an r-dimensional condition on the state of the corresponding dynamic system. The HOSM needs a large number of information, which may not always be at our disposal. Due to the low information requirement of SOSM, we choose it instead of HOSM. For our system, the switching function and its first time derivative are defined as in Eqs. (8.4) and (8.8), respectively. Therefore, the second time derivative of S is:

$$\begin{aligned} \ddot{S} &= e^{(3)} + \lambda\ddot{e} \\ &= \ddot{x}_2 - \ddot{x}_{2d} + \lambda\ddot{e} \\ &= \frac{d}{dt}[h(t)] + \frac{d}{dt}[\bar{g}]u(t) + \bar{g}\dot{u}(t) - \ddot{x}_{2d} + \lambda\ddot{e} \end{aligned} \tag{8.23}$$

and since $\bar{g}$ is a constant matrix, then $\frac{d}{dt}[\bar{g}] = 0$.

Now, let's define a new system formed by $z_1 = S$ and $z_2 = \dot{S}$, then:

$$\begin{cases} \dot{z}_1 = \dot{S} \\ \dot{z}_2 = \frac{d}{dt}[h(t)] + \bar{g}\dot{u}(t) - \ddot{x}_{2d} + \lambda\ddot{e} \end{cases} \tag{8.24}$$

In Eq. (8.24), the time derivative $\dot{u}(t)$ would be designed to act on the second order derivative of the sliding surface. Here, the time derivative, $\dot{u}(t)$ would be designed as a discontinuous signal, but its integral $u(t)$ would be continuous by eliminating the high frequency chattering.

To determine a high order sliding mode control, a novel sliding surface is defined for the system given in Eq. (8.24) as:

$$\sigma = z_2 + \beta z_1 \tag{8.25}$$

where $\beta = diag(\beta_1, \ldots, \beta_n)$ and β_i for $i = 1, \ldots, n$ are positive constants and σ satisfies:

$$\dot{\sigma} = -K\, sign(\sigma) \tag{8.26}$$

where $K = diag(k_1, \ldots, k_n)$ is the switching gain matrix with $k_i > 0$ for $i = 1, \cdots, n$.

Differentiating Eq. (8.25) and using Eqs. (8.24) and (8.26), $\dot{u}(t)$ is expressed as:

$$\dot{u}(t) = \bar{g}^{-1}\left(-\frac{d}{dt}[h(t)] + \ddot{x}_{2d} - \lambda\ddot{e} - \beta\dot{S} - K\, sign(\sigma)\right) \tag{8.27}$$

As $h(t)$ has unknown parts, then $\frac{d}{dt}[h(t)]$ has also unknown parts, the control law given by Eq. (8.26) can't be used for systems given by Eq. (8.3). Then, if **Assumption 2** given in Sect. 8.2 is verified, we estimate $\frac{d}{dt}[h(t)]$ by using a TDE as follows:

$$\begin{aligned} \frac{d}{dt}[\hat{h}(t)] &\cong \frac{d}{dt}[h(t-L)] \\ &= \ddot{x}_2(t-L) - \bar{g}\dot{u}(t-L) \end{aligned} \tag{8.28}$$

Finally, replacing $\frac{d}{dt}[h(t)]$ in Eq. (8.26) by the estimates $\frac{d}{dt}[\hat{h}(t)]$ given in Eq. (8.28), then, with sufficient condition given in the next Subsection, the SOSMTDC is obtained as:

$$\dot{u}(t) = \dot{u}(t-L) + \bar{g}^{-1}\left(-\ddot{x}_2(t-L) + \ddot{x}_{2d} - \lambda\ddot{e} - \beta\dot{S} - K\, sign(\sigma)\right) \tag{8.29}$$

where $\ddot{x}_2(t-L)$ is given by numerical differentiation as:

$$\ddot{x}_2(t-L) = \frac{1}{L}\left[\dot{x}_2(t) - \dot{x}_2(t-L)\right] \tag{8.30}$$

8.4.2 Stability Analysis

For the stability analysis of the overall system, we have to ensure that σ goes to zero, the method of Lyapunov is used and the Lyapunov function is given by:

$$V = \frac{1}{2}\sigma^T \sigma \tag{8.31}$$

Its time derivative is as follows:

$$\begin{aligned}\dot{V} &= \sigma^T \dot{\sigma} \\ &= \sigma^T \left(\frac{d}{dt}[h(t)] + \bar{g}\dot{u}(t) - \ddot{x}_{2d} + \lambda\ddot{e} + \beta\dot{S}\right)\end{aligned} \tag{8.32}$$

By replacing $\dot{u}(t)$ by its expression given in Eq. (8.29) in the time derivative of the Lyapunov function in Eq. (8.32), we obtain:

$$\begin{aligned}\dot{V} &= \sigma^T \left(\frac{d}{dt}[h(t)] - \frac{d}{dt}[h(t-L)] - K\, sign(\sigma)\right) \\ &= \sum_{i=1}^{n} [\sigma_i \Delta H_i - k_i\sigma_i sign(\sigma_i)] \\ &\leq \sum_{i=1}^{n} |\sigma_i|(|\Delta H_i| - k_i)\end{aligned} \tag{8.33}$$

To ensure $\dot{V}$ is a negative-definite function for Lyapunov stability, the following condition must be verified:

$$k_i > |\Delta H_i|,\ for\ i = 1, \ldots, n \tag{8.34}$$

where $\Delta H_i = \frac{d}{dt}[h_i(t)] - \frac{d}{dt}[h_i(t-L)]$ is the term due to TDE error.

Otherwise, as $h(t)$ is a Lipschitz function, then:

$$\begin{aligned}|\Delta H_i| &= |\frac{d}{dt}[h_i(t)] - \frac{d}{dt}[h_i(t-L)]| \\ &\leq l_i|(t) - (t-L)| \\ &\leq l_i L\end{aligned} \tag{8.35}$$

where $l_i > 0$ is the Lipschitz constant. Hence, for $k_i > l_i L$, the stability of the closed loop system is proven. The finite-time convergence of the tracking error can be shown by using that:

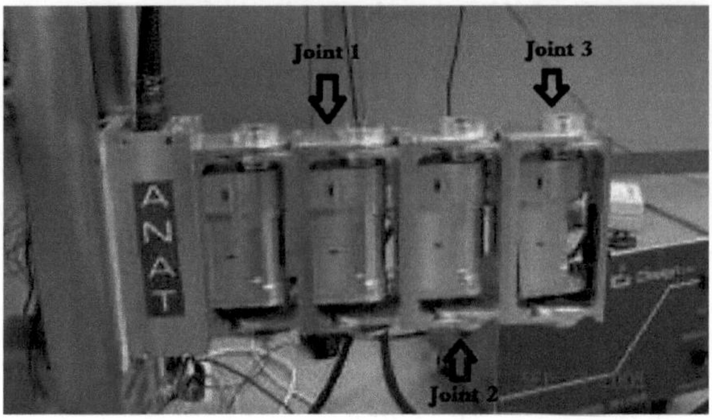

Fig. 8.1 ANAT robot arm

$$\sum_{i=1}^{n} \sigma_i \dot{\sigma}_i \leq \sum_{i=1}^{n} |\sigma_i|(|\Delta H_i| - k_i)$$
$$\leq \sum_{i=1}^{n} |\sigma_i|(l_i L - k_i) \tag{8.36}$$

Then, dividing by $|\sigma_i|$ and integrating both sides between 0 and t gives:

$$\int_0^t \frac{\sigma_i}{|\sigma_i|} \dot{\sigma}_i dt \leq \int_0^t (l_i L - k_i) dt$$
$$|\sigma_i(t)| - |\sigma_i(0)| \leq (l_i L - k_i)t \tag{8.37}$$

Hence, by considering again that $|\sigma_i(t_r)| = 0$, one has:

$$t_r \leq \frac{|\sigma_i(0)|}{k_i - l_i L}, \ for\ i = 1, \ldots, n \tag{8.38}$$

which guarantees the convergence of the tracking error in a time smaller than $|\sigma_i(0)|/(k_i - l_i L)$.

8.5 Simulation Examples

This section contains four simulations using the MATLAB software. The simulations use designs performed on the robot arm ANAT with 3-DOF (Fallaha et al. 2011) shown in Fig. 8.1 of conventional SMC and the presented theory given in Sects. 8.3.1, 8.3.3 and 8.4.1.

The dynamic of the robot arm ANAT with 3-DOF is given by:

$$\begin{cases} \dot{x}_1 = x_4 \\ \dot{x}_2 = x_5 \\ \dot{x}_3 = x_6 \\ \dot{x}_4 = h_1(t) + \sum_{i=1}^{3} \bar{g}_{1i} u_i \\ \dot{x}_5 = h_2(t) + \sum_{i=1}^{3} \bar{g}_{2i} u_i \\ \dot{x}_6 = h_3(t) + \sum_{i=1}^{3} \bar{g}_{3i} u_i \end{cases} \tag{8.39}$$

The disturbances are considered here as:

$$\tau_d = \begin{bmatrix} 0.3e^{-5(t-5)^2} \sin(4\pi t) \\ 0.15e^{-6(t-6)^2} \sin(5\pi t) \\ 0.09e^{-8(t-8)^2} \sin(6\pi t) \end{bmatrix}. \tag{8.40}$$

The model of the ANAT robot given in Eq. (8.35) satisfies all the assumptions given in Sect. 8.2. The parameter initiation vector and dynamic parameters of the robot, are mentioned in Table 8.1.

The objective is to ensure that the joint position x_1, x_2 and x_3 tracks respectively the desired trajectories x_{1d}, x_{2d} and x_{3d} even in the presence of unknown dynamics and external disturbances. To this end, let define the tracking error for each joint as $e_i = x_i - x_{id}$ for $i = 1, 2, 3$, where x_{id} is obtained with a smooth fifth-order polynomial as:

$$x_{id} = a_{xi5} t^5 + a_{xi4} t^4 + a_{xi3} t^3 + a_{xi2} t^2 + a_{xi1} t + a_{xi0} \tag{8.41}$$

where:

$$a_{xi5} = \frac{6(x_{ifd} - x_{i0d})}{t_1^5}, \; a_{xi4} = \frac{15(x_{ifd} - x_{i0d})}{t_1^4}, \; a_{xi3} = \frac{10(x_{ifd} - x_{i0d})}{t_1^3}. \tag{8.42}$$

and $a_{xi2} = a_{xi1} = a_{xi0} = 0$; x_{i0d} and x_{ifd} are respectively the desired initial and final position of link i, and t_1 is the time required for the reference trajectory to reach x_{ifd} starting from x_{i0d}. The desired trajectory parameters are given in Table 8.2.

Table 8.1 Joints initiations and parameters

Joint	Initial position (rad)	Mass (kg)	Length (m)
1	$x_1(0) = 0$	0.9	0.16
2	$x_2(0) = 0$	0.9	0.16
3	$x_3(0) = 0$	0.9	0.16

Table 8.2 Desired trajectory parameters

t_1	x_{10d}	x_{20d}	x_{30d}	x_{1fd}	x_{2fd}	x_{3fd}
5 s	0 rad	0 rad	0 rad	1.3 rad	−1.3 rad	1.3 rad

8.5.1 Conventional SMC Simulated on Robotic Arm

A complete study of conventional sliding-mode theory can be found in (Utkin 1992). The control law for the ANAT robot is given as:

$$u(t) = g(x,t)^{-1}\left(-f(x,t) + \dot{x}_{2d} - \lambda\dot{e} - K\,sign(S)\right) \tag{8.43}$$

where $g(x,t)$ is assumed to be smooth and the uncertain dynamics $f(x,t)$ is bounded for all x. That is $\|f(x,t)\| < F$, here $F \in R$ is an overestimated positive constant and the switching gain matrix $K = diag(k_1, k_2, k_3)$ is chosen such as $k_i > F$ for $i = 1, 2, 3$. The parameters used in the simulation are $\lambda = diag(10, 10, 10)$, $K = diag(15, 15, 15)$.

The simulation results are shown in Figs. 8.2 and 8.3.

8.5.2 FOSMTDC Simulated on Robotic Arm

The FOSMTDC simulated on the ANAT robot is given in Eq. (8.11) with $\lambda = diag(10, 10, 10)$, $K = diag(1, 1, 1)$ and $L = T_s = 0.03s$. While the FOSMTDC with saturation function applied to the robot is given in Eq. (8.11) with the same parameters.

The simulation results of FOSMTDC are shown in Figs. 8.4 and 8.5 while the results of FOSMTDC with saturation function are depicted in Figs. 8.6 and 8.7.

8.5.3 SOSMTDC Simulated on Robotic Arm

The SOSMTDC simulated on the ANAT robot is given in Eq. (8.29) with $\lambda = diag(10, 10, 10)$, $\beta = diag(5, 5, 5)$, $K = diag(1, 1, 1)$ and $L = T_s = 0.03s$. The desired trajectory is obtained by the same smooth fifth-order polynomial in Eq. (8.41).

The simulation results are shown in Figs. 8.8 and 8.9.

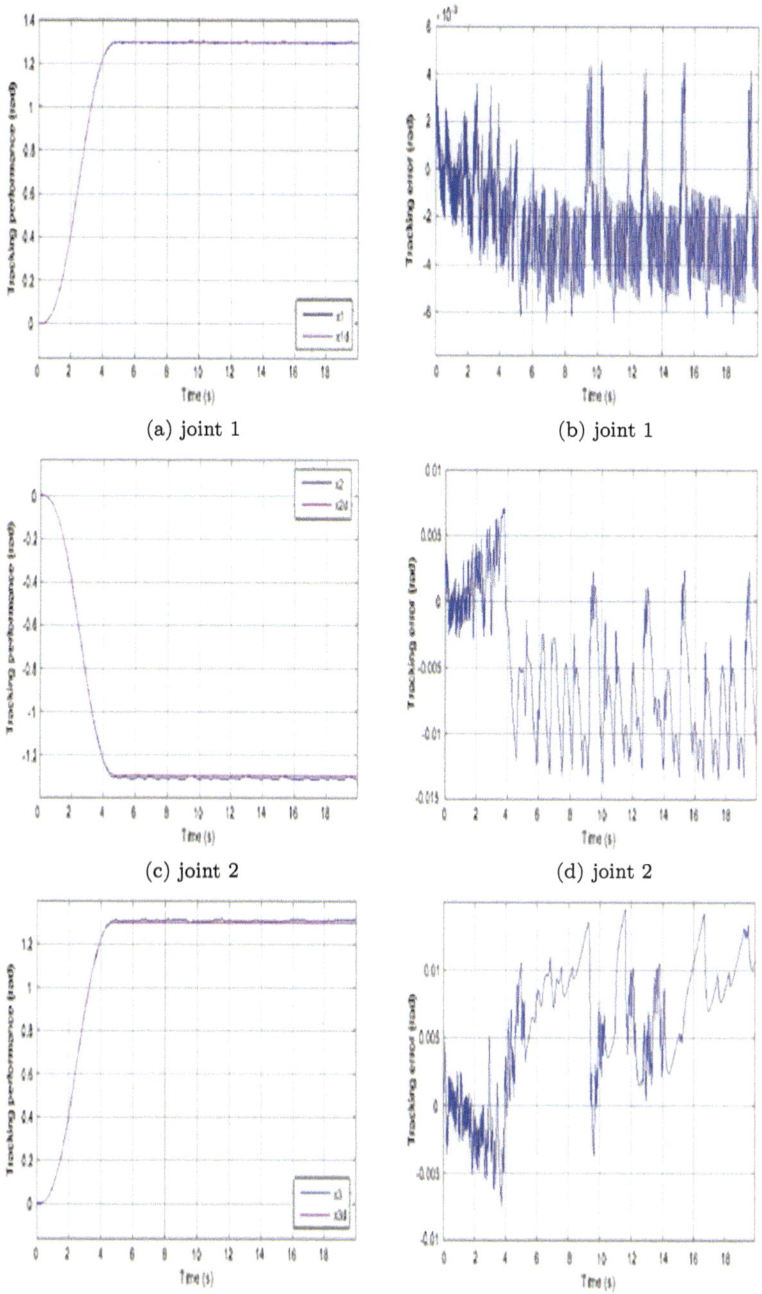

(a) joint 1 (b) joint 1

(c) joint 2 (d) joint 2

(e) joint 3 (f) joint 3

Fig. 8.2 Tracking position and error with SMC

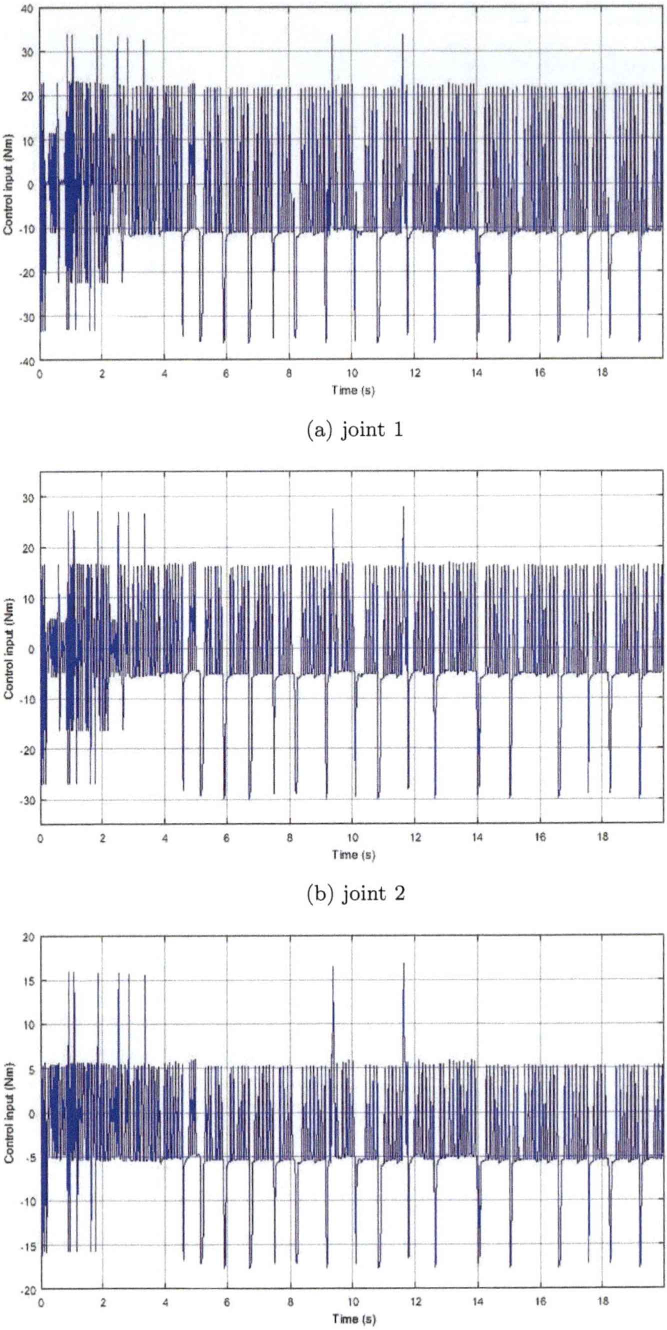

(a) joint 1

(b) joint 2

(c) joint 3

Fig. 8.3 Control input with SMC

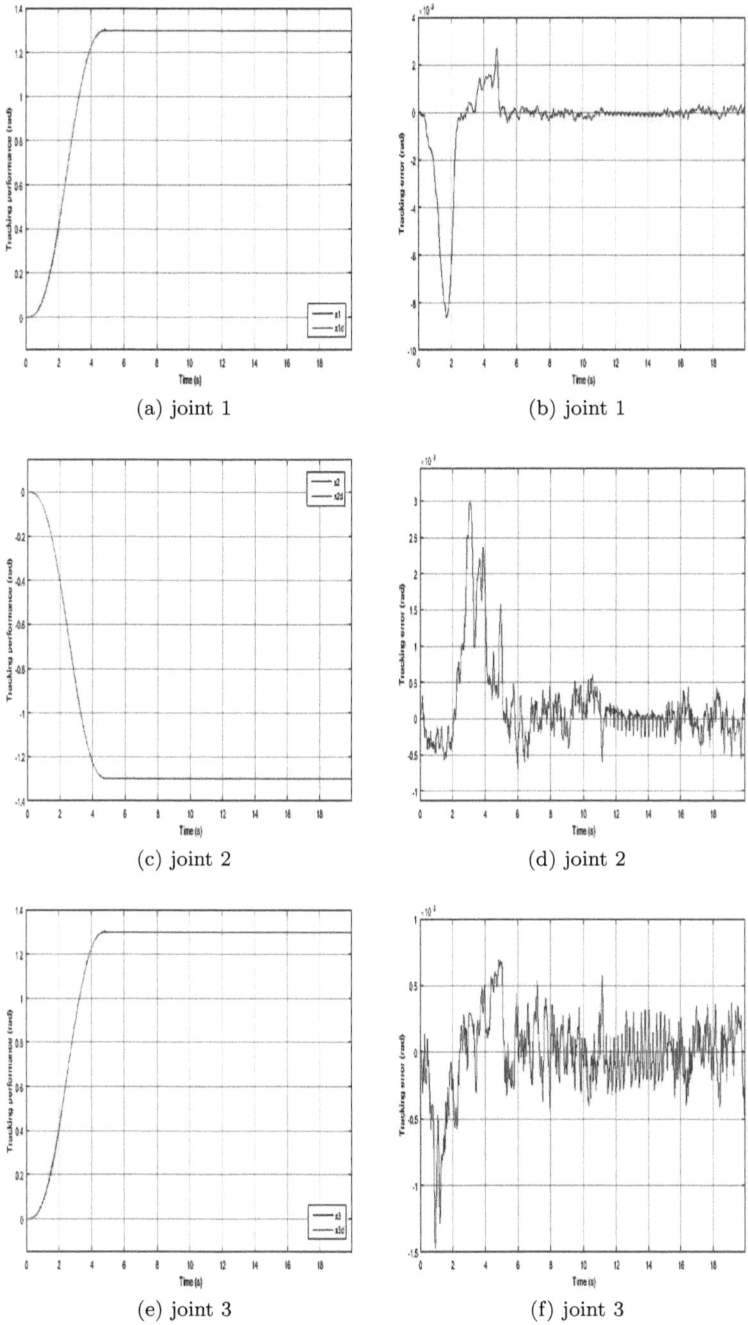

Fig. 8.4 Tracking position and error with FOSMTDC

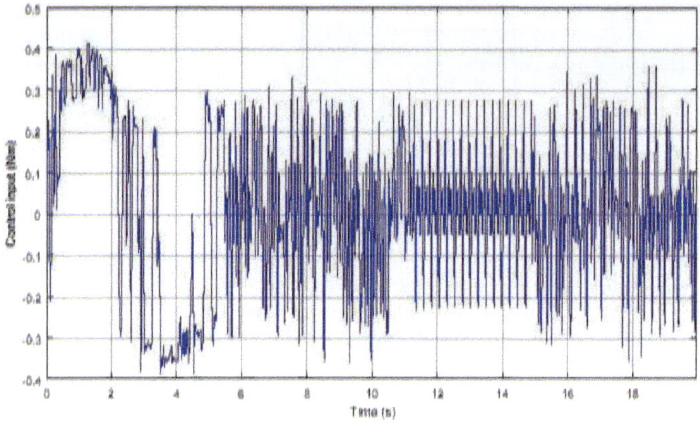

(a) joint 1

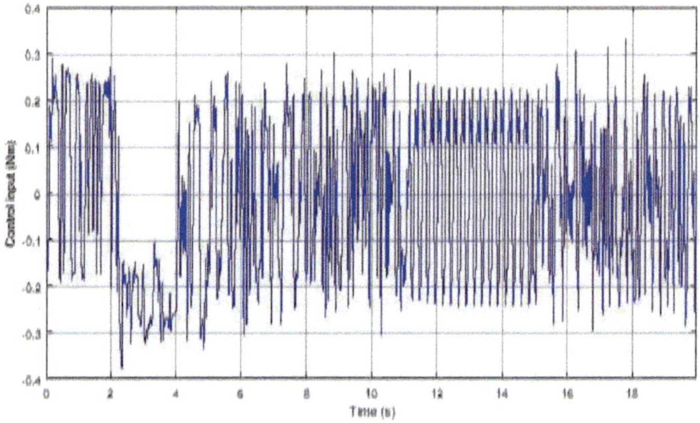

(b) joint 2

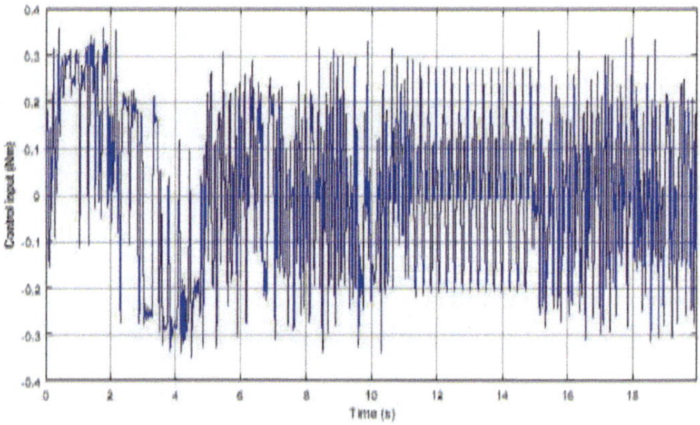

(c) joint 3

Fig. 8.5 Control input with FOSMTDC

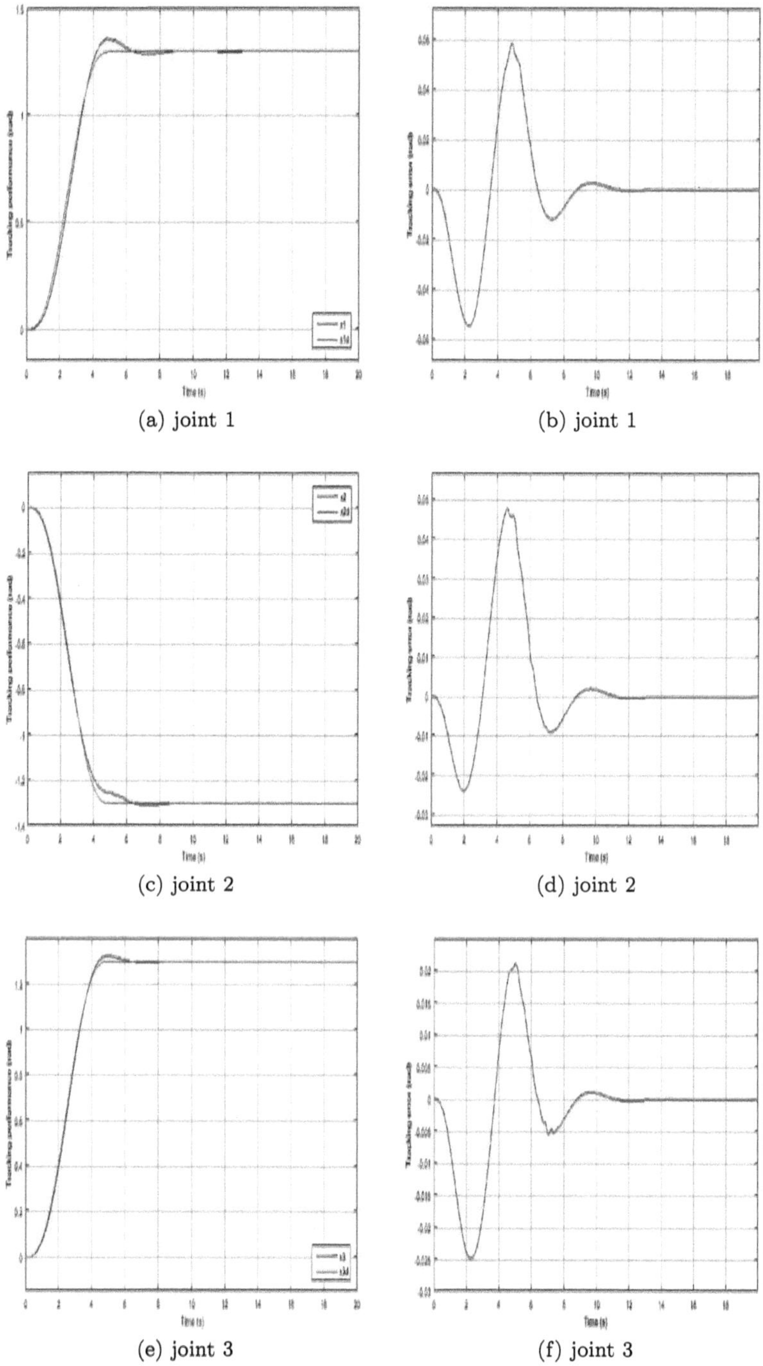

Fig. 8.6 Tracking position and error with FOSMTDC using (sat) function

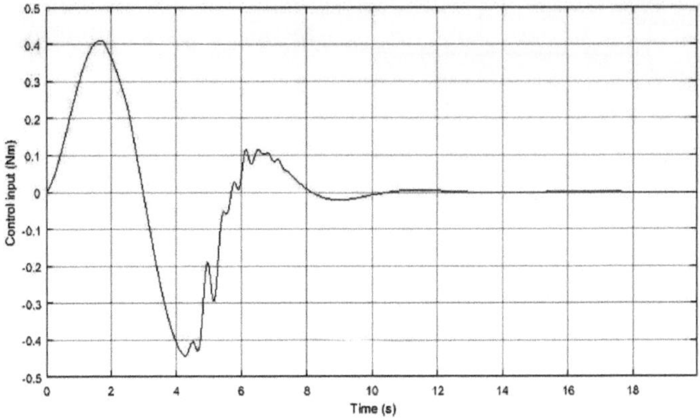

(a) joint 1

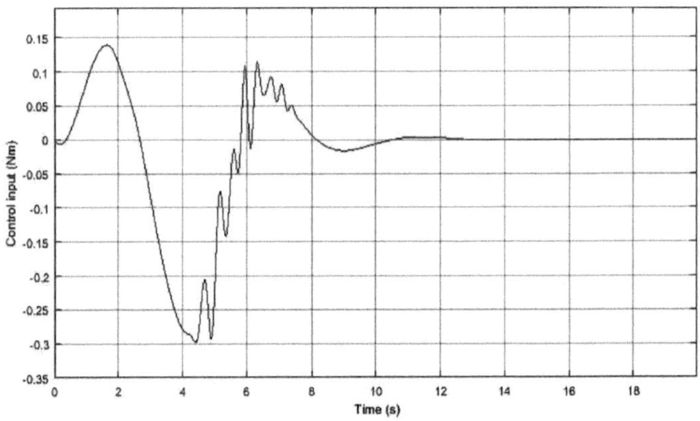

(b) joint 2

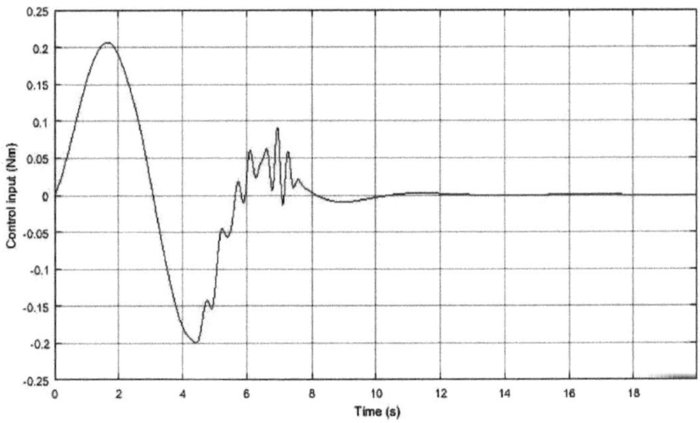

(c) joint 3

Fig. 8.7 Control input with FOSMTDC using (sat) function

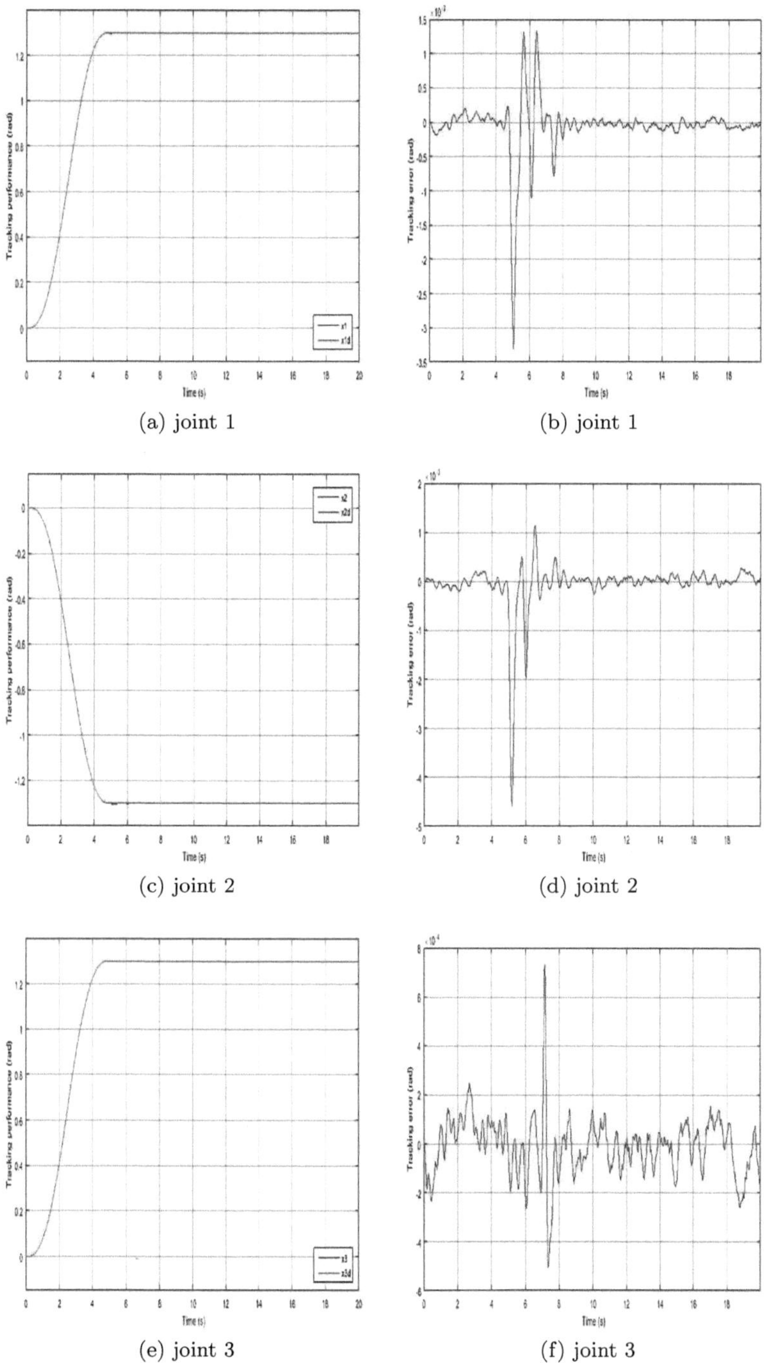

(a) joint 1 (b) joint 1

(c) joint 2 (d) joint 2

(e) joint 3 (f) joint 3

Fig. 8.8 Tracking position and error with SOSMTDC

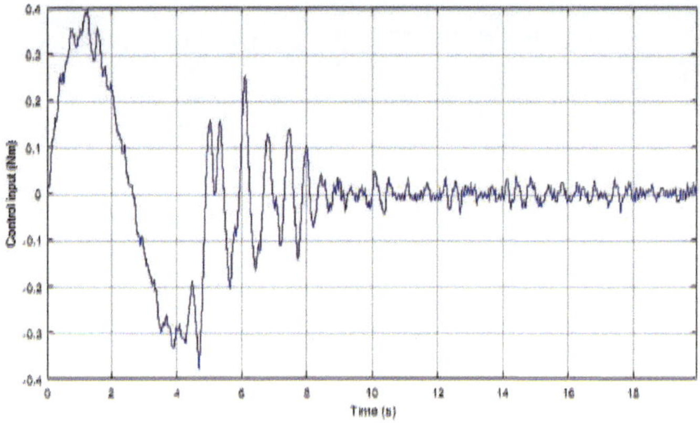

(a) joint 1

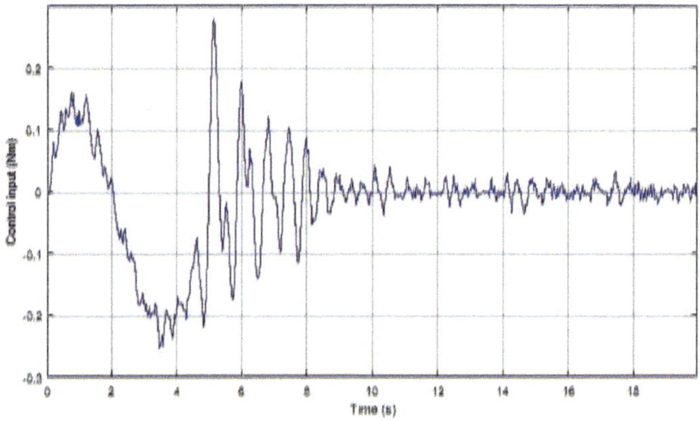

(b) joint 2

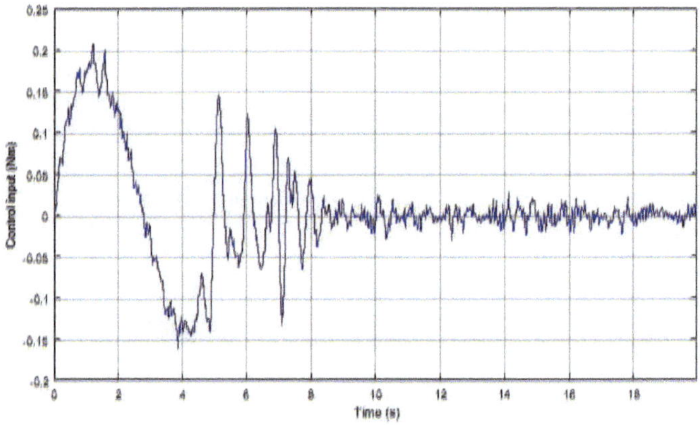

(c) joint 3

Fig. 8.9 Control input with SOSMTDC

Table 8.3 Comparative study

Control	Convergence time $t_r(s)$	Chattering
SMC	∞	excessive
FOSMTDC	5	still present
FOSMTDC with (sat) function	13	eliminate
SOSMTDC	1	reduced

8.5.4 Simulation Results and Discussion

Conventional SMC: The states x_1, x_2 and x_3 shown respectively in Fig. 8.2a, c, e did not follow their desired references properly. In addition, Fig. 8.3 shows that the chattering phenomenon is very important due to the choice of the switching gain, which is excessive and do not have any physical sense.

FOSMTDC: The positions depicted in Fig. 8.4a, c, e follow their desired references in a finite time even if the dynamics is completely unknown and the presence of disturbances. However, the chattering still present on control input as shown in Fig. 8.5, which let the tracking error oscillate around the origin.

FOSMTDC with saturation function: Figure 8.7 shows that the chattering is eliminated in control input, but the system deals with a large steady state error as depicted in Fig. 8.6b, d, f.

SOSMTDC: The simulation results show the effectiveness of this method, since the convergence of the states x_1 Fig. 8.8a, x_2 Fig. 8.8c and x_3 Fig. 8.8e in finite time, the chattering reduction as shown in Fig. 8.9.

A comparative study between the four techniques simulated is summarized in Table 8.3.

From the above comparative study, SOSMTDC shows the superiority such that the steady state error converges to zero in a finite time unlike with FOSMTDC with saturation function. The chattering is still present in FOSMTDC, which can damage the controlled robot, while is reduced in SOSMTDC and FOSMTDC with saturation function.

8.6 Conclusion

In this chapter, two controllers FOSMTDC and SOSMTDC for robot manipulators with unknown dynamics and external disturbances have been presented to achieve our control objective. According to Lyapunov stability, the resulting closed loop in both cases are stable. Furthermore, it has been proven that the tracking error converges in finite time. Due to the chattering effect on FOSMTDC as seen in simulation results, the second controller showed his superiority and seemed to be excellent for tracking control of robot manipulators with unknown dynamics and unexpected disturbances.

In the further research, we will implemented the controllers on real systems and we hope to make the second controller better in the way that high frequency uncertainties may cause complication.

Acknowledgments The authors is grateful to Jawhar Ghommam, Nabil Derbel and Quan Zhu for the opportunity to contribute to the Applications of Sliding Mode Control Lecture Notes.

References

Boiko, I., & Fridman, L. (2005). Analysis of chattering in continuous sliding-mode controllers. *IEEE Transactions on Automatic Control, 50*, 1442–1446.

Cherrid, H., M'Sirdi, N., Errahimi, F., & N-Gauthier, N. (2000). The second order sliding mode control for a bipedal walking robot. In *Third International Conference on Climbing and Walking Robots, CLAWAR, Madrid, Spain.*

Fallaha, C., Saad, M., Kanaan, H., & Al-Haddad, K. (2011). Sliding mode robot control with exponential reaching law. *IEEE Transactions on Industrial Electronics, 58*, 600–610.

Fridman, L. (2001). An averaging approach to chattering. *IEEE Transactions on Automatic Control, 46*, 1260–1265.

Fridman, L. & Levant, A. (2002). Higher order sliding mode. In *Systems and control book series.*

Hsia, T., & Gao, L. (1991). Robust independent joint controller design for industrial robot manipulators. *IEEE Transactions on Industrial Electronics, 38*, 21–25.

Jezernik, K., Rodic, M., Safaric, R., & Curk, B. (1997). Neural network sliding mode robot control. *Robotica, 15*, 23–30.

Kali, Y., Saad, M., Benjelloun, K., & Benbrahim, M. (2015). Sliding mode with time delay control for mimo nonlinear systems with unknown dynamics. In *International Workshop on Recent Advances in Sliding Modes, April* (pp. 9–11). Turkey: Istanbul.

Liu, J., & Wang, X. (2012). *Advanced sliding mode control for mechanical systems*. Heidelberg: Springer.

Palm, R. (1992). Sliding mode fuzzy control. In *IEEE International Conference on Fuzzy Systems* (pp. 519–526).

Park, C.-W., & Cho, Y.-W. (2007). Robust fuzzy feedback linearization controllers for takagi sugeno fuzzy models with parametric uncertainties. *Control Theory and Applications, IET, 1*, 1242–1254.

Sastry, S., & Bodson, M. (1994). *Adaptive control: Stability, convergence, and robustness*. New York: Prentice-Hall.

Shtessel, Y., Edwards, C., Fridman, L., & Levant, A. (2014). *Sliding mode control and observation*. New York: Springer.

Utkin, V. (1992). *Sliding mode in control and optimization*. Berlin: Springer.

Utkin, V., Guldner, J., & Shi, J. (1999). *Sliding mode control in electromechanical systems*. London: Taylor-Francis.

Youcef-Toumi, K., & Ito, O. (1990). A time delay controller for systems with unknown dynamics. *ASME Journal of Dynamic System, Measurement ans Control, 112*, 133–141.

Zhou, J., & Wen, C. (2008). *Adaptive backstepping control of uncertain systems*. Heidelberg: Springer.

Chapter 9
Kinematics and a Comparison Between Two SM Control Strategies for a 5DOF Serial Robot for Tele-Echography

Amina Jribi, Fatma Abdelhedi, Yassine Bouteraa and Nabil Derbel

Abstract This paper presents the kinematic structure and the design of a robust control strategy for a 5 Degree Of Freedom (DOF) serial robot for tele-echography. According to the manipulator kinematics, the dynamic model is formulated in the task space by using the Lagrangian formalism. Based on the system dynamics, a classical sliding mode control scheme based on the contour error is proposed to implement a performant trajectory tracking motion control. Additionally, a second order sliding mode control is applied to provide robustness in the face of uncertainties and to minimize the chattering effect caused by discontinuous control signals. The stability proof of the suggested scheme is analyzed in terms of the Lyapunov function which proves that, the system output and its second order derivative converge to the origin in a finite time. Simulation results indicate the effectiveness of the adopted controller and demonstrate satisfactory tracking performances compared to the conventional controller in the face of uncertain system parameters and external disturbances.

Keywords Dynamic model · Classical sliding mode control · Trajectory tracking · Second order sliding mode control · Lyapunov function

---

A. Jribi (✉) · F. Abdelhedi · Y. Bouteraa · N. Derbel
Control and Energy Management Laboratory (CEM-Lab), Sfax Engineering School, University of Sfax, Sfax, Tunisia
e-mail: aminajribimohamed@gmail.com

F. Abdelhedi
e-mail: fatma.abdelhedi@live.com

Y. Bouteraa
e-mail: yassinebouteraa@gmail.com

N. Derbel
e-mail: n.derbel@enis.rnu.tn

N. Derbel et al. (eds.), *Applications of Sliding Mode Control*,
Studies in Systems, Decision and Control 79, DOI 10.1007/978-981-10-2374-3_9

9.1 Introduction

Among several types of medical equipments, ultrasound tele-echography robots are widely used due to their important role in the medical field. The tele-echography system concept allows expert physicians to realize a remote echo examination so that peoples who live in isolated locations, far from medical centers and experts, can benefit from this review. The constraints of the remote tele-echography procedure led to the development of a 5-DOF serial robot (Protech). This robot is composed of two subsystems placed on two distant sites: the master site where the expert is located and the slave site where the robot is positioned and held by an assistant on the patient's body. The master subsystem is composed of a haptic device connected to a computer workstation, this device is manipulated by the expert as a virtual ultrasound probe to control the real one. In the other hand, the real probe is placed on the slave subsystem which is composed of the ultrasound system and of a robot, both connected to a second computer workstation. The slave robot executes orders sent from the master site. Both subsystems include an audio-video conference system. In this way, the expert can see the patient, communicate with him or with the non-expert operator which is located close to the patient, and then to visualize ultrasound images (Vilchis et al. 2001).

The Protech robot is used to perform tele-echography acts via a satellite link and to eliminate the problem of singularity which is faced by several tele-ultrasound robotic systems. Indeed, the presence of singular points can provide ultrasound return images that temporarily does not correspond to the set desired by the expert. Moreover, its rigidity and accuracy quickly deteriorate and its behavior becomes dangerous. In order to solve such problems and realize potential advantages of the Protech robot, an efficient control algorithm is introduced. Generally, classical proportional-differential (PD) and augmented PD controllers are the most used control schemes in actual manipulators, (Kelly et al. 2005). However, they cannot always ensure better controller performances because of the presence of high nonlinearity aspects in such complex systems, (Duchaine et al. 2007). For this reason, several advanced control algorithms have been recently developed and have led to efficient results. One of the most famous nonlinear robust control is the sliding mode controller. This theory was first proposed in the early 1950 by Emelyanov and several co-workers. Since then, it has been widely developed with the invention of high speed control devices, (Walker et al. 2005). The Classical sliding mode control has long proved its interests, that are presented in invariance to matched uncertainties, the relative simplicity of design and robustness against perturbations, (Iordanov and Surgenor 1997; Boiko et al. 2007), etc. However, this control scheme suffers from several problems, mainly from the chattering phenomenon that is presented as high frequency oscillations in the controller output, (Fridman 2001). This phenomenon may cause dangerous problems such as saturation and heat of mechanical parts of the robotic system. To overcome such difficulties, a second order sliding mode approach has been considered. Its basic idea is to change the dynamics in a small vicinity of the discontinuity surface in order to keep system states away from the real discontinuity and to maintain principal properties of the whole system.

The first step in the actual work is to present the mechanical structure of the Protech robot and to establish its dynamic and kinematics formulation. The second step is to develop the concept of the classical and that of the second order sliding mode controller. Obtained tracking performances of these controllers are then compared. Based on the control evaluation indices, simulation results demonstrate the superiority of the second order sliding mode approach compared to the classical control concept performances.

9.2 Preliminaries

9.2.1 Mechanical Structure of the Robot

In this study, simulations are based on a serial Protech robot manipulator having five degrees of freedom with three separate modules. Figure 9.1 shows the structure of this robot.

The first module has a degree of freedom corresponding to a prismatic joint allowing the end effector of the robot to move along the axis y_0. The second module has three degrees of freedom composed of revolute joints. The role of this module is to reproduce the orientation of the virtual probe which is manipulated by the expert. All the rotation axis (corresponding to each joint) are intersected in a same point, forming a remote center of rotation. So, they form a spherical wrist with concurrent axes. The angle α between joints is equal to 30°. Finally, the third module is presented as a prismatic joint. It allows a translation of the end effector along the rotation axis of the fourth joint. This translation keeps the ultrasonic probe in contact with the patient's skin and apply the same force as that practiced by the doctor on the virtual probe. Figure 9.2 shows the kinematics representation of the robot.

Fig. 9.1 Protech robot structure

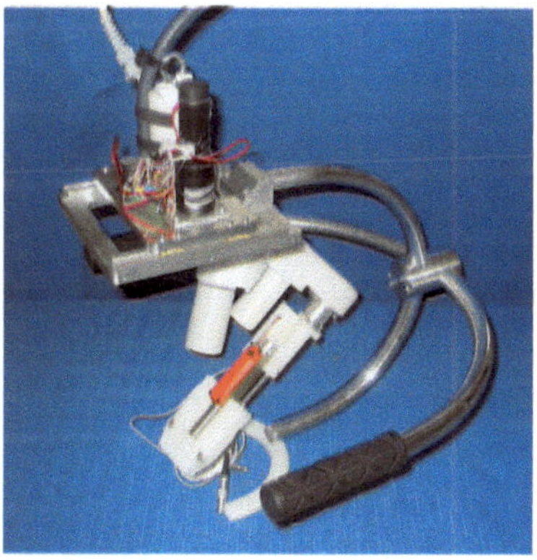

Fig. 9.2 Robot kinematics structure

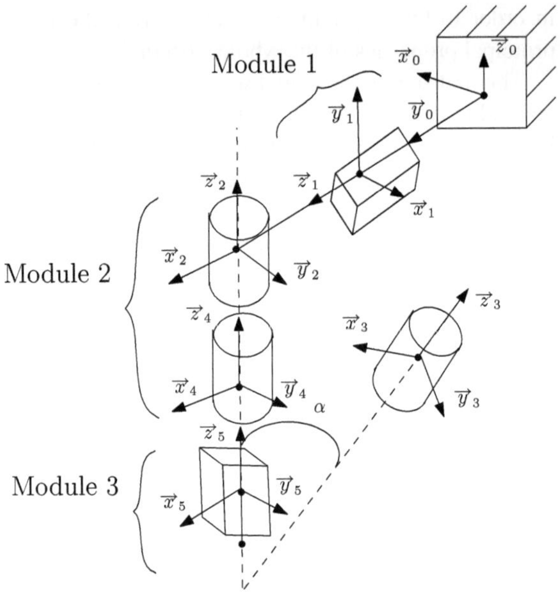

9.2.2 *Kinematics of the Protech Robot*

In order to analyse the kinematics of the Protech robot, several methods have been proposed. The most commonly used is the Denavit–Hartenberg method because of its flexibility and simplicity for modelling any number of joints and links in a manipulator, (Khalil and Kleinfinger 1986). This formalism allows the obtention of a homogeneous description for both structures, simple and complex, of articulated mechanical systems with a minimum number of parameters. Table 9.1 lists the Denavit–Hartenberg parameters of the considered robot whose axes normalization that is associated with each joint is shown in Fig. 9.2.

The homogeneous transformation matrix ${}^0T_5$, that represents the kinematic robot model is the product of transformation matrices:

$$ {}^0T_5 = {}^0T_1\,{}^1T_2\,{}^2T_3\,{}^3T_4\,{}^4T_5 \tag{9.1} $$

Table 9.1 Parameters of Denavit–Hartenberg of the robot

i	σ_i	α_i	d_i	θ_i	r_i
1	1	$\frac{\pi}{2}$	0	0	r_1
2	0	$-\frac{\pi}{2}$	0	φ_1	0
3	0	$-\alpha$	0	φ_2	0
4	0	α	0	φ_3	0
5	1	0	0	0	r_5

$$
{}^{0}T_5 = \begin{pmatrix} s_x & n_x & a_x & p_x \\ s_y & n_y & a_y & p_y \\ s_z & n_z & a_z & p_z \\ 0 & 0 & 0 & 1 \end{pmatrix} \tag{9.2}
$$

where:

$$
\begin{aligned}
s_x &= C_2C_{13} - C_\alpha S_2 S_{13} - S_1 S_\alpha^2 S_3(1 - C_2) \\
n_x &= -C_2 S_{13} - C_\alpha S_2 C_{13} - S_1 S_\alpha^2 C_3(1 - C_2) \\
a_x &= S_\alpha(C_1 S_2 - S_1 C_\alpha(1 - C_2)) \\
p_x &= a_x r_5 \\
s_y &= C_2 S_{13} + C_\alpha S_2 C_{13} + C_1 S_\alpha^2 S_3(1 - C_2) \\
n_y &= C_2 C_{13} - C_\alpha S_2 S_{13} + C_1 S_\alpha^2 C_3(1 - C_2) \\
a_y &= S_\alpha(S_1 S_2 + C_1 C_\alpha(1 - C_2)) \\
p_y &= a_y r_5 - r_1 \\
s_z &= S_\alpha(-S_2 C_3 + C_\alpha S_3(1 - C_2)) \\
n_z &= S_\alpha(S_2 S_3 + C_\alpha C_3(1 - C_2)) \\
a_z &= 1 - S_\alpha^2(1 - C_2) \\
p_z &= a_z r_5
\end{aligned} \tag{9.3}
$$

where $C_i = \cos\varphi_i,\ S_i = \sin\varphi_i,\ C_{ij} = \cos(\varphi_i + \varphi_j),\ S_{ij} = \sin(\varphi_i + \varphi_j),\ C_\alpha = \cos\alpha$ and $S_\alpha = \sin\alpha$ for $i, j = 1, 2, 3$.

The position vector of matrix ${}^{0}T_5$ is used to calculate cartesian coordinates of the robot end effector in the original reference R_0 in terms of its joint coordinates. The orientation matrix ${}^{0}A_5$ connects the base (x_5, y_5, z_5) which is related to the robot end effector, to the base (x_0, y_0, z_0) that is connected to the support of the robot, as follows:

$$
(x_0, y_0, z_0)^T = {}^{0}A_5.(x_5, y_5, z_5)^T \tag{9.4}
$$

The expression of the robot joint coordinates that we want to achieve are calculated by the following equations:

$$
\begin{aligned}
\cos\varphi_2 &= \frac{a_z - \cos^2\alpha}{\sin^2\alpha} \\
\cos\varphi_1 &= \frac{\sin^2\alpha \sin\phi_2.a_x + \cos\alpha(1 - a_z).a_y}{\sin\alpha(1 - a_z^2)} \\
\sin\varphi_1 &= \frac{-\cos\alpha(1 - a_z).a_x + \sin^2\alpha \sin\phi_2.a_y}{\sin\alpha(1 - a_z^2)} \\
\cos\varphi_3 &= \frac{-\sin^2\alpha \sin\phi_2.s_z + \cos\alpha(1 - a_z).n_z}{\sin\alpha(1 - a_z^2)}
\end{aligned}
$$

$$\sin\varphi_3 = \frac{\cos\alpha(1-a_z).s_z + \sin^2\alpha\sin\phi_2.n_z}{\sin\alpha(1-a_z^2)}$$
$$r_5 = \frac{P_z}{a_z}$$
$$r_1 = a_y r_5 - P_y \tag{9.5}$$

These equations show two singularities for $a_z = \pm 1$.

- The case $a_z = -1$ corresponds to the angle $\theta = \pi$ which is an unreachable mechanical configuration.
- The case $a_z = 1$ corresponds to a joint angle $\varphi_2 = 0$. The axis z_1 and z_3 are confused. This indicates that rotations φ_1 and φ_3 are carried on the same axis. So it is impossible to calculate independently its two angles. The following equations are obtained when $\varphi_2 = 0$:

$$\cos(\varphi_1 + \varphi_3) = s_x$$
$$\sin(\varphi_1 + \varphi_3) = s_y \tag{9.6}$$

9.2.3 Dynamic of the Protech Robot

The dynamic analysis has a double role in robotics, in such a way that it can be used both in modeling and control of robot mechanisms, (Groover et al. 2008). The dynamic model of the robotic system has been derived using the Lagrange formulation, motion equations can be then derived in a systematic way independently of the reference coordinate frame (Al Bassit 2005). The dynamic equation of the proposed manipulator can be generally written as:

$$\Gamma_i = \frac{d}{dt}\left(\frac{\partial L}{\partial \dot{q}_i}\right) - \frac{\partial L}{\partial q_i} \tag{9.7}$$

where Γ_i is the generalized force associated with the joint variables q_i, and L is the Lagrangian of the mechanical system. The Lagrangian expression can be defined as:

$$L = U + T \tag{9.8}$$

with T and U are the kinetic and the potential energies of the mechanical system, there are expressed as:

$$T = \sum \frac{1}{2}(\Omega_i^T I_{G_i}\Omega_i + m_i V_{G_i}^T V_{G_i}) \ , \quad i = 1..5$$
$$U = \sum m_i g z_i \ , \quad i = 1..5 \tag{9.9}$$

where:

- Ω_i represents the angular velocity of the ith body.
- I_{G_i} is the inertia matrix of the ith body, relatively to a reference whose origin is the center of gravity of the body. The inertia tensor is constant seen that it is expressed in a reference related to the rigid body. In the case of symmetric bodies, it is expressed by:

$$I_{G_i}/R_i = \begin{pmatrix} I_{i1} & 0 & 0 \\ 0 & I_{i2} & 0 \\ 0 & 0 & I_{i3} \end{pmatrix} \tag{9.10}$$

- m_i is the mass of the ith body.
- V_{G_i} represents the velocity of the gravity center of the ith body which is defined as:

$$V_{G_i} = \frac{d\overrightarrow{O_0 G_i}}{dt} \tag{9.11}$$

Further calculations of the kinetic and the potential energies of the system, we find the following result:

$$\begin{aligned} U &= a\dot{r_1}^2 + b\dot{\varphi_1}^2 + c\dot{\varphi_2}^2 + d\dot{\varphi_3}^2 + e\dot{r_5}^2 + f\dot{\varphi_1}\dot{\varphi_2} \\ &\quad + g\dot{\varphi_1}\dot{r_5} + h\dot{\varphi_2}\dot{r_5} + i\dot{\varphi_1}\dot{r_1} + j\dot{\varphi_2}\dot{r_1} + k\dot{r_1}\dot{r_5} \\ &\quad + l\dot{\varphi_3}\dot{\varphi_1} + m\dot{\varphi_3}\dot{\varphi_2} + n\dot{\varphi_3}\dot{r_5} \\ T &= l_1 + l_2 \sin\varphi_2 + l_3 \cos\varphi_2 + L_4 r_5 + l_5 r_5 \cos\varphi_2 \end{aligned} \tag{9.12}$$

Coefficients of the kinetic energy $(a, b, c, \ldots, n)$ are expressed in terms of the joint coordinates. The resulting torque equation yields the following equation:

$$M(q)\ddot{q} + C(q, \dot{q})\dot{q} + G(q) = \Gamma \tag{9.13}$$

where:

- $q, \dot{q}, \ddot{q} \in \mathbb{R}^n$ are joint positions, velocities and accelerations, respectively.
- $M(q)$ is a positive definite symmetric matrix of inertial accelerations.
- $C(q, \dot{q})$ is the matrix of forces / torques due to the Coriolis and centrifuged accelerations.
- $G(q)$ represents the vector of pairs of joints which equilibrates the gravity forces when the robot is at rest and in the absence of forces.
- Γ is the equivalent generalized forces/torques applied by the probe on the patient's skin.

The parameter values corresponding to the adopted Protech robot manipulator system are given in Table 9.2.

Table 9.2 Parameters of the robot

Link i	$m_i(kg)$	$L_i(m)$	$l_i(m)$	$R_i(m)$
Link 1	0.4	0.5	0.2	0.1
Link 2	0.4	0.5	0.2	0.1
Link 3	0.4	0.5	0.2	0.1
Link 4	0.4	0.5	0.2	0.1
Link 5	0.4	0.5	0.2	0.1

9.3 Control Design

9.3.1 Classical Sliding Mode Control

Sliding Mode Control is known as one of the most powerful nonlinear conventional controller due to its robustness properties, relative simplicity of design and stability. The object of sliding mode control is to restrict the system trajectories to reach and stay on a given sliding surface in the state space after a finite time, (Emelyanov et al. 1986). When confined with the sliding surface, the system dynamic is connected to the discontinuous part which describes an optimum sliding motion and denotes the controlled system behavior, (Vecchio 2008).

Let us consider a nonlinear system described by the following state equation:

$$\dot{x} = f(x) + g(x)u \tag{9.14}$$

where x and u are the state and the control variables, and $g(x)$ has a full rank. Define the sliding surface as:

$$s(x) = h(x) - h(x_d) \tag{9.15}$$

with x_d is a desired trajectory. The sliding mode controller for this system could be written as follow:

$$u = u_{eq} + \delta u \tag{9.16}$$

where the equivalent control u_{eq} represents the required control to reach and to remain on the sliding surface, (Utkin 1992). The corrected term δu is required to guarantee the remaining on the surface $s(x) = 0$. The equivalent control can be designed as follows: when the system remains on the sliding surface, we have $s(x) = 0$, then $\dot{s}(x) = 0$. Consequently:

$$
\begin{aligned}
\dot{s}(x) &= \frac{dh}{dx}\Big[f(x) + g(x)u\Big] - \frac{dh}{dx_d}\dot{x}_d \\
&= A(x, x_d) + B(x)u
\end{aligned}
\tag{9.17}
$$

This yields the following expression of the equivalent control:

$$
\begin{aligned}
u_{eq} &= \left[\frac{dh}{dx}g(x)\right]^{-1}\left[\frac{dh}{dx_d}\dot{x}_d - \frac{dh}{dx}f(x)\right] \\
&= -[B(x)]^{-1}A(x, x_d)
\end{aligned}
\tag{9.18}
$$

The term δu can be expressed as:

$$
\delta u = -u_0 \text{sign}\big[B^T(x)s(x)\big] \tag{9.19}
$$

In order to prove the stability of the proposed controller, we consider the Lyapunov function $V(x) = \frac{1}{2}s^T s$. Its differential with respect to time is expressed as:

$$
\begin{aligned}
\dot{V} &= s^T\dot{s} \\
&= -s^T B(x)\text{sign}\big[B^T(x)s(x)\big] < 0
\end{aligned}
\tag{9.20}
$$

Thus, according to the Lyapunov theory, the sliding mode control law is able to guarantee the stability of the closed loop control system.

9.3.2 Second Order Sliding Mode Control

The classical SMC law presents oscillations at high frequencies leading to the appearance of a chattering phenomenon, which reflects the discontinuous nature of the control. In order to eliminate the chattering action, the sign function can be replaced by an approximation such as *sat* or *tanh*, however this method may present steady state errors and deteriorate system performances, (Slotine and Li 1991). In this study, we are interested in the second order sliding mode control approach, (Bartolini et al. 1998). This methodology consists in considering the derivatives of the sliding variable s in order to eliminate or to minimize oscillations around the sliding surface, (Levant 1993). Then, the control action guarantees the finite time convergence of the state to the sliding manifold $s = \dot{s} = 0$, (Levant and Fridman 1996). To design the adequate controller, it is useful to observe the expression of the first sliding surface derivative $\dot{s} = \sigma$, that gives the following control law:

$$u = -[B(x)]^{-1}A(x, x_d) + [B(x)]^{-1}\sigma$$
$$= u_{eq} + \left[\frac{dh}{dx}g(x)\right]^{-1}\sigma \quad (9.21)$$

The adequate system representation becomes:

$$\begin{cases} \dot{s} = \sigma \\ \dot{\sigma} = -h_0\, s - h_1\, \sigma + v \end{cases} \quad (9.22)$$

where v is a control variable, h_0 and h_1 are positive scalars that can be deduced from the following Hurwitz polynomial (its roots have negative real parts) $H(p)$:

$$H(p) = p^2 + h_1\, p + h_0$$
$$= (p + \eta)^2 \quad (9.23)$$

System (9.22) can be then represented by the following form:

$$\dot{S} = \psi S + \Lambda\, v \quad (9.24)$$

where:

$$S = \begin{bmatrix} s \\ \sigma \end{bmatrix},\ \psi = \begin{bmatrix} 0 & 1 \\ -h_0 & -h_1 \end{bmatrix},\ \Lambda = \begin{bmatrix} 0 \\ 1 \end{bmatrix}$$

The characteristic equation of matrix ψ is $H(p) = 0$. Accordingly, all eigenvalues of matrix ψ have negative real parts. For that reason, ψ represents a Hurwitz matrix. Consequently, there exists two positive definite matrices P and Q such that:

$$P\psi + \psi^T P = -Q \quad (9.25)$$

We recommend the following adequate choices:

$$P = \begin{bmatrix} 3\eta^2 & \eta \\ \eta & 1 \end{bmatrix},\ Q = 2\,\eta \begin{bmatrix} \eta^2 & 0 \\ 0 & 1 \end{bmatrix} \quad (9.26)$$

The control term v is defined by:

$$v = -v_0 \text{sign}(\Lambda^T P S) \quad (9.27)$$

- **Stability analysis**:

In order to prove the stability of the proposed controller, the following stability theorem is considered:

Theorem: Consider a nonlinear uncertain dynamical system represented by (9.13). If the second order sliding mode control shown in (9.21) is applied, asymptotic robust stability of the closed-loop system is guaranteed.

Proof: Let's consider the positive Lyapunov function candidate:

$$V = \frac{1}{2} S^T P S \tag{9.28}$$

Differentiating the previous Lyapunov expression (9.28) with respect to time, gives:

$$\begin{aligned} \dot{V} &= \dot{S}^T P S + S^T P \dot{S} \\ &= S^T (\psi^T P + P\psi) S + 2v\Lambda^T P S \\ &= -S^T Q S - 2v_0|w| < 0 \end{aligned} \tag{9.29}$$

Thus, the second order sliding mode control law guarantees the stability of the closed loop control system.

9.4 Simulation Results

In this section, in order to verify the performance comparison of the classical sliding mode and the second order sliding mode strategies, simulations of the trajectory tracking task performed by a 5 DOF Protech robot are carried out in two cases of simulation: with or without external disturbances.

We have selected the vector of reference trajectories in the joint space which is noted $q_d(t)$ as follows:

$$q_d(t) = \begin{bmatrix} 0.1 \operatorname{sign}(1-t) \\ \frac{\pi}{6} \cos(\pi t) \\ \frac{\pi}{8} \sin(2\pi t) \\ \frac{\pi}{4} \cos(\frac{\pi}{2} t) \\ 0.2 \operatorname{sign}(1-t) \end{bmatrix} \tag{9.30}$$

Initial conditions of the robot are given in Table 9.3.

Control parameter values have been chosen as indicated in Table 9.4.

Simulation results concerning the classical SMC and the second order SMC without external disturbances are shown in Figs. 9.3, 9.4, 9.5 and 9.6.

Table 9.3 Initial conditions of the robot

$q_1(0)$	$q_2(0)$	$q_3(0)$	$q_4(0)$	$q_5(0)$	$\dot{q}_1(0)$	$\dot{q}_2(0)$	$\dot{q}_3(0)$	$\dot{q}_4(0)$	$\dot{q}_5(0)$
−0.3	−0.2	−0.1	0	0	0	0	0	0	0

Table 9.4 Control parameters

Control parameters	Values
λ	10
v_0	10
η	5

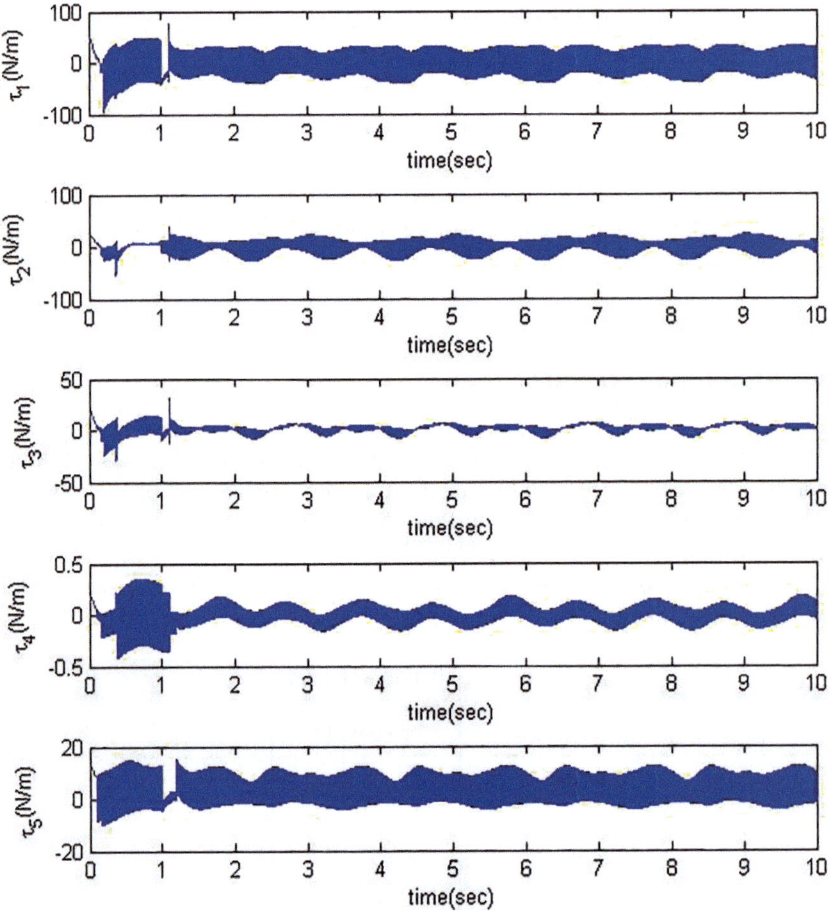

Fig. 9.3 Simple SMC torques

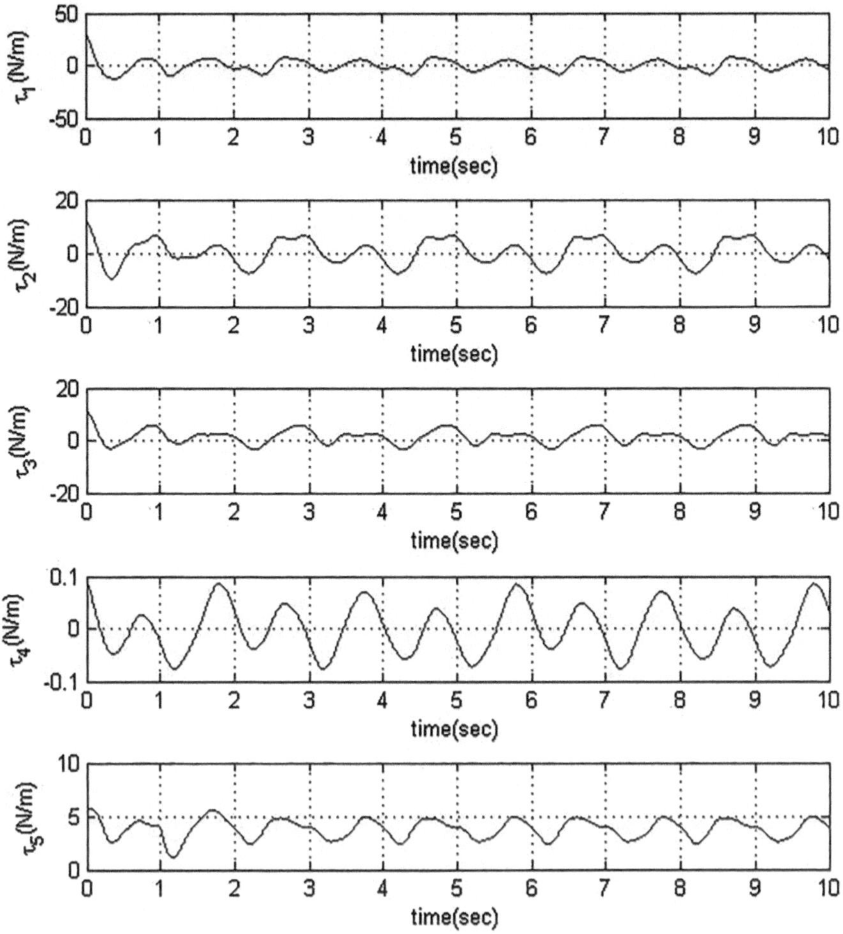

Fig. 9.4 Second order SMC torques

By making a torque comparison between Figs. 9.3 and 9.4, it is obvious that the second order SMC can considerably reduce the chattering phenomenon.

In fact, Fig. 9.4 shows a reduction of the commutation frequencies level in the second order sliding mode control input compared to that produced by the classical SMC. Besides, the evolution of the trajectory tracking motion with respect to time of the classical SMC and the second order SMC are shown in Fig. 9.5.

Compared to the simpler controller, an improved tracking behavior which is performed by the second order SMC can be clearly observed. Indeed, the higher order controller follows the desired trajectory in the absence of disturbing fluctuations in the control torques while realizing a rapid and efficient trajectory tracking task. Tracking errors are demonstrated through Fig. 9.6, where obtained curves rapidly converge to zero.

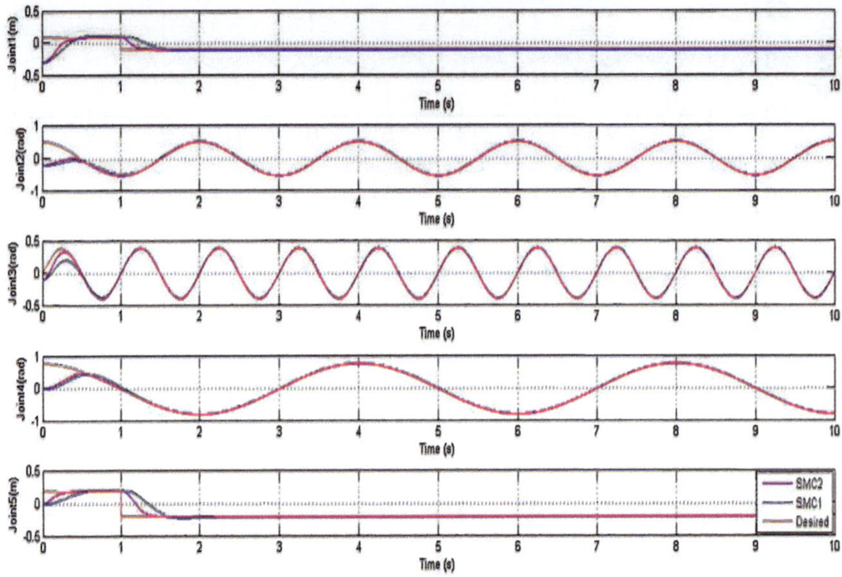

Fig. 9.5 Reference and actual trajectories

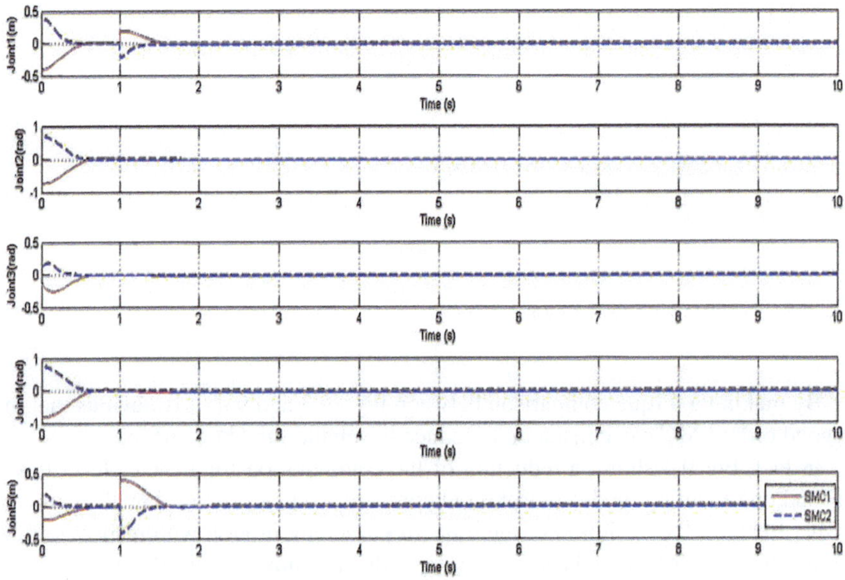

Fig. 9.6 Tracking error

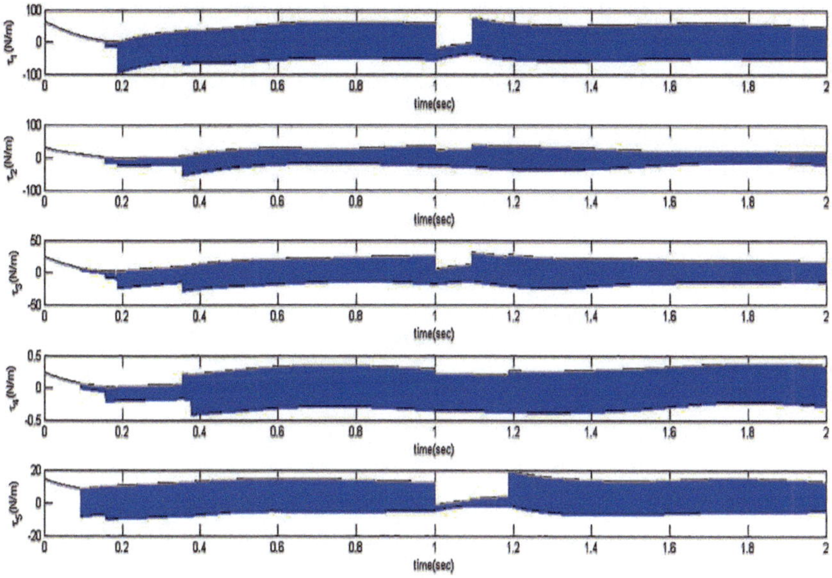

Fig. 9.7 Simple SMC torques in presence of external disturbances

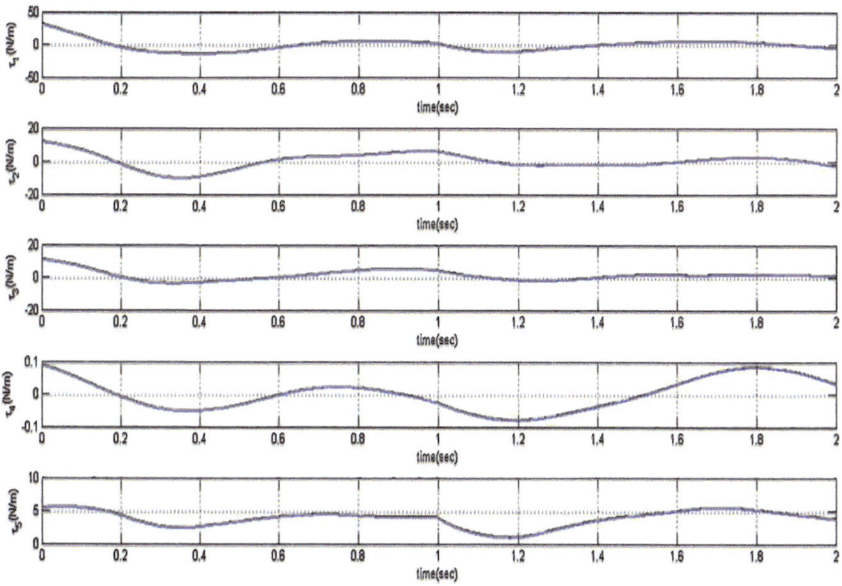

Fig. 9.8 Second order SMC torques in presence of external disturbances

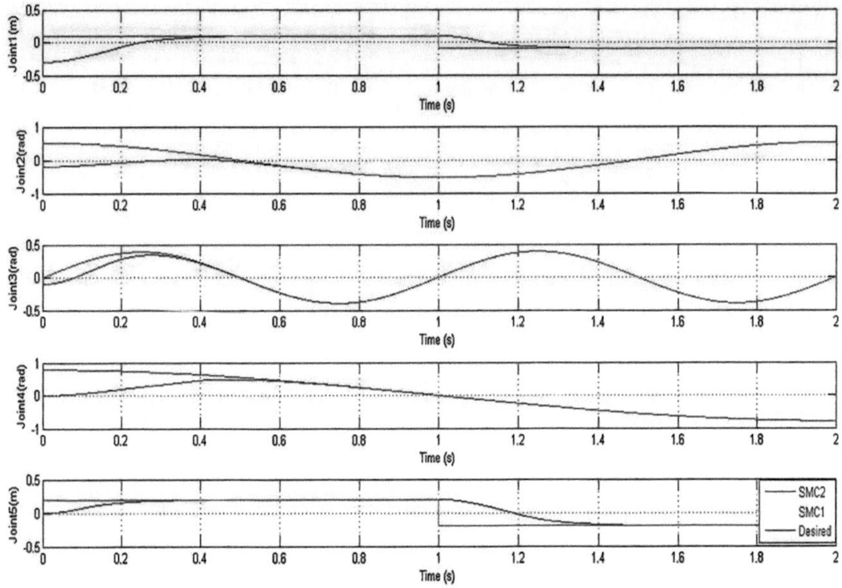

Fig. 9.9 Reference and actual trajectories in presence of external disturbances

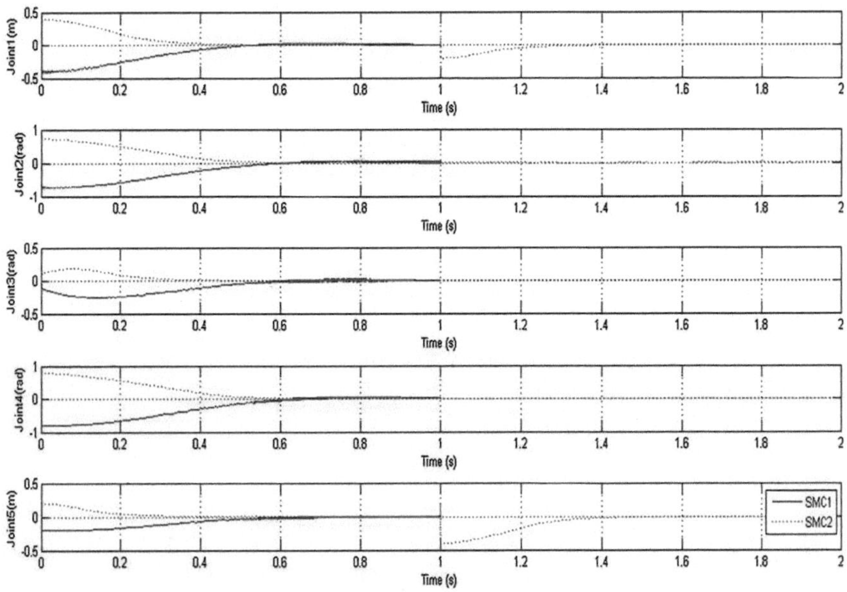

Fig. 9.10 Tracking error in presence of external disturbances

Simulation results concerning the classical SMC and the second order SMC in the presence of external disturbances are shown in Figs. 9.7, 9.8, 9.9 and 9.10. The position state vector is afflicted by 10 % measured errors.

Figure 9.7 shows that the classical SMC in presence of external disturbances causes high level commutation frequency in the control signal. This important switching frequency appears through the strong disturbance oscillations. Concerning the second order SMC torques in Fig. 9.8, we noted that the behavior of this controller has not changed despite the appearance of external disturbances.

So, it is clear that the proposed second order SMC provides good robustness properties. Figure 9.9 shows that the trajectory tracking motion keeps same behaviors of those presented without external disturbances. Consequently, the trajectory tracking task of the classical SMC and the second order SMC are indifferent to any change of environmental conditions. Figure 9.10 shows that the tracking error of the classical SMC and the second order SMC present small parasitic effects in the form of small disturbance oscillations.

Therefore, we can deduce that the second order SMC is the best approach to considerably reduce the chattering effect relatively to the classical SMC. The second order SMC improves the system response quality and provides a superior tracking performance of the closed-loop control system.

9.5 Conclusion

This paper illustrates the contribution of the second order sliding mode control as a stabilizing control law for nonlinear systems. The objective of this controller consists to minimize the oscillations in the output of the actuators by reducing considerably the level of the control and the chattering phenomenon. The stability analysis of the first order and the second order sliding mode controllers is proved via Lyapunov methodology. Performances of the second order SMC have been compared with a first order SMC, applied to the nonlinear model of a five link robot. We conclude that the second order SMC gives the best simulation results and shows a great robustness, best performances and higher accuracy even in presence of external disturbances.

References

Al Bassit, L. (2005). *Structures mécaniques á modules sphériques optimisées pour un robot médical de télé-échographie mobile*. Doctoral Thesis, Université d'Orléans.

Bartolini, G., Ferrara, A., & Usai, E. (1998). Chattering avoidance by second-order sliding mode control. *IEEE Transactions on Automatic Control*, *43*(2), 241–246.

Boiko, I., Fridman, L., Pisano, A., & Usai, E. (2007). Analysis of chattering in systems with second-order sliding modes. *IEEE Transactions on Automatic Control*, *52*(11), 2085–2102.

Duchaine, V., Bouchard, S., & Gosselin, C. M. (2007). Computationally efficient predictive robot control. *IEEE/ASME Transactions on Mechatronics*, *12*(5), 570–578.

Emelyanov, S. V., Korovin, S. V., & Levantovskii, L. V. (1986). Higher order sliding modes in the binary control systems. *Soviet Physics, Doklady, Journal Magazine*, *31*(4), 291–293.
Fridman, L. (2001). An averaging approach to chattering. *IEEE Transactions on Automatic Control*, *46*(8), 1260–1265.
Groover, M. P., Weiss, M., Nagel, R. N., & Odrey, N. G. (2008). *Industrial robotics - Technology, programming and applications*. Tata McGraw-Hill Publishing.
Iordanov, H. N., & Surgenor, B. W. (1997). Experimental evaluation of the robustness of discrete sliding mode control versus linear quadratic control. *IEEE Transactions on Control System Technology*, *5*(2), 254–260.
Kelly, R., Santibáñez, V., & Loría, A. (2005). *Control of robot manipulators in joint space*. USA: Springer.
Khalil, W., & Kleinfinger, J.-F. (1986). A new geometric notation for open and closed-looped robots. *Proceedings of the IEEE International Conference on Robotics and Automation* (pp. 1174–1180). San Francisco.
Levant, A. (1993). Sliding order and sliding accuracy in sliding mode control. *International Journal of Control*, *58*(6), 1247–1263.
Levant, A., & Fridman, L. (1996). Higher order sliding modes as a natural phenomenon in control theory. In F. Garofalo & L. Glielmo (Eds.), *Robust control via variable structure and Lyapunov techniques* (pp. 107–133). London: Springer.
Slotine, J. E., & Li, W. (1991). *Applied nonlinear control*. Englewood Cliffs: Prentice-Hall.
Utkin, V. I. (1992). *Sliding modes in control optimization*. Berlin: Springer.
Vecchio, C. (2008). *Sliding mode control: Theoretical developments and applications to uncertain mechanical systems*. Universitá degli Studi di Pavia.
Vilchis, A., Cinquin, P., Troccaz, J., Guerraz, A., Hennion, B., Pellissier, F., et al. (2001). *TER: A system for robotic TeleEchography* (pp. 326–334). Lectures notes in computer science, medical image computing and computer-assisted intervention (MICCAI'01).
Walker, I. D., Dawson, D., Flash, T., Grasso, F., Hanlon, R., Hochner, B., et al. (2005). Continuum robot arms inspired by Cephalopods. *Proceedings SPIE conference on unmanned ground vehicle technology VII* (pp. 303–314). Orlando, FL.

Chapter 10
Estimated Model-Based Sliding Mode Controller for an Active Exoskeleton Robot

Yassine Bouteraa and Ismail Ben Abdallah

Abstract This paper presents a new design of a robot elbow rehabilitation: Wireless Remote Control Arm exoskeleton (WRCAE). The robot is designed for the upper limb therapy. The developed system is an exoskeleton with two degrees of freedom that can be used for the treatment, evaluation and reinforcement. The exoskeleton actuates the both movements: flexion/extension for the elbow and pronation/supination for the forearm. Angles limits (max and min) should be introduced by the physiotherapist through a Human-Machine Interface (HMI). Desired angles of the both movements (elbow flexion/extension or forearm pronation/supination) are sent remotely via the ZigBee protocol (xbee-pro communication). A kinematic model has been developed based on Denavit–Hartenberg approach to make first tests. A sliding mode robust law control has been implemented. A kinect camera was built to detect different measures of flexion/extension and send feedback to the controller. The Lyapunov-based approach has been used to establish the system asymptotic stability. Experimental results are provided to demonstrate performances of the developed robot of upper limb remote rehabilitation.

Keywords Rehabilitation · Upper limb exoskeleton · Robust control · xbee communication · Human-machine interface · Kinect camera

Y. Bouteraa (✉) · I. Ben Abdallah
Control and Energy Management Laboratory (CEM-Lab), Sfax Engineering School, University of Sfax, Sfax, Tunisia
e-mail: yassinebouteraa@gmail.com

N. Derbel et al. (eds.), *Applications of Sliding Mode Control*,
Studies in Systems, Decision and Control 79, DOI 10.1007/978-981-10-2374-3_10

10.1 Introduction

10.1.1 General Introduction

Ability to move is so necessary to perform basic activities of daily living. Motor disability reduces significantly the quality of the patient life. A disabled member specifically lost its independence. Motorization repetitive exercises of the human joint is proven as an effective solution for the lost motion recovery. Recent technological advances have focused on the functional movement recovering of the upper and lower extremities. Patient population needing physical rehabilitation of the upper extremity was dramatically increased. This phenomenon has automatically led to an increase in the number of therapists and caregivers assisting physically disabled individuals at home, which can become serious problem in the near future. Rehabilitation robotic systems can be considered a highly effective solution for these kind of problem. However, the Availability of these devices is still limited, a field to work more. Provide robotic solutions for the upper limb rehabilitation by trying to improve the existing, is the main purpose of this work.

10.1.2 Motivation

Indeed, having a positive outcome of physical rehabilitation, mainly in the case of neurologically based disorders, is in strong relationship with the patient's health condition (Patton et al. 2008), its resistance, treatment duration, exercises intensity and orientation trainings (Feys et al. 2004). However Intense repetitions of coordinated motor activities is almost the most effective solution to improve and restore lost functions (Platz 2003; Feys et al. 2004; Patton et al. 2008). However this intense repetition is a huge burden for physiotherapists. Automate the motor activities can be a very effective solution that can lighten the physical doctors tasks. The main Goal is to develop a robotic system able to move the upper limb in repetitive actions for both motions of the elbow and the forearm. This robotic system can also be a platform with pre programmed scenarios for rehabilitation movements which will be applied according to the state of the patient in question. The developed robotic assistive rehabilitation device would be used to provide repeated motor practice in an effort to promote neurological recovery and improve functional use of the upper extremity. The proposed exoskeleton can be used to support upper extremity rehabilitation in individuals who sustain neurological impairments such as cervical level spinal cord injuries (SCI), acquired brain injuries (ABIs) or stroke. Have such systems will not only produce a more easy rehabilitation process, fast and efficient but also have an impact on the good management in health facilities. Indeed, these devices are affordable and may be for domestic use. A procedure that will greatly alleviate rehabilitation hospital stays. Even in health facilities, through these kind of equipment, physical therapists can take care of several patients at once, which significantly decreases the waiting time at the physical rehabilitation service.

10.1.3 Previous Works

Several approaches to restore the functionality of the upper extremity. Orthotics can be a solution to solve the problems in the upper limbs disability. Electrical stimulation is also among the techniques used for the treatment and rehabilitation. Physical therapy is still among the most effective techniques to deal with disability problems. Various works have been focused on the upper limb rehabilitation (Masiero et al. 2011; Varalta et al. 2013; Gijbels et al. 2011; Bovolenta et al. 2011; Morales et al. 2011; Culmer et al. 2011; Pignolo et al. 2012; Klein et al. 2010). Further researches have been specialized only in the rehabilitation of the elbow (Cozens 1999; Cheng et al. 2003; Mavroidis et al. 2005; Sulzer et al. 2007; Song et al. 2008; Oda et al. 2009; Vanderniepen et al. 2009; Pylatiuk et al. 2009; Rosen et al. 2001; Stein et al. 2007; Stein 2009) or hand (Sale et al. 2012; Vanoglio et al. 2013; Palsbo and Hood-Szivek 2012; Parrinello et al. 2013; Ho et al. 2011). In order to overcome the problems of sensory processing deficit, an innovative integrated setup which provides the user with an EMG-based visual-haptic biofeedback, is presented in Casellato et al. (2013). In Sale et al. (2012), authors propose a clinical evidence about the robotic contribution in the hand motor recovery improvement for acute stroke patients. A new design of a rehabilitation exoskeleton with multidisciplinary support and specific rehabilitation exercises dedicated to the paretic upper limb after stroke is developed in ARAMIS (Automatic Arm Mobility Integrated System) project (Pignolo et al. 2012). Robotic technology efficacy is explored in improving handwriting in children cases having impaired motor skills (Palsbo and Hood-Szivek 2012). In Cheng et al. (2003) authors present an assistive torque system with an homogeneic surface electromyogram (EMG) signals in order to improve the elbow torque capability of stroke patients. An EMG-based robotic hand device is conceived to provide training on impaired hand for person after stroke (Ho et al. 2011). Supinator Extender (SUE), a 2 DOF serial pneumatically actuated robot is a developed to measure and assist forearm supination-pronation and wrist flexion-extension (MasAllingtoniero et al. 2011). A clinical experience brief report about the improvement provided by robot aided therapy for upper limbs dedicated to patients with stroke-related lesion is presented in Parrinello et al. (2013). Modelling, design and control of 2DOF exoskeleton robot (ExoRob) to rehabilitate the elbow and forearm movements of physically disabled individuals with impaired upper-limb function is developed in Rahman et al. (2010, 2011, 2012). Based on servo motor to apply torque about the elbow, a simple system was developed by Cozens in Cozens (1999), proved that the mean range of active extension flexion was increased for every patient of a group of ten. Authors in Mavroidis et al. (2005) introduce a new design of smart and portable rehabilitation devices with real time monitoring where the recovery process could be dramatically improved.

10.1.4 Comparison and Contribution

Control point of view, compared to recent studies showing elbow rehabilitation prototypes (Vanderniepen et al. 2009; Oda et al. 2009; Pylatiuk et al. 2009) with an on-off control strategy, the proposed exoskeleton takes into consideration the system dynamics and presents a robust nonlinear control law which ensures the device stability. In contrast of similar studies using DC motors as actuators (Mali and Munih 2006; Mavroidis et al. 2005; Oda et al. 2009), the developed system is closed loop controlled. Compared to recent works in the field (Vanderniepen et al. 2009; Mavroidis et al. 2005; Mali and Munih 2006), the developed solution provides the velocity control during operation of the rehabilitation process operation. Some works control the system dynamics (Rahman et al. 2011, 2012). Compared these architectures, the developed strategy incorporates a robust velocity observer. Compared to all cited works, it can be considered as a first prototype that incorporates Kinect solution for the feedback information. The Kinect sensor has shown great efficiency and great accuracy in measuring of the human joints. Also in relation to previous works, including the most recent (Casellato et al. 2013; Rahman et al. 2012; Vanoglio et al. 2013; Varalta et al. 2013; Pignolo et al. 2012; Klein et al. 2010; Gijbels et al. 2011; Bovolenta et al. 2011; Morales et al. 2011; Masiero et al. 2011), we present in this work a wireless remote controlled rehabilitation robot. We can summarize the major contributions of this work in the following points:

- The proposed work, offers a remotely controlled exoskeleton.
- The developed communication solution is a wireless technology and does not consume energy (battery life: years).
- The mechatronics design brings to a low cost production.
- The system software part presents an extensible Human-Machine Interface that facilitates the work of physiotherapist.
- The rehabilitation system of the device incorporates Kinect technology to have a very effective measure of elbow angles.
- This work also focuses on the modeling and control of the proposed system device. Indeed, a kinematic model has been developed based on Denavit–Hartenberg approach.
- The exoskeleton control system is equipped by a nonlinear robust controller.

The present paper is organized as follows: In Sect. 10.2, we present the rehabilitation robotic device mechatronic design. The control law and the system stability study are detailed in Sect. 10.3. In Sect. 10.4, Human-Machine Interface is presented as well as experiments. Finally, we conclude with a work summary.

10.2 Mechatronic Design

The design of the elbow-forearm exoskeleton should be have the same human arm shape and mainly should not compromise the natural arm motion and the operator workspace. The basic kinematic structure of the Wireless Remote Control Arm

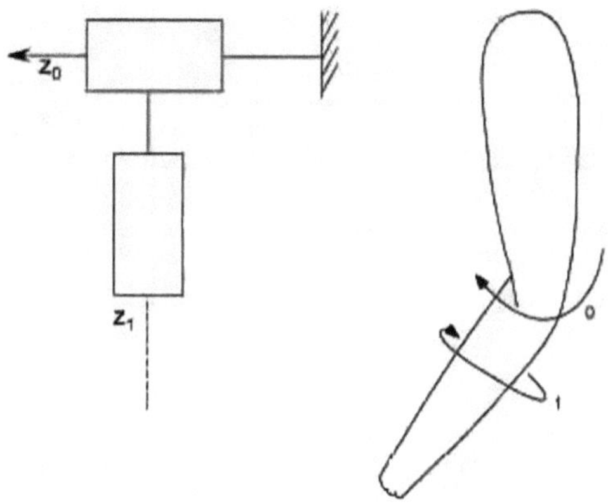

Fig. 10.1 Kinematic model of the elbow-forearm movement

Table 10.1 Control and coupling gains

Joint	Torque limits (N · m)	Workspace limits (degree)
Elbow	55	*Flex* : 90 *Ext* : 0
Forearm	5.08	*Supinat* : 90 *Pronat* : 90

exoskeleton is depicted in Fig. 10.1. The exoskeleton is comprised of a revolute joint at the elbow and a revolute joint for forearm rotation. Axes 0 and 1 represent elbow rotation and forearm rotation respectively.

The device should also have torque capabilities to match and enhance human abilities. The Elbow-Forearm rehabilitation system workspace and torque capabilities are shown in Table 10.1.

The WRCAE architecture system basically consists of 4 components (see Fig. 10.2): First, The exoskeleton, with its structure, sensors and actuators. Second, The control board is the interface between the CPU and actuators. Third, the CPU containing the control and supervision interface. This Human Machine Interface (HMI) allows robot control and feedbacks visualization (numerical values of the angles, movement curve). The last, the kinect sensor supposed to detect the angles of each joint and communicate information to the supervision interface (Fig. 10.4).

From a mechanical standpoint, the rehabilitation robot is composed essentially by the following parts:

Upper-limb support: It is an exoskeleton fitting part on the human arm. The hold of this piece guaranteed the proper functioning of the entire robot. Indeed, it should be fixed either on the fixed support or directly to the triceps with scratches on one side and metal plates to the other side. Having an arc structure, the stainless steel plates take the triceps shape.

Wrist Handle: Forming a sliding connection with the spindle, this piece fits with the size of the patient's arms which gives some flexibility to the system. In addition,

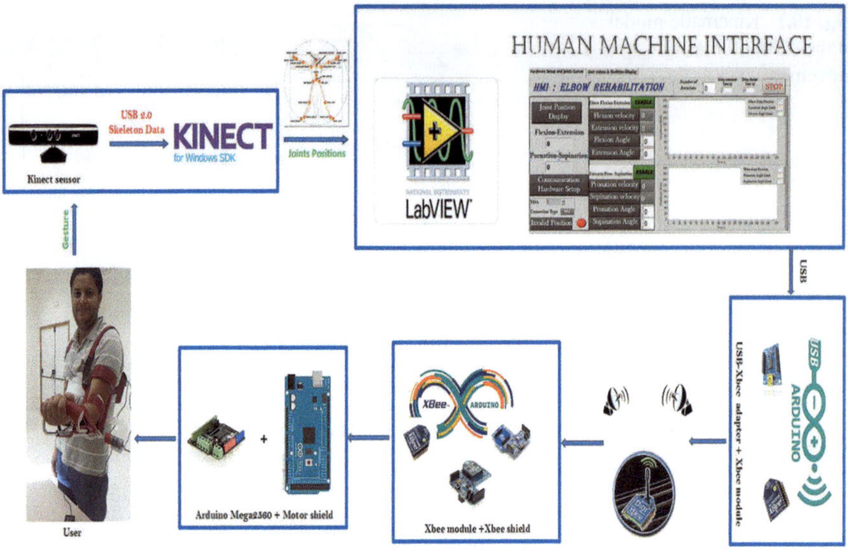

Fig. 10.2 WRCAE architecture system

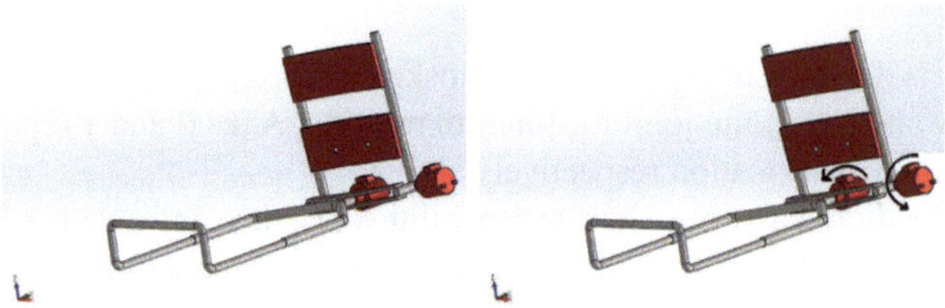

Fig. 10.3 3D design of Elbow-Forearm rehabilitation device

Fig. 10.4 Microsoft kinect sensor

wrist handle is necessary to perform the movement of pronation-supination. Patient's hand should be hold on the wrist handle.

Connecting Spindle is the main component of pronation-supination exercise. Indeed, attached directly to the motor shaft, the spindle provides the second degree of freedom: the pronation-supination joint movement. Moreover, this part provides the connection between Intermediate metal plate link and the Wrist Handle.

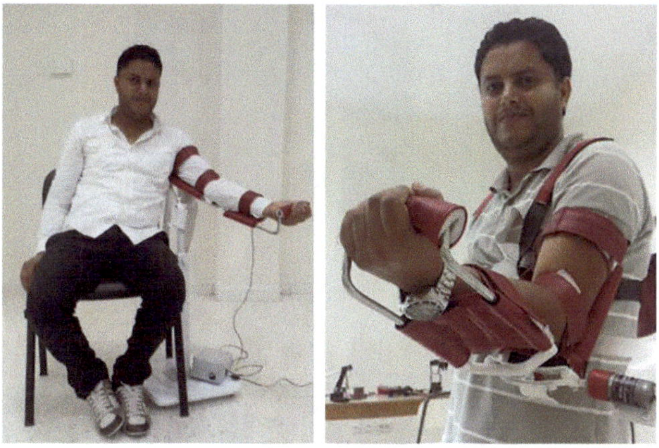

Fig. 10.5 Appropriate use of rehabilitation robot

Intermediate metal plate: The main function of this portion is the binding of the two supports: upper limb support and forearm support, providing a rotation movement between the both parts (flexion/extension).

How to use: We have more than mode of using (Fig. 10.5): 1. Portable mode with a bag on his back. The bag encompasses the electronic control board and it is connected to the charger (power from battery is possible as a solution).
2. Sitting position: user sitting on chair and the rehabilitation exoskeleton is fixed on a supporting holder with a flexible length. A small box placed on the supporting base containing the PCB connected to the electrical outlet.

10.3 Kinematic Measurement

Over the last years, controlling systems using gestures has become a common practice in our daily lives. Various works in literature deal with gesture recognition from human body using video cameras (Weinland et al. 2011; Turaga et al. 2008). However, the cameras-based control approaches was found to be significantly difficult and subsequently they are used mainly in certain well-dedicated applications. Recently, the judgment that fear the difficulty of the cameras-based solutions development becomes obsolete with the wide availability of new 3D depth cameras, such as Microsoft Kinect (see Fig. 10.4). Indeed, Microsoft Kinect as a 3D depth camera promotes the development of natural interaction applications in many domains among much larger audiences (Ibanez et al. 2014). Composed of a Red-Green-Blue camera coupled with a Depth Sensor and a processing module, kinect sensor is able to estimate various body parts movements (Fig. 10.7). The 3D position of body joints, RGB-D camera and microphones are accessible in real time thanks to Several

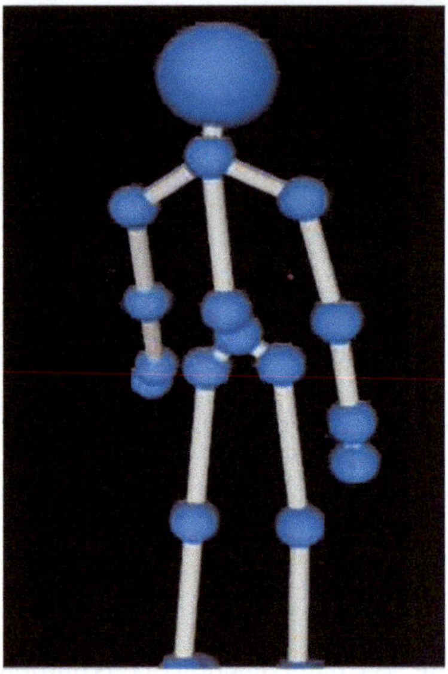

Fig. 10.6 3D skeleton display

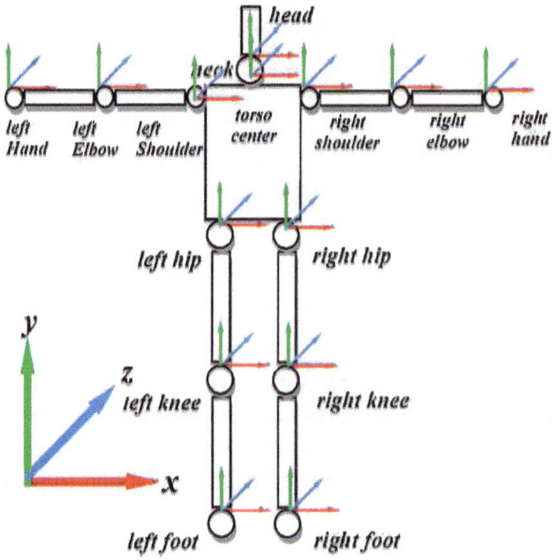

Fig. 10.7 Joints detected by kinect

SDKs (Software Development Kits), such as Microsoft Kinect SDK2, OpenNI 3 and OpenKinect. Indeed, The Kinect for Windows Software Development Kit (SDK) enables developers to create applications that support gesture and voice recognition. Kinect depth camera, called RGB-D sensor, couples RGB images with depth data.

Since 2010 (first release of Microsoft Kinect depth camera), kinect sensor gathered the scientific community attention as a vision system with a low price and a high precision (Fig. 10.4).

10.4 Control Design

Based on Lagrangian formulation, the exoskeleton robot dynamic is described as follow

$$M(q)\ddot{q} + C(q, \dot{q})\dot{q} + g(q) = \tau \tag{10.1}$$

where $q = [q_1, q_2]^T$; $\tau = [\tau_1, \tau_2]^T$. Let $q_d(t)$ denoted the desired trajectory. The tracking error is defined as: $e = q_d(t) - q(t)$.

Define

$$\dot{q}_r = q_d + \lambda(q_d - q) \tag{10.2}$$

where λ is definite positive matrix.

According to the linear characteristic of robotics, we obtain

$$\hat{M}(q)\dot{q}_r + \hat{C}(q, \dot{q})\dot{q}_r + \hat{g}(q) = Y(q, \dot{q}, \dot{q}_r, \ddot{q}_r)\hat{P} \tag{10.3}$$

where P is a robot estimated parameters vector and $Y(q, \dot{q}, \dot{q}_r, \ddot{q}_r)$ is a regressor matrix.

Define $\tilde{M}(q) = M(q) - \hat{M}(q)$; $\tilde{C}(q, q) = C(q, q) - \hat{C}(q, q)$; $\tilde{g}(q) = g(q) - \hat{g}(q)$;

Let's define the sliding variable as:

$$s = \dot{e} + \lambda e \tag{10.4}$$

We design the sliding mode controller as the following

$$\tau = \hat{M}(q)\ddot{q}_r + \hat{C}(q, \dot{q})\dot{q}_r + \hat{g}(q) + \tau_s \tag{10.5}$$

where τ_s is the robustness component.

Define: $\bar{\bar{\tilde{P}}}_i$ such $\forall i \mid \tilde{P}_i \mid \leq \bar{\bar{\tilde{P}}}_i$; Yij such $\forall i \mid \tilde{Y}_{ij} \mid \leq \bar{\bar{\tilde{Y}}}_{ij}$.

Theorem *The proposed controller (10.5) guarantees the asymptotic stability of the system (10.1).*

Proof Select the LFC as:

$$V = \frac{1}{2}s^T M(q)s \tag{10.6}$$

Differentiating with respect to time, yields:

$$\dot{V} = s^T M(q)\dot{s} + \frac{1}{2}s^T \dot{M}(q)s \tag{10.7}$$

$$= s^T M(q)\dot{s} + s^T C(\dot{q}, q)s$$

$$= s^T \left[M(q)(\ddot{q}_r - \ddot{q}) + C(\dot{q}, q)(\dot{q}_r - \dot{q})\right]$$

$$= s^T \left[M(q)\ddot{q}_r + C(\dot{q}, q)\dot{q}_r + G(q) - \tau\right] \tag{10.8}$$

From Eqs. 10.5 and 10.7, the time derivative of the Lyapunov-candidate-function can be written as:

$$\dot{V} = s^T \left[M(q)\ddot{q}_r + C(\dot{q}, q)\dot{q}_r + G(q) - \hat{M}(q)\ddot{q}_r - \hat{C}(\dot{q}, q)\dot{q}_r - \hat{G}(q) - \tau_s\right]$$

$$= s^T \left[\tilde{M}(q)\ddot{q}_r + \tilde{C}(\dot{q}, q)\dot{q}_r + \tilde{G}(q) - \tau_s\right]$$

$$= s^T \left[Y(q, \dot{q}, \dot{q}_r, \ddot{q}_r)P - \tau_s\right] \tag{10.9}$$

Let's select the robustness element τ_s as

$$\tau_s = k\text{sign}(s) + s$$

$$\tau_s = \begin{bmatrix} k_1\text{sign}(s) \\ k_2\text{sign}(s) \end{bmatrix} \tag{10.10}$$

where $k_i = \sum_{j=1}^{4} \bar{Y}_{ij}\bar{\tilde{P}}_j$, $i = 1, 2$. Further calculation, yields

$$\begin{aligned}
\dot{V}(t) &= \sum_{i=1}^{2}\sum_{j=1}^{4} s_i Y_{ij}\tilde{P}_j - \sum_{i=1}^{2} s_i k_i \text{sign}(s_i) - \sum_{i=1}^{2} s_i^2 \\
&= \sum_{i=1}^{2}\sum_{j=1}^{4} s_i Y_{ij}\tilde{P}_j - \sum_{i=1}^{2} \mid s_i \mid \bar{Y}_{ij}\bar{\tilde{P}}_j - \sum_{i=1}^{2} s_i^2 \\
&\le -\sum_{i=1}^{2} s_i^2 \le 0
\end{aligned} \tag{10.11}$$

which proves the asymptotic system stability.

10.5 Experimental Results

We further carried out experiments on the designed robot (Fig. 10.8). The user interface called HMI Elbow Rehabilitation is developed on the Labview IDE. Actually this control software is developed to serve the physiotherapist. The Human-Robot Interface contains two screens. The first screen (see Fig. 10.10) shows the user video and the skeleton display. The second screen (see Fig. 10.11) represents the hardware setup and joints curves. The user behaviors and its skeleton image, both are transmitted in real time to the physician interface. Before starting the rehabilitation session, the physiotherapist can choose the desired iteration member in each exercise. He has also the opportunity to precise the delay extension/flexion time. Some hardware setup are already set like the kind of connection and communication. The numerical

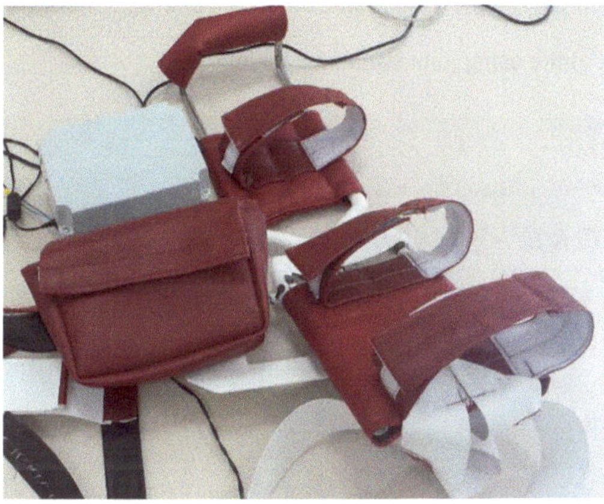

Fig. 10.8 The designed robot: portable version

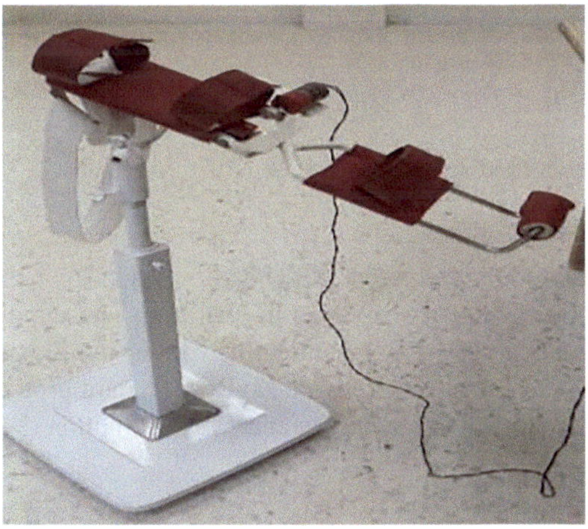

Fig. 10.9 The designed robot: sitting version

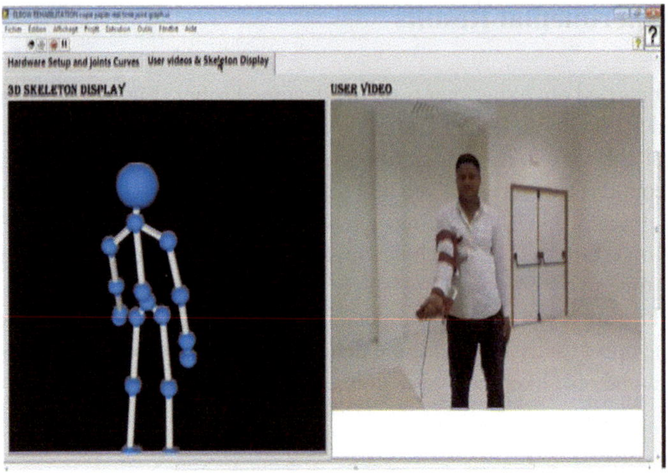

Fig. 10.10 Motion capture using joint skeleton tracking

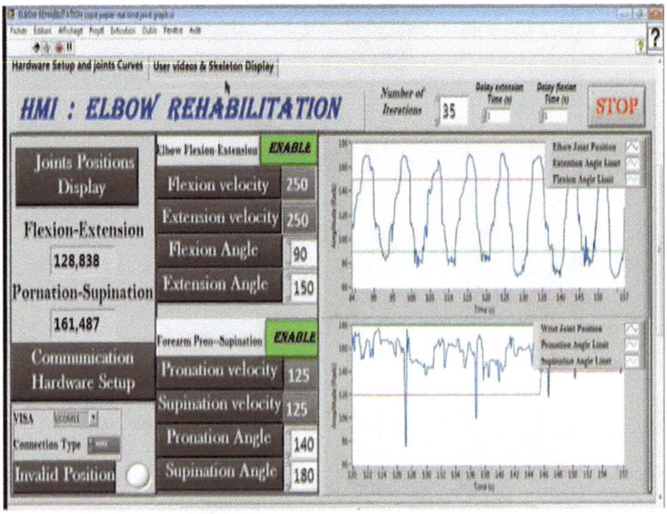

Fig. 10.11 User interface and experimental results

data about the acquired measurement of the flexion/extension behaviors and pronation/supination movements are displayed in real time. The same acquired data are graphically displayed as curves representing the various joints as a function of time. The robot is wireless controlled through a ZigBee standard protocol. The choice of the xbee solution is motivated by the low energy consumption (Fig. 10.9).

10.6 Conclusion

The stroke is a disease that leads to a large disability and even death, according to the world health organization. One of the main benefits of the physical therapy automation process is the relief and release of physiotherapist from such annoying physical tasks including repeatability, continuously monitoring and patient guide. In this context, our project focuses on the design and development of upper limb rehabilitation robot. The elbow rehabilitation robot is remotely controlled through a ZigBee protocol. The pronation/supination angles limits and the flexion/extension joints bornes, the both are remotely sent by the physiotherapist through a wireless communication. The angles made by the subject are picked up by a kinect camera and sent to the doctor interface for display and recording.

References

Allington, J., Spencer, S. J., Klein, J., Buell, M., Reinkensmeyer, D. J., & Bobrow, J. (2011). Supinator Extender (SUE): a pneumatically actuated robot for forearm/wrist rehabilitation after stroke. *Conference Proceedings of the IEEE Engineering in Medicine and Biology Society* (pp. 1579–1582).

Bovolenta, F., Sale, P., Dall-Armi, V., Clerici, P., & Franceschini, M. (2011). Robot-aided therapy for upper limbs in patients with stroke-related lesions. Brief report of a clinical experience. *Journal of NeuroEngineering and Rehabilitation*, *8*, 18.

Casellato, C., Pedrocchi, A., Zorzi, G., Vernisse, L., Ferrigno, G., & Nardocci, N. (2013). EMG-based visual-haptic biofeedback: A tool to improve motor control in children with primary dystonia. *IEEE Transactions on Neural Systems and Rehabilitation Engineering*, *21*(3), 474–480.

Cheng, H. S., Ju, M. S., & Lin, C. C. K. (2003). Improving elbow torque output of stroke patients with assistive torque controlled by EMG signals. *Journal of Biomechanical Engineering*, *125*(6), 881–886.

Cozens, J. A. (1999). Robotic assistance of an active upper limb exercise in neurologically impaired patients. *IEEE Transactions on Rehabilitation Engineering*, *7*(2), 254–256.

Culmer, P. R., Jackson, A. E., Makower, S. G., Cozens, J. A., Levesley, M. C., Mon-Williams, M., et al. (2011). A novel robotic system for quantifying arm kinematics and kinetics: Description and evaluation in therapist-assisted passive arm movements post-stroke. *Journal of Neuroscience Methods*, *197*(2), 259–269.

Feys, H., Weerdt, W. D., Verbeke, G., Steck, G. C., Capiau, C., Kiekens, C., et al. (2004). Early and repetitive stimulation of the arm can substantially improve the long-term outcome after stroke: A 5-year follow-up study of a randomized trial. *Stroke*, *35*(4), 924–929.

Gijbels, D., Lamers, I., Kerkhofs, L., Alders, G., Knippenberg, E., & Feys, P. (2011). The Armeo Spring as training tool to improve upper limb functionality in multiple sclerosis: A pilot study. *Journal of NeuroEngineering and Rehabilitation*, *8*, 5.

Ho, N. S. K., Tong, K. Y., Hu, X. L., Fung, K. L., Wei, X. J., Rong, W., & Susanto, E. A. (2011). An EMG-driven exoskeleton hand robotic training device on chronic stroke subjects: task training system for stroke rehabilitation. *IEEE International Conference on Rehabilitation Robot Boston, MA* (p. 5975340)

Ibanez, R., Soria, L., Teyseyre, A., & Campo, M. (2014). Easy gesture recognition for Kinect. *Advances in Engineering Software*, *76*, 171–180.

Klein, J., Spencer, S., Allington, J., Bobrow, J. E., & Reinkensmeyer, D. J. (2010). Optimization of a parallel shoulder mechanism to achieve a high-force, low-mass, robotic-arm exoskeleton. *IEEE Transactions on Robotics*, *26*(4), 710–715.

Mali, U., & Munih, M. (2006). HIFE-haptic interface for finger exercise. *IEEE/ASME Transactions on Mechatronics*, *11*, 93–102.

Masiero, S., Armani, M., & Rosati, G. (2011). Upper-limb robot-assisted therapy in rehabilitation of acute stroke patients: Focused review and results of new randomized controlled trial. *Journal of Rehabilitation Research and Development*, *48*(4), 355–366.

Mavroidis, C., Nikitczuk, J., Weinberg, B., Danaher, G., Jensen, K., Pelletier, P., et al. (2005). Smart portable rehabilitation devices. *Journal of NeuroEngineering and Rehabilitation*, *2*, 18.

Morales, R., Badesa, F. J., Garca-Aracil, N., & Sabater, J. M. (2011). Pneumatic robotic systems for upper limb rehabilitation. *Medical and Biological Engineering and Computing*, *49*(10), 1145–1156.

Oda, K., Isozumi, S., Ohyama, Y., Tamida, K., Kikuchi, T., & Furusho, J. (2009). Development of isokinetic and iso-contractile exercise machine MEM-MRB using MR brake. *IEEE International Conference on Rehabilitation Robotics (ICORR), Kyoto, Japan* (pp. 6–11).

Palsbo, S. E., & Hood-Szivek, P. (2012). Effect of robotic-assisted three-dimensional repetitive motion to improve hand motor function and control in children with handwriting deficits: A nonrandomized phase 2 device trial. *American Journal of Occupational Therapy*, *66*(6), 682–690.

Parrinello, I., Faletti, S., & Santus, G. (2013). Use of a continuous passive motion device for hand rehabilitation: clinical trial on neurological patients. *41 National Congress of Italian Society of Medicine and Physical Rehabilitation*, Rome-Italy.

Patton, J., Small, S. L., & Rymer, W. Z. (2008). Functional restoration for the stroke survivor: Informing the efforts of engineers. *Topics in Stroke Rehabilitation*, *15*(6), 521–541.

Pignolo, L., Dolce, G., Basta, G., Lucca, L. F., Serra, S., & Sannita, W. G. (2012). Upper limb rehabilitation after stroke, ARAMIS a robot-mechatronic innovative approach and prototype. *4th IEEE RAS and EMBS International Conference on Biomedical Robotics and Biomechatronics (BioRob), Rome-Italy* (pp. 1410–1414).

Platz, T. (2003). Evidence-based arm rehabilitation: A systematic review of the literature. *Nervenarzt*, *74*(10), 841–849.

Pylatiuk, C., Kargov, A., Gaiser, I., Werner, T., Schulz, S., & Bretthauer, G. (2009). Design of a flexible fluidic actuation system for a hybrid elbow orthosis. *IEEE International Conference on Rehabilitation Robotics (ICORR), Kyoto, Japan* (pp. 167–171).

Rahman, M. H., Saad, M., Kenn, J. P., & Archambault P. S. (2010). Exoskeleton robot for rehabilitation of elbow and forearm movements. *18th Mediterranean Conference on Control and Automation, Congress Palace Hotel, Marrakech, Morocco*, June 23–25, 2010.

Rahman, M. H., Thierry Ouimet, K., Maarouf, S., Kenn, J. P., Philippe, S., & Archambault, M. A. (2011). Robot assisted rehabilitation for elbow and forearm movements. *International Journal of Biomechatronics and Biomedical Robotics*, *1*(4).

Rahman, M. H., Thierry Ouimet, K., Maarouf, S., Kenn, J. P., Philippe, S., & Archambault, M. A. (2012). Development of a 4DoFs exoskeleton robot for passive arm movement assistance, *International Journal of Mechatronics and Automation*, *2*(1).

Rosen, J., Brand, M., Fuchs, M. B., & Arcan, M. (2001). A myosignal-based powered exoskeleton system. *IEEE Transactions on Systems, Man, and Cybernetics, Part A*, *31*(3), 210–222.

Sale, P., Lombardi, V., Franceschini, M. (2012). Hand robotics rehabilitation: Feasibility and preliminary results of a robotic treatment in patients with hemiparesis. *Stroke Research and Treatment* (p. 820931).

Song, R., Yu Tong, K., & Hu, X. (2008). Assistive control system using continuous myoelectric signal in robot-aided arm training for patients after stroke. *IEEE Transactions on Neural Systems and Rehabilitation Engineering*, *16*(4), 371–379.

Stein, J., Narendran, K., McBean, J., Krebs, K., & Hughes, R. (2007). Electromyography-controlled exoskeletal upper-limb-powered orthosis for exercise training after stroke. *American Journal of Physical Medicine and Rehabilitation*, *86*(4), 255–261.

Stein, J. (2009). e100 NeuroRobotic system. *Expert Review of Medical Devices*, *6*, 15–19.

Sulzer, J. S., Peshkin, M. A., & Patton, J. L. (2007). Design of a mobile, inexpensive device for upper extremity rehabilitation at home. *10th IEEE International Conference on Rehabilitation Robotics (ICORR), Noordwijk, Netherlands* (pp. 933–937).

Turaga, P., Chellappa, R., Subrahmanian, V. S., & Udrea, O. (2008). Machine recognition of human activities: A survey. *IEEE Transactions on Circuits and Systems for Video Technology*, *18*(11), 1473–1488.

Vanderniepen, I., Van Ham, R., Van Damme, M., Versluys, R., Lefeber, D. (2009). Orthopaedic rehabilitation: A powered elbow orthosis using compliant actuation. *IEEE International Conference on Rehabilitation Robotics (ICORR), Kyoto, Japan* (pp. 172–177).

Vanoglio, F., Luisa, A., Garofali, F., & Mora, C. (2013). Evaluation of the effectiveness of Gloreha (Hand Rehabilitation Glove) on hemiplegic patients. Pilot study. *XIII Congress of Italian Society of Neurorehabilitation*, Bari-Italy.

Varalta, V., Smania, N., Geroin, C., Fonte, C., Gandolfi, M., & Picelli, A., et al. (2013). Effects of passive rehabilitation of the upper limb with robotic device Gloreha on visual-spatial and attentive exploration capacities of patients with stroke issues. *XIII Congress of Italian Society of Neurorehabilitation, Bari-Italy.*

Weinland, D., Ronfard, R., & Boyer, E. (2011). A survey of vision-based methods for action representation, segmentation and recognition. *Computer Vision and Image Understanding*, *115*(2), 224–241.

Chapter 11
An Adaptive Finite-Time Consensus Control for Higher-Order Nonlinear Multi-agent Systems

Sanjoy Mondal, Jawhar Ghommam and Maarouf Saad

Abstract This chapter presents a finite-time consensus problem of higher-order nonlinear multi-agent systems (MAS) in the presence of bounded disturbances. The nominal control is designed by homogeneous finite-time technique to track the desired target trajectories. The chattering is mitigated by designing an integral sliding surface using adaptive super twisting algorithm (STA). The design parameters of super twisting controller are estimated adaptively without knowing the bounds a priori. The finite time convergence of the consensus protocol for the higher-order MAS is presented using Lyapunov analysis. Simulation results shows the effectiveness of the proposed homogeneous adaptive sliding mode control for the MAS.

Keywords Higher-order sliding mode · Multi-gent system · Finite-time convergence · Reference tracking · Matched uncertainty · Adaptive super twisting algorithm

11.1 Introduction

Recently, network cooperative control design of multi-agent systems (MAS) has received significant attention from various scientific and research communities and emerged as a challenging new research area. In coordination control a common agreement is achieved by interacting with each other. Thus the application of the network consensus is widely used in satellite formation flying (Li et al. 2010; Canale

S. Mondal (✉)
School of Electrical and Electronic Engineering, Nanyang Technological University, Singapore 639798, Singapore
e-mail: mondals@ntu.edu.sg; mondal.sanjoy@gmail.com

J. Ghommam · M. Saad
Department of Electrical Engineering, Ecole de Technologie Superieure, Montreal, QC, Canada
e-mail: jawhar.ghommam@etsmtl.ca

M. Saad
e-mail: maarouf.saad@etsmtl.ca

N. Derbel et al. (eds.), *Applications of Sliding Mode Control*,
Studies in Systems, Decision and Control 79, DOI 10.1007/978-981-10-2374-3_11

et al. 2015), cooperative control of unmanned air vehicles (Pack et al. 2009; Ghommam and Saad 2014), congestion control of communication networks (Li and Zhang 2010), robotic systems (Mehrjerdi et al. 2011), synchronized control of mobile robots (Zhao et al. 2011), air traffic control and urban traffic control (Weigang et al. 2008), autonomous underwater vehicles (Yoon and Qiao 2011) and so on.

Among many other formation control the leader-follower technique for MAS is explicitly studied due to its vast applications in engineering background, physical meaning and simplicity (Dimarogonas et al. 2009; Hong et al. 2007; Lin and Jia 2009; Ghommam et al. 2013). In this formation, the leader is usually independent of its followers, but can affect the follower's behaviours (Consolini et al. 2008). While the leader's behaviour can easily be controlled to achieve the desire control objective. This type of consensus protocol can simplify the implementation, system design and reduce control energy and cost (Guo et al. 2011). The past decade has witnessed the dramatic progresses in the endeavor of asymptotic consensus problems for various agent dynamics and graph topologies (Canale et al. 2013; Ghommam et al. 2013; Zhang and Wang 2015). Compared with the asymptotical control method, a finite-time control approach demonstrates higher control accuracy, stronger robustness and disturbance rejection properties besides a faster convergence rate (Moulay and Perruquetti 2008; Ghasemi and Nersesov 2014; Ghasemi et al. 2014b). Therefore, many results related to finite-time consensus of MAS, have been developed in recent years from different perspectives.

Among other robust control strategies, the sliding mode control technique is an effective and popular robust control strategy for controlling systems affected by uncertainties and external disturbances. The major drawback of sliding mode control is chattering,the high frequency switching in the control input. Which makes it very difficult to implement in real time applications. Sliding mode control method combined with Finite-time consensus problem have been adopted in many cases (Khoo et al. 2009; Ghasemi et al. 2014a, b; Ghasemi and Nersesov 2014; He et al. 2014; Yu and Long 2015). In Khoo et al. (2009), He et al. (2014), Ghasemi et al. (2014b) the use of sliding mode control is discussed in formation control to achieve the finite time convergence of the agents. But these methods are restricted to the second order agent dynamics only, suffers significantly with the high frequency chattering. As the control input is in the direct influence of the switching function. So the use of terminal sliding mode in consensus control is also affected with chattering. Which is again a main drawback while considering it in practical applications. As it can easily damage the actuators.

The high-order sliding mode (HOSM) control is developed with higher accuracy, robustness and reduction of the chattering phenomenon (Levant 1993, 2001).

In Zuo (2015) a fixed-time consensus tracking problem for second-order MAS is developed to estimate the convergence time off-line. An integral sliding mode control for double integrator MAS is recently reported in Yu and Long (2015). All these new consensus control techniques can only be used when the agent dynamics are of second order,i.e. double integrators only. In many practical cases, this assumption can be very restrictive as the agent dynamics can be of any order and can have dynamics also (Zhang and Wang 2015). So designing a more general finite time controller for higher order agent dynamics is still a considerable challenge. To the best of the authors' knowledge, the sliding mode technique has not been addressed for more general nonlinear high order MAS.

In this chapter we proposed an integral adaptive super twisting control technique for a more general nonlinear higher order MAS. The use of adaptive law gives us the flexibility of tuning the parameter automatically without knowing the bounds of the disturbances a priori. Thus the proposed controller is more generalized can be effectively implemented to any higher order linear or nonlinear agent dynamics. Therefore the original contribution of this paper is to propose a finite-time control technique for more general higher order nonlinear agent dynamics. A new adaptive homogeneous integral sliding surface is used to guarantee the finite time convergence and chattering elimination for the MAS. The use of HOSM guarantees the higher accuracy and robustness to uncertainties and external disturbances.

The main contributions are listed as follows:

- In coordination control the agent dynamics are generally considered as second order i.e. the chain of integrators only (Khoo et al. 2009; Zhao et al. 2011; Ghasemi et al. 2014a; He et al. 2014; Yu and Long 2015; Zuo 2015). Proposed technique can easily be extended to a much more complex higher order nonlinear agent dynamics.
- The consensus control is proposed for higher order agent dynamics with higher relative degree more or equal to 2.
- The distributed consensus control law for each agent, as well as the over all control protocol using the Laplacian matrix is proposed for the MAS.
- Integral sliding surface with adaptive super twisting algorithm is proposed for the consensus protocol.
- The control protocol is more suitable to alleviate the chattering problem. Hence it is more effective for electromechanical applications.

The rest of the paper is organized as follows. Sections 11.2 and 11.3 includes the mathematical preliminaries and problem formulation. The finite time consensus control are explained in Sect. 11.4. Section 11.5 provides the illustrative examples. The conclusions are given in Sect. 11.6.

11.2 Mathematical Preliminaries

In this section, we present some preliminary notations on graph theory, finite-time stability and homogeneous finite-time consensus of MAS to be used throughout the paper, and then formulate the finite-time consensus tracking problem of higher-order MAS with external disturbances.

Lemma 1 *Consider a system described with $\dot{x} = f(x)$ with $f(0) = 0$ and there exists a continuous differential positive-definite function $V(x) : D \rightarrow \Re$, and $\beta > 0$, $0 < \eta < 1$, $p, \eta \in \Re$ such that*

$$\dot{V}(x) + \beta V^{\eta}(x) \leq 0 \ \ \forall x \in D \tag{11.1}$$

Then, the origin of the system is a locally finite-time stable (Moulay and Perruquetti 2008), with the settling time depending on the initial state $x(0) = x_0$, satisfies

$$T(x_0) \leq \frac{V^{1-\eta}(x_0)}{\beta(1-\eta)} \tag{11.2}$$

Graph Theory

Consider a multi-agent system consisting of one leader and N followers. The communication topology, i.e. the information exchange between agents, is modeled by a weighted directed graph $\mathcal{G} = \{\nu, \varepsilon, \mathcal{A}\}$, where $\nu = 0, 1, 2, \ldots, N$ is the vertex set of the agents, node i represents the ith agent, $\varepsilon = \{(i, j) \subseteq \nu \times \nu\}$, where ν is the set of vertices while ε represents the set of edges respectively. The weight adjacency matrix $\mathcal{A} = (a_{ij} \geq 0) \in \Re^{(N\times 1)\times(N\times 1)}$. $(i, j) \in \varepsilon$ means that there is an directed information flow between agent i and agent j. In other way an edge denoted by an unordered pair of agents $(i, j) \in \varepsilon$ if and only if there is a communication link between i and j and there is no self edge in the graph i.e. $(i, i) \notin \varepsilon$. The adjacency elements are associated with the edges are positive, i.e. $(i, j) \in \varepsilon \Leftrightarrow a_{ij} = a_{ji} > 0$, otherwise $a_{ij} = a_{ji} = 0$, $a_{ii} = 0, \ \forall i \in \nu$ because of $(i, i) \notin \varepsilon$. For the leader-follower MAS, another graph $\bar{\mathcal{G}}$ can be considered to associate the system consisting of N followers with the leader. The leader adjacency matrix is defined as $\bar{\mathcal{B}} = [b_1, b_2, \ldots, b_N]^T \in \Re^N$ with the adjacency element $b_i > 0$ if agent i is a neighbor of the leader, otherwise $b_i = 0$. The followers can receive information from the leader, but cannot send information to the leader. Let $\mathcal{D} = \text{diag}(d_0, d_1, \ldots, d_N) \in \mathbb{R}^{N\times N}$ be a diagonal matrix, where $d_i = \sum_{j=0}^{n} a_{ij}$ for $i = 0, 1, \ldots, N$. Laplacian of the weighted graph can be defined as

$$\mathcal{L} = \mathcal{D} - \mathcal{A} \tag{11.3}$$

Lemma 2 *In a directed graph $\mathcal{G} = \{\nu, \varepsilon, \mathcal{A}\}$, the Laplacian matrix $\mathcal{L}$, has at least one zero eigenvalue and all of the non-zero eigenvalues are in the open right-half plane. Along with $\mathcal{L}$ has exactly one zero eigenvalue if and only if $\mathcal{G}$ has a rooted spanning tree. Furthermore, Rank($\mathcal{L}$) $= N$ if and only if $\mathcal{L}$ has a simple zero eigenvalue (Ren and Beard 2005).*

Let us consider $\bar{\mathcal{G}} = \{\bar{\nu}, \bar{\varepsilon}, \bar{\mathcal{A}}\}$ be the subgraph, which is formed by N followers

$$\bar{\mathcal{A}} = \begin{pmatrix} a_{11} & a_{12} & \cdots & a_{1n} \\ \vdots & \vdots & \ddots & \vdots \\ a_{N1} & a_{N2} & \cdots & a_{Nn} \end{pmatrix} \tag{11.4}$$

Let $\bar{\mathcal{D}} = diag(\bar{d}_1, \bar{d}_2, \ldots, \bar{d}_N)$ be a diagonal matrix, $\bar{d}_i = \sum_{j=1}^{N} a_{ij}, \forall i = 1, \ldots, N$. Then the Laplacian of graph $\bar{\mathcal{L}}$ is defined as

$$\bar{\mathcal{L}} = \bar{\mathcal{D}} - \bar{\mathcal{A}} \tag{11.5}$$

The Laplacian matrix $\bar{\mathcal{L}}$ is having the same properties of $\mathcal{L}$ i.e.

$$a_{ij} = \begin{cases} 1, \ \textit{if}\,(j, i) \in \varepsilon \\ 0, \ \textit{otherwise} \end{cases} \tag{11.6}$$

The connection weight between agent i and leader is represented by

$$b_i = \begin{cases} 1, \ \textit{if agent i is connected to the leader} \\ 0, \ \textit{otherwise} \end{cases} \tag{11.7}$$

11.3 Problem Formulation

Suppose the ith follower is governed by nth order dynamics as,

$$\begin{aligned} \dot{x}_i &= F_i(x_i, t) + G_i(x_i, t)u_i \quad i = 1, 2, \ldots, N \\ y_i &= h_i(x_i) \end{aligned} \tag{11.8}$$

where $x_i \in \Re^n$, $u_i \in \Re$, and $F_i(x_i, t) = F_0(x_i, t) + \Delta F_i(x_i, t)$ and $G_i(x_i, t) = G_0(x_i, t) + \Delta G_i(x_i, t)$ are n dimensional vector fields. It is assumed that $F_0(x_i, t)$ and $G_0(x_i, t)$ are known functions, while $\Delta F_i(x_i, t)$ and $\Delta G_i(x_i, t)$ are unknown bounded uncertainties.

The Lie derivative of the output function $h_i(x_i)$ with respect to the vector field $F_i(x_i, t)$ can be obtained as follows:

$$L_{F_i} h_i(x_i) = \frac{\partial h}{\partial x} F_i(x_i, t) \tag{11.9}$$

Also the Lie derivative of $L_{F_i} h_i(x_i)$ with respect to the vector field $G_i(x_i, t)$ can be defined as:

$$L_{G_i} L_{F_i} h_i(x_i) = \frac{\partial}{\partial x}(L_{F_i} h_i(x_i)) G_i(x_i, t) \tag{11.10}$$

Since the follower (11.8) has a relative degree, $r = n$, therefore one can easily obtain:

$$\begin{aligned} &L_{G_i} L_{F_i}^{k-1} h_i(x_i) = 0 \quad \forall k = 1, 2 \ldots, n-1 \\ &L_{G_i} L_{F_i}^{n-1} h_i(x_i) \neq 0 \end{aligned} \tag{11.11}$$

Using above, the *nth* derivative of the output can be obtained as:

$$y_i^n = L_{F_i}^n h_i(x_i) + L_{G_i} L_{F_i}^{n-1} h_i(x_i) u_i \tag{11.12}$$

So the ith follower (11.8) can be transformed as:

$$\begin{aligned}
\dot{x}_{i1} &= x_{i2} \\
\dot{x}_{i2} &= x_{i3} \\
&\vdots \\
\dot{x}_{in} &= f_i(x_i) + g_i(x_i)u_i \quad i = 1, 2, \ldots, N \\
&= f_{i0}(x_i, t) + g_{i0}(x_i, t)u_i + \underbrace{\Delta f_i(x_i, t) + \Delta g_i(x_i, t)u_i}_{\nabla_i} \\
y_i &= x_{i1}
\end{aligned} \tag{11.13}$$

where $f_i = L_{F_i}^n h_i$ and $g_i = L_{G_i} L_{F_i}^{n-1} h_i \neq 0$ are the Lie derivatives. It is also assumed that, $f_i(x_i, t) = f_{i0}(x_i, t) + \Delta f_i(x_i, t)$ and $g_i(x_i, t) = g_{i0}(x_i, t) + \Delta g_i(x_i, t)$. $\Delta f_i(x_i, t)$ and $\Delta g_i(x_i, t)$ are unknown uncertainties and external disturbances, whose upper bound is known. $f_{i0}(x_i, t)$ and $g_{i0}(x_i, t)$ are known nominal functions and available for feedback. $g_{i0}(x_i, t)$ is invertible.

Now $x_{i1} \in \Re$ is the output of the ith agent, $x_i = [x_{i1}, x_{i2}, \ldots x_{in}]^T \in \Re^n$ are the state variables, $u_i \in \Re$ is the control input for the ith agent. $\nabla_i = \Delta f_i(x_i, t) + \Delta g_i(x_i, t)u_i$ is the unknown disturbances and uncertainties of the ith agent. For further analysis we assume, the disturbance is bounded and continuously differentiable.

The leader dynamics can be modeled as

$$\begin{aligned}
\dot{x}_{01} &= x_{02} \\
&\vdots \\
\dot{x}_{0n} &= u_0 \\
y_0 &= x_{01}
\end{aligned} \tag{11.14}$$

where $x_0 = [x_{01}, \ldots, x_{0n}]^T \in \Re^n$ are the state vectors of the leader, $y_0 = x_{01} \in \Re$ is the output. $u_0 \in \Re$ is the control input. Here the leader dynamics is considered as a chain of integrators only, as the leader will act as a signal generators for the followers.

The homogenous MAS (11.13) is said to reach consensus in finite time, if for any initial condition, there exists a finite-time such that $lim_{t \to T_0}(x_{i1} - x_{01}) = 0$.

In this section, the finite-time consensus algorithm for nth order MAS with one leader under the directed network topology is proposed. The dynamics of the followers and the single leader are given by (11.13) and (11.14). Let $\chi_1 = [e_1^1, \ldots, e_N^1]^T$, $\chi_2 = [e_1^2, \ldots, e_N^2]^T$ and $\chi_n = [e_1^n, \ldots, e_N^n]^T$. Here for each follower, define the consensus tracking errors

$$e_i^1 = \sum_{j=1, j \neq i}^{N} a_{ij}(x_{i1} - x_{j1}) + b_i(x_{i1} - x_{01})$$

$$e_i^2 = \sum_{j=1, j\neq i}^{N} a_{ij}(x_{i2} - x_{j2}) + b_i(x_{i2} - x_{02})$$

$$\vdots$$

$$e_i^n = \sum_{j=1, j\neq i}^{N} a_{ij}(x_{in} - x_{jn}) + b_i(x_{in} - x_{0n}) \tag{11.15}$$

The error dynamics can be written as,

$$\begin{aligned}
\dot{\chi}_1 &= \chi_2 \\
\dot{\chi}_2 &= \chi_3 \\
&\vdots \\
\dot{\chi}_n &= (\bar{\mathcal{L}} + \bar{\mathcal{B}})f(\chi) + (\bar{\mathcal{L}} + \bar{\mathcal{B}})g(\chi)u + \underbrace{(\bar{\mathcal{L}} + \bar{\mathcal{B}})\nabla}_{d_o} - \bar{\mathcal{B}}\bar{1}u_0
\end{aligned} \tag{11.16}$$

where $f(\chi) = [f_{10}(x_1), \ldots, f_{N0}(x_N)]^T$, $g(\chi) = diag[g_{10}(x_1), \ldots, g_{N0}(x_N)]$, $u = [u_1, u_2, \ldots, u_N]^T$, $\bar{1} = [1, \ldots, 1]^T \in \Re^N$ and $\nabla = [\nabla_1, \nabla_2, \ldots, \nabla_N]^T$ and d_o is the over all uncertainty.

Let us consider the leader-follower system given by (11.13) and (11.14). Also let the directed graph $\mathcal{G}$ has a directed spanning tree and $\chi_1 = 0$, $\chi_2 = 0$ and $\chi_n = 0$, then

$$[x_{11} \ldots x_{N1}]^T = \bar{1}x_{01} \tag{11.17}$$

$$[x_{1n} \ldots x_{Nn}]^T = \bar{1}x_{0n} \tag{11.18}$$

Now $e_i^1 = \sum_{j=1, j\neq i}^{N} a_{ij}(x_{i1} - x_{j1}) + b_i(x_{i1} - x_{01})$, when $\chi_1 = 0$ it is easy to see that

$$(\bar{\mathcal{L}} + \bar{\mathcal{B}})\begin{pmatrix} x_{11} \\ \vdots \\ x_{N1} \end{pmatrix} = \bar{\mathcal{B}}\bar{1}x_{01} \tag{11.19}$$

Since $\bar{\mathcal{L}}\bar{1} = 0$, we have

$$(\bar{\mathcal{L}} + \bar{\mathcal{B}})\begin{pmatrix} x_{11} \\ \vdots \\ x_{N1} \end{pmatrix} = (\bar{\mathcal{L}}\bar{1} + \bar{\mathcal{B}}\bar{1})x_{01} = (\bar{\mathcal{L}} + \bar{\mathcal{B}})\bar{1}x_{01} \tag{11.20}$$

We know the matrix $(\bar{\mathcal{L}} + \bar{\mathcal{B}})$ is invertible (Khoo et al. 2009; Zhao et al. 2011; He et al. 2014). Hence one can easily write

$$\begin{pmatrix} x_{11} \\ \vdots \\ x_{N1} \end{pmatrix} = \bar{1}x_{01} \tag{11.21}$$

Remark 1 Sometime fixed distances are expected to be maintained between the leader and the followers. Then the consensus error can be defined by following way:

$$e_i^1 = \sum_{j=1, j\neq i}^{N} a_{ij}(x_{i1} + \Delta_i - x_{j1} - \Delta_j) + b_i(x_{i1} + \Delta_i - x_{01} - \Delta_0) \tag{11.22}$$

where Δ_i represents the distance between the agent i and the leader in different directions, $i = 1, 2, \ldots, N$ (Khoo et al. 2009; Zhao et al. 2011).

Remark 2 The missing derivatives of e_i^j, $j = 1, \ldots, N$ *and* $i = 1, \ldots, N$, can be estimated by means of the robust exact finite time convergent differentiator (Levant 2003).

11.4 Homogeneous Finite-Time Consensus Control with Integral Sliding Mode Control

With the notations above, the control objectives can be summarized as: to develop an adaptive control approach for nth order nonlinear MAS (11.13) with corresponding stability analysis, and to make the leader-follower system finite-time stable. To converge the error dynamics (11.16) in finite-time, let us consider an integral sliding surface (Feng et al. 2014) is given by

$$\begin{aligned} s = \chi_n + \int_{t_0}^{t} \{ & k_n|\chi_n|^{\alpha_n} sign(\chi_n) + k_{n-1}|\chi_{n-1}|^{\alpha_{n-1}} sign(\chi_{n-1}) \\ & + \cdots + k_1|\chi_1|^{\alpha_1} sign(\chi_1)\} d\tau \end{aligned} \tag{11.23}$$

where k_l and $\alpha_l (l = 1, 2, \ldots, n)$ are constants. k_l can be found such $\psi \in \Re$ the polynomial

$$\psi^n + k_n\psi^{n-1} + \cdots + k_2\psi + k_1 \tag{11.24}$$

is Hurwitz, α_l can be obtained satisfying the following conditions

$$\alpha_1 = \alpha, \quad n = 1$$
$$\alpha_{l-1} = \frac{\alpha_l \alpha_{l+1}}{2\alpha_{l+1} - \alpha_l}, \quad l = 2, \ldots, n \quad \forall n \geq 2 \tag{11.25}$$

where $\alpha_{n+1} = 1$, $\alpha_n = \alpha$, $\alpha \in (1-\epsilon, 1)$, $\epsilon \in (0, 1)$.

When the ideal sliding-mode $\dot{s} = 0$ is reached, one can easily get

$$\begin{aligned}
\dot{\chi}_n &+ k_n|\chi_n|^{\alpha_n} sign(\chi_n) + k_{n-1}|\chi_{n-1}|^{\alpha_{n-1}} sign(\chi_{n-1}) \\
&+ \cdots + k_1|\chi_1|^{\alpha_1} sign(\chi_1) = 0 \\
\dot{\chi}_n &= -k_n|\chi_n|^{\alpha_n} sign(\chi_n) - k_{n-1}|\chi_{n-1}|^{\alpha_{n-1}} sign(\chi_{n-1}) \\
&- \cdots - k_1|\chi_1|^{\alpha_1} sign(\chi_1)
\end{aligned} \tag{11.26}$$

Hence the error dynamics (11.16) can be expressed as,

$$\begin{aligned}
\dot{\chi}_1 &= \chi_2 \\
&\vdots \\
\dot{\chi}_n &= -k_n|\chi_n|^{\alpha_n} sign(\chi_n) - k_{n-1}|\chi_{n-1}|^{\alpha_{n-1}} sign(\chi_{n-1}) \\
&- \cdots - k_1|\chi_1|^{\alpha_1} sign(\chi_1)
\end{aligned} \tag{11.27}$$

which represents the establishment of the ideal sliding-mode $s = 0$, for system (11.16). It can converge to its equilibrium point from any initial condition in finite-time (Bhat and Bernstein 2005).

Theorem 1 *The error dynamics (11.16) will converge to equilibrium in finite-time along $s = 0$, if the sliding surface s is selected as (11.23) and the control law is designed as follows:*

$$\begin{aligned}
u &= [g(\chi)(\bar{\mathcal{L}} + \bar{\mathcal{B}})]^{-1}[u^{nom} + u^{sw}] \\
u^{nom} &= \bar{\mathcal{B}}\bar{1}u_0 - (\bar{\mathcal{L}} + \bar{\mathcal{B}})f(\chi) - k_n|\chi_n|^{\alpha_n} sign(\chi_n) - k_{n-1}|\chi_{n-1}|^{\alpha_{n-1}} sign(\chi_{n-1}) \\
&- \cdots - k_1|\chi_1|^{\alpha_1} sign(\chi_1) \\
u^{sw} &= -\lambda|s|^{1/2} sign(s) - \kappa \int_0^t sign(s) d\tau
\end{aligned} \tag{11.28}$$

where λ and κ are positive constants and and the inverse of $g(\chi)(\bar{\mathcal{L}} + \bar{\mathcal{B}})$ exists.

Distributed Format of the Control Law

For the *ith* agent (11.13) with error dynamics (11.15), the *ith* sliding variable can be expressed as

$$\begin{aligned}
s_i = e_i^n &+ \int_{t_0}^t \{k_{in}|e_i^n|^{\alpha_{in}} sign(e_i^n) + k_{i(n-1)}|e_i^{n-1}|^{\alpha_{i(n-1)}} sign(e_i^{n-1}) \\
&+ \cdots + k_{i1}|e_i^1|^{\alpha_{i1}} sign(e_i^1)\} d\tau
\end{aligned} \tag{11.29}$$

where k_{il} and α_{il} ($\forall i = 1, 2, \ldots, N$ and $l = 1, \ldots, n$) are constants. k_{il} can be found such $\psi \in \Re$ the polynomial

$$\psi^n + k_{in}\psi^{n-1} + \cdots + k_{i2}\psi + k_{i1} \tag{11.30}$$

is Hurwitz, α_{il} $\forall\, l = 1, \ldots n$ can be obtained satisfying the following conditions (Defoort et al. 2009)

$$\begin{aligned} \alpha_{il} &= \alpha, \quad n = 1, i = 1 \ldots N \\ \alpha_{i(l-1)} &= \frac{\alpha_{il}\alpha_{i(l+1)}}{2\alpha_{i(l+1)} - \alpha_{il}}, \quad l = 2, \ldots, n \quad \forall n \geq 2, \quad i = 1, \ldots, N \end{aligned} \tag{11.31}$$

where $\alpha_{i(n+1)} = 1$, $\alpha_{il} = \alpha$, $\alpha \in (1 - \epsilon, 1)$, $\epsilon \in (0, 1)$.

Matrix $\bar{\mathcal{L}} + \bar{\mathcal{B}}$ has been proved invertible (Khoo et al. 2009; Zhao et al. 2011). Note that $\bar{\mathcal{L}} + \bar{\mathcal{B}} = \bar{\mathcal{D}} + \bar{\mathcal{B}} - \bar{\mathcal{A}}$, therefore error dynamics (11.16) can be written as

$$\begin{aligned} \dot{\chi}_1 &= \chi_2 \\ \dot{\chi}_2 &= \chi_3 \\ &\vdots \\ \dot{\chi}_n &= (\bar{\mathcal{D}} + \bar{\mathcal{B}})f(\chi) - \bar{\mathcal{A}}f(\chi) + (\bar{\mathcal{D}} + \bar{\mathcal{B}})g(\chi)u \\ &\quad - \bar{\mathcal{A}}g(\chi)u + (\bar{\mathcal{D}} + \bar{\mathcal{B}} - \bar{\mathcal{A}})\nabla - \bar{\mathcal{B}}\bar{1}u_0 \end{aligned} \tag{11.32}$$

The error dynamics (11.32) of ith agent will converge to equilibrium in finite-time along $s_i = 0$ within, if the sliding-mode surface s_i is selected as (11.23) and the control is designed as follows:

$$\begin{aligned} u &= [g(\chi)(\bar{\mathcal{D}} + \bar{\mathcal{B}})]^{-1}[u^{nom} + u^{sw}] \\ u^{nom} &= \bar{\mathcal{B}}\bar{1}u_0 + \bar{\mathcal{A}}g(\chi)u + \bar{\mathcal{A}}f(\chi) - (\bar{\mathcal{D}} + \bar{\mathcal{B}})f(\chi) - k_n|\chi_n|^{\alpha_n} sign(\chi_n) \\ &\quad - k_{n-1}|\chi_{n-1}|^{\alpha_{n-1}} sign(\chi_{n-1}) - \cdots - k_1|\chi_1|^{\alpha_1} sign(\chi_1) \\ u^{sw} &= -\lambda|s|^{1/2} sign(s) - \kappa \int_0^t sign(s)d\tau \end{aligned} \tag{11.33}$$

where the inverse of $[g(\chi)(\bar{\mathcal{D}} + \bar{\mathcal{B}})]$ exists.

So in distributed format the control law can be written as

$$\begin{aligned} u_i &= \left\{\left(\sum_{j=1, j\neq i}^{N} a_{ij} + b_i\right) g_{i0}\right\}^{-1} [u_i^{nom} + u_i^{sw}] \\ u_i^{nom} &= b_i u_0 + \sum_{j=1, j\neq i}^{N} a_{ij}(f_{j0}(x_j) + g_{j0}(x_j)u_j) - \left(\sum_{j=1, j\neq i}^{N} a_{ij} + b_i\right) f_{i0}(x_i) \end{aligned}$$

$$
\begin{aligned}
&-k_{in}|e_i^n|^{\alpha_{in}} sign(e_i^n) - k_{i(n-1)}|e_i^{n-1}|^{\alpha_{i(n-1)}} sign(e_i^{n-1}) \\
&- \cdots - k_{i1}|e_i^1|^{\alpha_{i1}} sign(e_i^1) \\
u_i^{sw} &= -\lambda_i |s_i|^{1/2} sign(s_i) - \kappa_i \int_0^t sign(s_i) d\tau
\end{aligned} \tag{11.34}
$$

where λ_i and κ_i are positive constants and $(\sum_{j=1, j\neq i}^{N} a_{ij} + b_i) g_{i0}$ is invertible. u_j is the control information of the *jth* agent.

In the above control law the values of λ_i and κ_i must be known a priori.

For the error system (11.16) the sliding surface can be written as,

$$
\begin{aligned}
s &= \chi_n + \int_{t_0}^{t} \{k_n|\chi_n|^{\alpha_n} sign(\chi_n) + k_{n-1}|\chi_{n-1}|^{\alpha_{n-1}} sign(\chi_{n-1}) \\
&\quad + \cdots + k_1|\chi_1|^{\alpha_1} sign(\chi_1)\} d\tau \\
\dot{s} &= (\bar{\mathcal{L}} + \bar{\mathcal{B}}) f(\chi) + (\bar{\mathcal{L}} + \bar{\mathcal{B}}) g(\chi) u + \underbrace{(\bar{\mathcal{L}} + \bar{\mathcal{B}})\nabla - \bar{\mathcal{B}}\bar{1}u_0}_{d_o} + k_n|\chi_n|^{\alpha_n} sign(\chi_n) \\
&\quad + k_{n-1}|\chi_{n-1}|^{\alpha_{n-1}} sign(\chi_{n-1}) + \cdots + k_1|\chi_1|^{\alpha_1} sign(\chi_1)
\end{aligned} \tag{11.35}
$$

Using (11.28) one can easily obtain

$$
\dot{s} = u^{sw} + d_o \tag{11.36}
$$

In the control design (11.28) the values of λ and κ must be known a priori. In this work the proposed consensus algorithm, the values of λ and κ are chosen adaptively to reject the perturbations and convergence the sliding surface in finite time.

Hence the switching control in (11.28) can be written as

$$
u^{sw} = -\hat{\lambda}\phi_1(s) - \int_0^t \hat{\kappa}\phi_2(s) d\tau \tag{11.37}
$$

where $\hat{\lambda}$ and $\hat{\kappa}$ are unknown adaptive parameters. $\phi_1(s)$ and $\phi_2(s)$ are related as,

$$
\begin{aligned}
\phi_1(s) &= \lambda_3 s + |s|^{1/2} sign(s) \\
\phi_2(s) &= \lambda_3^2 s + 1/2 sign(s) + 3/2\lambda_3 |s|^{1/2} sign(s), \ \lambda_3 > 0.
\end{aligned} \tag{11.38}
$$

When $\lambda_3 = 0$ and $\hat{\lambda} = \lambda$ and $\hat{\kappa} = \kappa$ are constants, it is just like a simple super twisting algorithm. Let us consider the disturbance is bounded and satisfies the following equalities (Rath et al. 2015),

$$
\begin{aligned}
d_o(x,t) &= d_{1o}(x,t) + d_{2o}(x,t) \\
|d_{1o}(x,t)| \leq \varrho_1 |s| &\ \text{and} \ |\dot{d}_{2o}(x,t)| \leq \varrho_2
\end{aligned} \tag{11.39}
$$

where ϱ_1 and ϱ_2 are positive constants.

The term λ_3 is tuned in order to withstand the exciting perturbation terms.

Theorem 2 *Considering the system (11.36) satisfying the conditions* $s(0) > \mu_1$ *and* $|s| \leq \bar{\eta}_1$ *and* $|\dot{s}| \leq \bar{\eta}_2$, *is established if the adaptive gains are calculated as (Shtessel et al. 2010, 2012),*

$$\dot{\hat{\lambda}} = \begin{cases} \omega_1\sqrt{\frac{\delta_1}{2}}sign(|s| - \mu_1), & if \ \hat{\lambda}_1 \geq \delta_m \\ \zeta, & \hat{\lambda}_1 < \delta_m \end{cases}$$
$$\hat{\kappa} = 2\epsilon_* \hat{\lambda} \tag{11.40}$$

where $\epsilon_*, \delta_1, \omega_1, \mu_1, \delta_m$ *are arbitrary positive constants and and* $\bar{\eta}_1 \geq \mu_1, \bar{\eta}_2 > 0$.

Proof For finite time convergence the sliding surface dynamics can be simplified as,

$$\begin{aligned} \dot{s} &= \rho - \hat{\lambda}\phi_1(s) + d_{1o}(x,t) \\ \dot{\rho} &= -\hat{\kappa}\phi_2(s) + \dot{d}_{2o}(x,t) \end{aligned} \tag{11.41}$$

We consider a new state vector as,

$$\gamma = \begin{bmatrix} \gamma_1 \\ \gamma_2 \end{bmatrix} = \begin{bmatrix} \phi_1(s) \\ \rho \end{bmatrix} \tag{11.42}$$

Its derivative can be written as,

$$\begin{aligned} \dot{\gamma} &= \begin{bmatrix} (\lambda_3 + 1/2|s|^{-1/2})(\rho - \hat{\lambda}\phi_1(s)) + d_{10} \\ -\hat{\kappa}\phi_2(s) + \dot{d}_{2o} \end{bmatrix} \\ &= \left(\lambda_3 + 1/2|s|^{-1/2}\right)\left(A_0\gamma + \begin{bmatrix} d_{1o}(s) \\ 0 \end{bmatrix}\right) + \begin{bmatrix} 0 \\ \dot{d}_{2o} \end{bmatrix} \end{aligned} \tag{11.43}$$

where A_o is a Hurwitz matrix selected as,

$$A_0 = \begin{bmatrix} -\hat{\lambda} & 1 \\ -\hat{\kappa} & 0 \end{bmatrix} \tag{11.44}$$

To analyze stability of the system, a Lyapunov function chosen as,

$$V_1(\gamma, \lambda, \theta) = V_\gamma + \frac{1}{2\delta_1}\tilde{\lambda}^2 + \frac{1}{2\delta_2}\tilde{\kappa}^2 \tag{11.45}$$

where $\tilde{\lambda} = \hat{\lambda} - \lambda$ and $\tilde{\kappa} = \hat{\kappa} - \kappa$, i.e. estimated value minus the actual value. The function V_γ is given by,

$$V_\gamma = \gamma^T P \gamma \tag{11.46}$$

$$P = \begin{bmatrix} \mu + 4\epsilon_*^2 & -2\epsilon_* \\ -2\epsilon_* & 1 \end{bmatrix}. \tag{11.47}$$

where P is positive definite, for any $\mu > 0$ and $\epsilon_* > 0$.

Where the time derivative of V_γ can be obtained as,

$$\dot{V}_\gamma = \left(\lambda_3 + 1/2|s|^{-1/2}\right)\left(\gamma^T(A_0^T P + PA_0)\gamma + 2\gamma^T P \begin{bmatrix} d_{1o}(s) \\ 0 \end{bmatrix}\right) + 2\gamma^T P \begin{bmatrix} 0 \\ \dot{d}_{2o} \end{bmatrix} \tag{11.48}$$

Now $\dot{d}_{20} = \rho_2 \phi_2$. Thus it can be simplified as,

$$\dot{V}_\gamma = \left(\lambda_3 + 1/2|s|^{-1/2}\right)\left(\gamma^T(A_0^T P + PA_0)\gamma + 2\gamma^T P \begin{bmatrix} \rho_1\gamma \\ 0 \end{bmatrix}\right) + \left(\lambda_3 + 1/2|s|^{-1/2}\right) 2\gamma^T P \begin{bmatrix} 0 \\ \rho_2\gamma \end{bmatrix} \tag{11.49}$$

Thus it can be simplified as,

$$\dot{V}_\gamma = -\left(\lambda_3 + 1/2|s|^{-1/2}\right)\gamma^T Q \gamma < 0 \tag{11.50}$$

where P and Q are related by

$$A^T P + PA = -Q \tag{11.51}$$

The symmetric matrix Q_1 can be given as,

$$Q = \begin{bmatrix} 2(\hat{\lambda} - \rho_1)(\mu + 4\epsilon_*^2) - 4\epsilon_*(\hat{\kappa} - \rho_3) & * \\ \hat{\kappa} - 2\hat{\lambda}\epsilon_* + 2\epsilon_*\rho_1 - \rho_3 - (\mu + 4\epsilon_*^2) & 4\epsilon_* \end{bmatrix}$$

If we choose

$$\hat{\kappa} = 2\hat{\lambda}\epsilon_* \tag{11.52}$$

and

$$\hat{\lambda} > \frac{(2\epsilon_*\rho_1 + \rho_3 - (\mu + 4\epsilon_*^2))^2}{8\epsilon_*(\mu + 4\epsilon_*^2) - 32\epsilon_*^3} + \frac{8\epsilon_*\rho_1(\mu + 4\epsilon_*^2) - 16\epsilon_*^2\rho_3}{8\epsilon_*(\mu + 4\epsilon_*^2) - 32\epsilon_*^3} \tag{11.53}$$

then Q is positive definite. Since

$$\xi_{min}\{P\}||\gamma||^2 \leq \gamma^T P\gamma \leq \xi_{max}\{P\}||\gamma||^2 \tag{11.54}$$

From (11.42) we have,

$$||\gamma||^2 = \gamma_1^2 + \gamma_2^2 \geq \lambda_3^2|s| \tag{11.55}$$

and

$$|\gamma_1| \leq ||\gamma|| \leq \frac{V^{1/2}}{\xi_{min}^{1/2}\{P\}} \tag{11.56}$$

Thus we can conclude that,

$$\dot{V}_\gamma \leq r_1 V_\gamma^{1/2} - r_2 V_\gamma \tag{11.57}$$

where $r_1 = \nu\frac{\xi_{min}(Q)}{\xi_{max}^{1/2}(P)}\frac{\lambda_3^2}{2}$ and $r_2 = \nu\frac{\xi_{min}(Q)}{\xi_{max}(P)}$. where $0 < \nu < 1$ is a positive constant. Taking the derivative of Now in view of the above analysis derivative of Lyapunov function can be obtained as,

$$\begin{aligned}
&\dot{V}_1(\gamma, \hat{\lambda}, \hat{\kappa}) = \dot{\gamma}^T P\gamma + \gamma^T P\dot{\gamma} + \frac{1}{\delta_1}(\hat{\lambda} - \lambda)\dot{\hat{\lambda}} \\
&\frac{1}{\delta_2}(\hat{\kappa} - \kappa)\dot{\hat{\kappa}} \leq -rV_\gamma^{1/2} - r_2 V_\gamma - \frac{\omega_1}{\sqrt{2\delta_1}}|(\hat{\lambda} - \lambda)| \\
&-\frac{\omega_2}{\sqrt{2\delta_2}}|(\hat{\kappa} - \kappa)| + \frac{1}{\delta_1}(\hat{\lambda} - \lambda)\dot{\hat{\lambda}} + \frac{1}{\delta_2}(\hat{\kappa} - \kappa)\dot{\hat{\kappa}} \\
&+\frac{\omega_1}{\sqrt{2\delta_1}}|(\hat{\lambda} - \lambda)| + \frac{\omega_2}{\sqrt{2\delta_2}}|(\hat{\kappa} - \kappa)|
\end{aligned} \tag{11.58}$$

Using Lemma 1 one can easily write

$$\begin{aligned}
&-r_1 V_\gamma^{1/2} - \frac{\omega_1}{\sqrt{2\delta_1}}|(\hat{\lambda} - \lambda)| - \frac{\omega_2}{\sqrt{2\delta_2}}|(\hat{\kappa} - \kappa)| \leq \\
&-\eta_1[V_1(\gamma, \hat{\lambda}, \hat{\kappa})]^{1/2}
\end{aligned} \tag{11.59}$$

where $\eta_1 = min(r, \omega_1, \omega_2)$ Using the above inequities one can easily write,

$$\begin{aligned}
&\dot{V}_1(\gamma, \hat{\lambda}, \hat{\kappa}) \leq -\eta_1[V_1(\gamma, \hat{\lambda}, \hat{\kappa})]^{1/2} - r_2 V_\gamma \\
&+\frac{1}{\delta_1}(\hat{\lambda} - \lambda)\dot{\hat{\lambda}} + \frac{1}{\delta_2}(\hat{\kappa} - \kappa)\dot{\hat{\kappa}} \\
&+\frac{\omega_1}{\sqrt{2\delta_1}}|(\hat{\lambda} - \lambda)| + \frac{\omega_2}{\sqrt{2\delta_2}}|(\hat{\kappa} - \kappa)|
\end{aligned} \tag{11.60}$$

Now let us assume that the adaptation law makes the adaptive gains bounded i.e. $(\hat{\lambda}-\lambda)<0$ and $(\hat{\kappa}-\kappa)<0$. From the above assumptions one can write that

$$\begin{aligned}&\dot{V}_1(\gamma,\hat{\lambda},\hat{\kappa})\leq-\eta_1[V_1(\gamma,\hat{\lambda},\hat{\kappa})]^{1/2}\\&-|(\hat{\lambda}-\lambda)|(\dot{\hat{\lambda}}/\delta_1-\omega_1/\sqrt{2\delta_1})\\&-|(\hat{\kappa}-\kappa)|(\dot{\hat{\kappa}}/\delta_2-\omega_2/\sqrt{2\delta_2})\end{aligned}\tag{11.61}$$

By selecting $\dot{\hat{\lambda}}=\omega_1\sqrt{\frac{\delta_1}{2}}$ and $\dot{\hat{\kappa}}=\omega_2\sqrt{\frac{\delta_2}{2}}$, one can easily get, $\epsilon_*=\frac{\omega_2}{\omega_1}\sqrt{\frac{\delta_2}{\delta_1}}$. Thus it follows that,

$$\dot{V}_1(\gamma_1,\hat{\lambda},\hat{\kappa})\leq-\eta_1[V_1(\gamma_1,\hat{\lambda},\hat{\kappa})]^{1/2}-r_2V_\gamma\tag{11.62}$$

In order to achieve the finite time convergence it is necessary that $\hat{\lambda}$ satisfies inequality (11.53). It implies that $\hat{\lambda}$ increases in accordance with $\dot{\hat{\lambda}}=\omega_1\sqrt{\frac{\delta_1}{2}}$ until (11.53) is met which guarantees the positive definiteness of matrix Q.

In the condition when $|s|<\delta$

$$\dot{\hat{\lambda}}=\begin{cases}-\omega_1\sqrt{\frac{\delta_1}{2}} & if\ \hat{\lambda}\geq\delta_m\\ \zeta, & \hat{\lambda}<\delta_m\end{cases}\tag{11.63}$$

$$\begin{aligned}&\dot{V}_1(\gamma,\hat{\lambda},\hat{\kappa})\leq-\eta_1[V_1(\gamma,\hat{\lambda},\hat{\kappa})]^{1/2}\\&-2|(\hat{\lambda}-\lambda)|(1-\frac{1}{\sqrt{2\delta_1}})\end{aligned}\tag{11.64}$$

By selecting $\dot{\hat{\lambda}}=-\omega_1\sqrt{\frac{\delta_1}{2}}$ It can be deduced that second equation in (11.64) is valid for a finite duration. As soon as λ becomes less than or equal to δ_m Then the equation in (11.64) is valid. The derivative of the Lyapunov function becomes sign indefinite. During the adaptation process the sliding variable reaches the domain in finite time. It is however guaranteed that it always stays in a larger domain in the real sliding mode.

11.5 Results

Example 1 **Second order MAS** A leader-follower system composed of five agents was considered (Khoo et al. 2009; Zhao et al. 2011), where the leader was indexed by 0, and the four followers were indexed by 1, 2, 3 and 4, respectively. The communication topology graph is shown in Fig. 11.1.

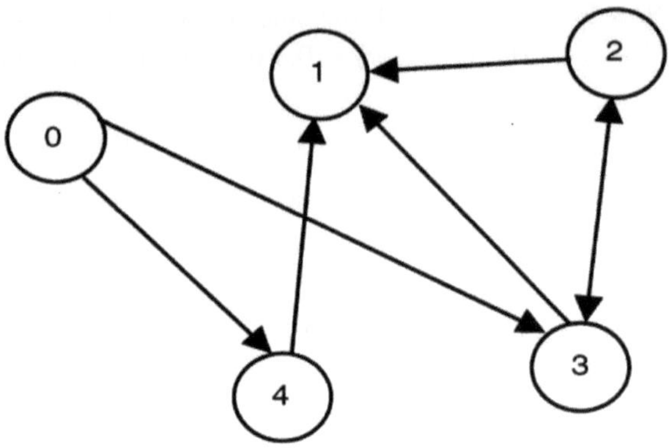

Fig. 11.1 The communication graph

In this section, we consider the agent dynamics are of second order. As compared to Khoo et al. (2009), Zhao et al. (2011) where agents dynamics were considered to be simple double integrators only, here we added some input dynamics in the agent

$$\begin{aligned} \dot{x}_{i1} &= x_{i2} \\ \dot{x}_{i2} &= x_{i1}^2 + x_{i2}\cos(x_{i2}) + u_i + 0.01\sin(x_{i1}) \end{aligned} \tag{11.65}$$

where $i = 1, \ldots 4$. The term $d_i = 0.01\,\sin(x_{i1})$ represents the uncertainty. The initial conditions of the agents are taken as $[x_{11}^0\ \ x_{12}^0]^T = [1\ \ 0]^T$,$[x_{21}^0\ \ x_{22}^0]^T = [1.2\ \ 0]^T$, $[x_{31}^0\ \ x_{32}^0]^T = [2\ \ 0]^T$, $[x_{41}^0\ \ x_{41}^0]^T = [-1.2\ \ 0]^T$.

The control objective is to track the leader whose dynamic equation is given by Khoo et al. (2009), Zhao et al. (2011)

$$\begin{aligned} \dot{x}_{01} &= \dot{x}_{02} \\ \dot{x}_{02} &= -\sin(x_{01})/(1 + e^{-t}) \end{aligned} \tag{11.66}$$

The initial conditions of the leader is $[x_{01}^0\ \ x_{02}^0]^T = [\pi/2\ \ 0]^T$.

In order to obtain the finite time convergence and to mitigate the chattering, we design an integral sliding mode controller using adaptive super twisting algorithm. The sliding surface for the agents can be chosen as

$$s = \chi_2 + \int_{t_0}^{t} \{k_2|\chi_2|^{\alpha_2} sign(\chi_2) + k_1|\chi_1|^{\alpha_1} sign(\chi_1)\} d\tau \tag{11.67}$$

$k_1 = \text{diag}([0.3, 0.3, 0.3, 0.3])$, $k_2 = \text{diag}([0.45, 0.45, 0.45, 0.45])$, $\alpha_2 = 2/3$, $\alpha_1 = 1/2$. The control law for each agent can be obtained from (11.28), where the design parameters for the adaptive STA controller are selected as, $\omega_1 = 100$, $\zeta = 2$, $\delta_m = 2$, $\delta = 2$, $\mu_1 = 2$ and $\epsilon = 0.02$.

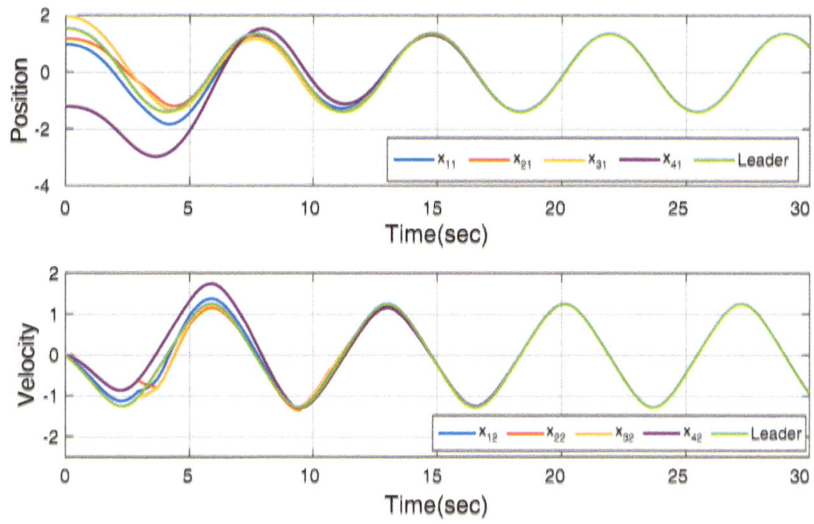

Fig. 11.2 Consensus tracking

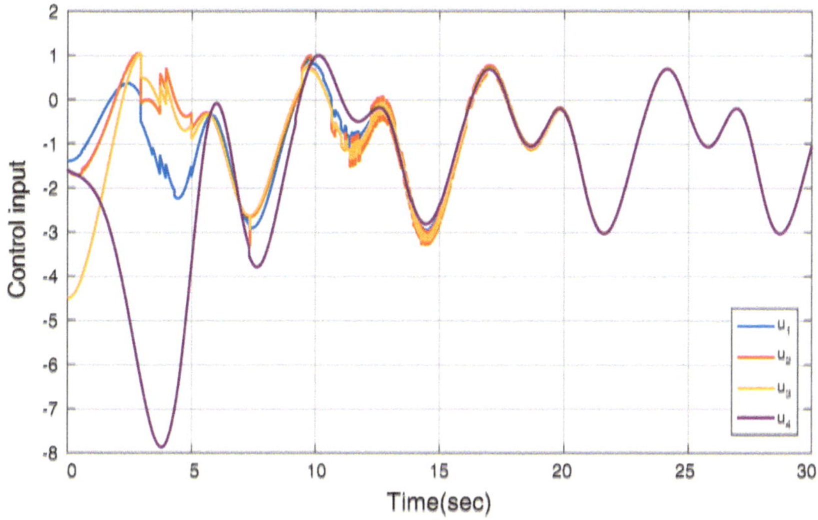

Fig. 11.3 Control signal

Consensus tracking, control signal sliding surface and estimated adaptive parameters using the proposed method (11.28) are shown in Figs. 11.2, 11.3, 11.4 and 11.5. It is clear that, the outputs of the followers track accurately that of the leader.

The control input of each follower is provided in Fig. 11.3. Since the adaptive super twisting algorithm is used hence the control signal is much more smooth and it is more superior in eliminating chattering problem in the MAS.

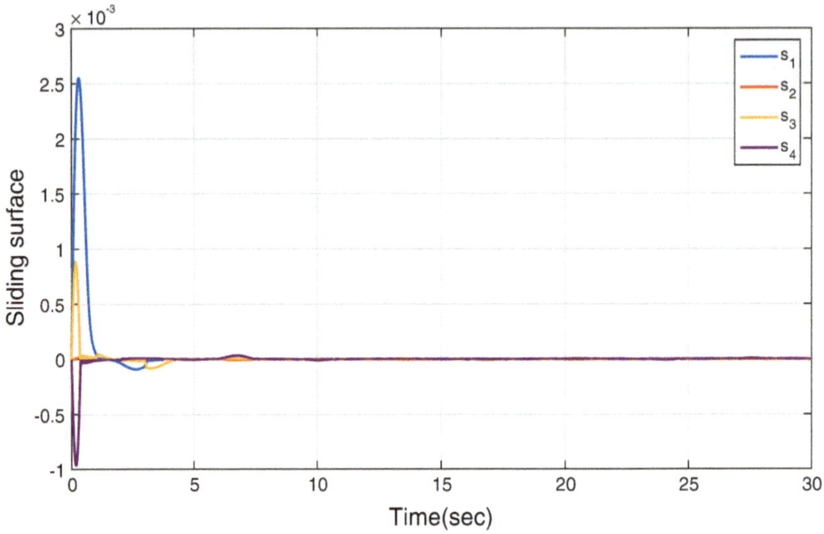

Fig. 11.4 Sliding surface

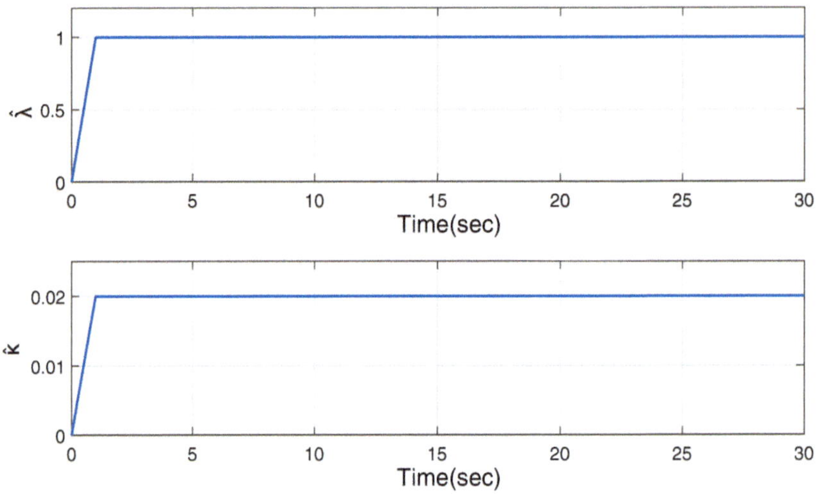

Fig. 11.5 Estimated parameters

The convergence of sliding surfaces and estimated parameters are shown in Figs. 11.4 and 11.5 respectively.

Example 2 **Third order MAS** The effectiveness of the proposed control technique for higher-order MAS, the following directed topologies Fig. 11.6, with five followers and one leader are selected for the simulation.

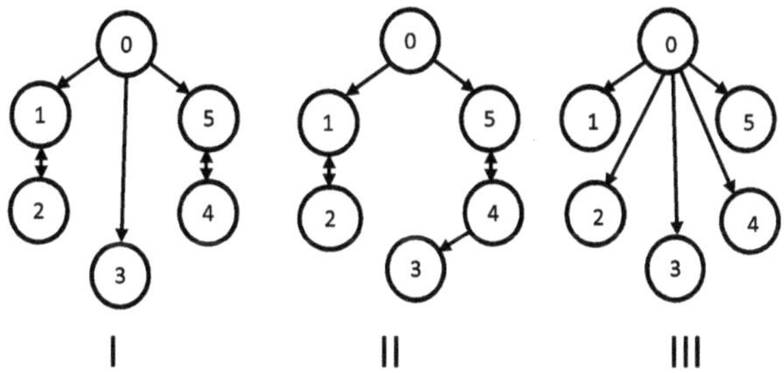

Fig. 11.6 The communication graph

Now we consider the agent dynamics are of third order

$$\begin{aligned}\dot{x}_{i1} &= x_{i2}\\ \dot{x}_{i2} &= x_{i3}\\ \dot{x}_{i3} &= x_{i2}^2 + x_{i2}x_{i3} + u_i + d_i.\ \ i = 1, \ldots n.\end{aligned} \tag{11.68}$$

where $d_i \neq 0$ are the uncertainties, chosen as: $d_1{=}1.15\sin(0.3t + pi/3)$, $d_2 = 1.6\sin(0.1t + pi/6)$, $d_3{=}1.25\sin(0.4t + pi/3)$, $d_4 = \sin(0.5t + pi/2)$, $d_5 = 1.5\sin(0.6t + pi/2)$. Control objective is to design a controller such that the agents follow the path of leader. The output of the leader is considered as $y_0 = t$. The initial conditions of the agents are taken as $[x_{11}^0\ x_{12}^0\ x_{13}^0]^T = [-2\ 0\ 0]^T$, $[x_{21}^0\ x_{22}^0\ x_{23}^0]^T = [6\ 3\ -3]^T$, $[x_{31}^0\ x_{32}^0\ x_{33}^0]^T = [3\ 0\ 3]^T$, $[x_{41}^0\ x_{42}^0\ x_{43}^0]^T = [-5\ -5\ 0]^T$, $[x_{51}^0\ x_{52}^0\ x_{53}^0]^T = [5\ 0\ 0]^T$. The sliding surface chosen from (11.23) for the agents are taken as

$$s = \chi_3 + \int_{t0}^{t} \{k_3|\chi_3|^{\alpha_3} sign(\chi_3) + k_2|\chi_2|^{\alpha_2} sign(\chi_2) + k_1|\chi_1|^{\alpha_1} sign(\chi_{i1})\} d\tau$$

where $\alpha_3 = 3/4, \alpha_2 = 3/5, \alpha_1 = 1/2$, $k_1 = 1 \times diag([2, 2, 2, 2, 2])$, $k_2 = 1 \times diag([3, 3, 3, 3, 3])$, $k_3 = 1 \times diag([4, 4, 4, 4, 4])$. The control law for each agent can be obtained from (11.28) where $\lambda = 2.2 \times diag([2, 2, 2, 2, 2])$ and $\theta = 2.2 \times diag([3, 3, 3, 3, 3])$.

The design parameters for the adaptive STA controller are selected as, $\omega_1 = 100$, $\zeta = 2$, $\delta_m = 2$, $\delta = 2$, $\mu_1 = 2$ and $\epsilon = 0.02$.

Simulation results are shown in the Figs. 11.7, 11.8 and 11.9.

The consensus tracking performance of the MAS under the proposed consensus protocol with different topologies are shown in Figs. 11.7, 11.8 and 11.9, from which it can be observed that the leader is accurately followed by the followers. From the above two examples, it can be observed that the proposed method can realize the globally finite-time stability of the nth-order MAS using a higher-order sliding surface. In addition, the control signals are smooth, as shown in Fig. 11.3, which means that the chattering problem in sliding mode control has been further attenuated.

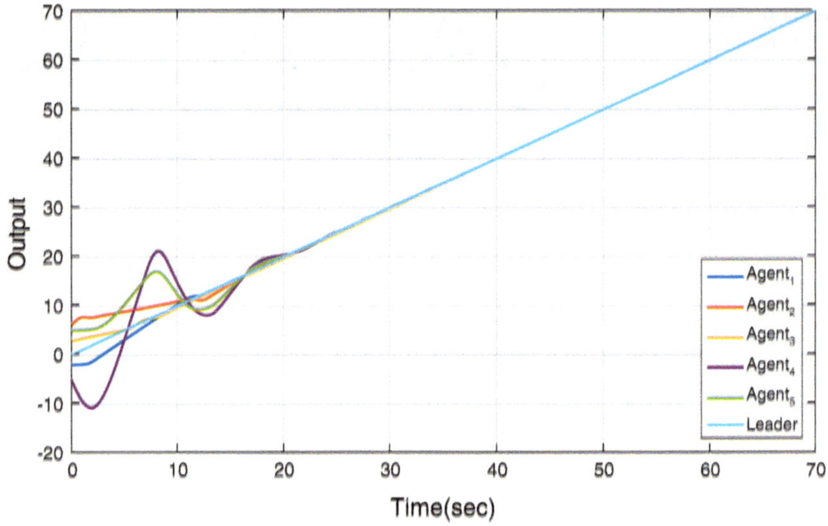

Fig. 11.7 Consensus tracking using the topology I

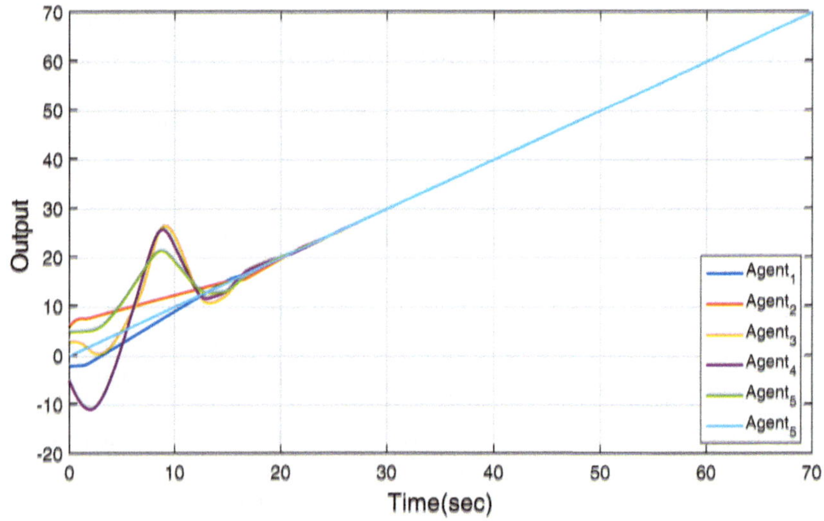

Fig. 11.8 Consensus tracking using the topology II

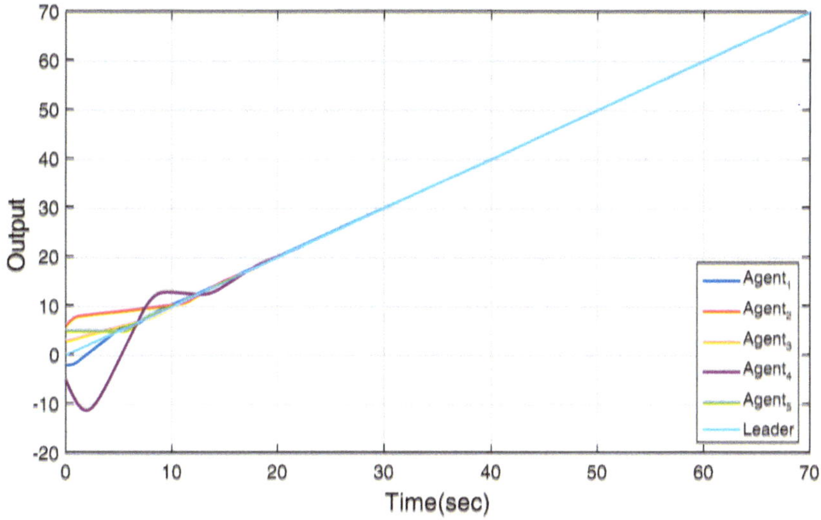

Fig. 11.9 Consensus tracking using the topology III

11.6 Conclusion

An adaptive finite time consensus protocol is developed for higher-order nonlinear multi-agent systems(MAS) using higher-order integral sliding mode. The proposed method shows the finite-time tracking of the agents. The consensus control law generalizing the results to the systems of relative degree more or equal than two. The switching control is obtained by using adaptive super twisting algorithm, h thus it alleviates the chattering effectively. The adaptive tuning law estimates the bounds of the uncertainties without knowing it a priori. The finite-time convergence analysis of higher-order MAS is obtained by using strict Lyapunov function.

References

Bhat, S. P., & Bernstein, D. S. (2005). Geometric homogeneity with applications to finite-time stability. *Mathematics of Control Signals Systems*, *17*, 101–127.

Canale, E., Dalmao, F., Mordecki, E., & Souza, M. O. (2015). Robustness of cucker-smale flocking model. *IET Control Theory and Applications*, *9*(3), 346–350.

Cao, Y., Yu, W., Ren, W., & Chen, G. (2013). An overview of recent progress in the study of distributed multi-agent. *An Overview of Recent Progress in the Study of Distributed Multi-agent*, *9*(1), 427–438.

Consolini, L., Morbidi, F., & Tosques, D. P. M. (2008). Leaderfollower formation control of non-holonomic mobile robots with input constraints. *Automatica*, *44*(5), 1343–1349.

Defoort, M., Floquet, T., Kokosy, A., & Perruquetti, W. (2009). A novel higher order sliding mode control scheme. *System and Control Letters*, *58*, 102–108.

Dimarogonas, D. V., Tsiotras, P., & Kyriakopoulos, K. J. (2009). Leader-follower cooperative attitude control of multiple rigid bodies. *Systems and Control Letters, 58*, 429–435.
Feng, Y., Han, F., & Yu, X. (2014). Chattering free full-order sliding-mode control. *Automatica, 50*, 1310–1314.
Ghasemi, M., & Nersesov, S. G. (2014). Finite-time coordination in multiagent systems using sliding mode control approach. *Automatica, 50*, 1209–1216.
Ghasemi, M., Nersesov, S. G., Clayton, G., & Ashrafiuon, H. (2014a). Sliding mode coordination control for multiagent systems with underactuated agent dynamics. *International Journal of Control, 87*(12), 2612–2633.
Ghasemi, M., Nersesovn, S. G., & Clayton, G. (2014b). Finite-time tracking using sliding mode control. *Journal of the Franklin Institute, 351*(5), 2966–2990.
Ghommam, J., Mehrjerdi, H., & Saad, M. (2013). Robust formation control without velocity measurement of the leader robot. *Control Engineering Practice, 21*, 1143–1156.
Ghommam, J., & Saad, M. (2014). Backstepping-based coop- erative and adaptive tracking control design for a group of underactuated AUVs in horizontal plan. *International Journal of Control, 87*(5), 1076–1093.
Guo, W., Lü, J., Chen, S., & Yu, X. (2011). Second-order tracking control for leader-follower multi-agent flocking in directed graphs with switching topology. *Systems and Control Letters, 60*, 1051–1058.
He, X., Wang, Q., & Yu, W. (2014). Finite-time containment control for second-order multiagent systems under directed topology. *IEEE Transactions on Circuit and Systems-II, 61*(8), 619–623.
Hong, Y., Gao, L., Cheng, D., & Hu, J. (2007). Lyapunov-based approach to multiagent systems with switching jointly connected interconnection. *IEEE Transactions on Automatic Control, 52*(5), 943–948.
Khoo, S., Xie, L., & Man, Z. (2009). Robust finite-time consensus tracking algorithm for multirobot systems. *IEEE/ASME Transactions on Mechatronics, 14*(2), 219–228.
Levant, A. (1993). Sliding order and sliding accuracy in sliding mode control. *International Journal of Control, 58*(6), 1247–1263.
Levant, A. (2001). Universal single-input single-output (SISO) sliding-mode controllers with finite-time convergence. *IEEE Transactions on Automatic Control, 46*(1), 1447–1451.
Levant, A. (2003). Higher-order sliding modes, differentiation and output-feedback control. *International Journal of Control, 76*(9/10), 924–941.
Li, T., & Zhang, J.-F. (2010). Consensus conditions of multi-agent systems with time-varying topologies and stochastic communication noises. *IEEE Transactions on Automatic Control, 55*(9), 2043–2057.
Li, Z., Duan, Z., Chen, G., & Huang, L. (2010). Consensus of multiagent systems and synchronization of complex networks: A unified viewpoint. *IEEE Transactions on Circuits and Systems -I, 57*(1), 213–224.
Lin, P., & Jia, Y. (2009). Consensus of second-order discrete-time multiagent systems with nonuniform time-delays and dynamically changing topologies. *Automatica, 45*(9), 2154–2158.
Mehrjerdi, H., Saad, M., & Ghommam, J. (2011). Hierarchical fuzzy cooperative control and path following for a team of mobile robots. *IEEE/ASME Transactions on Mechatronics, 16*(5), 907–917.
Moulay, E., & Perruquetti, W. (2008). Finite time stability conditions for non autonomous continuous systems. *International Journal of Control, 81*(5), 797–803.
Pack, D. J., DeLima, P., Toussaint, G. J., & York, G. (2009). Cooperative control of UAVs for localization of intermittently emitting mobile targets. *IEEE Transactions on Systems Man and Cybernatics-Part B: 39*(4), 959–970.
Rath, J., Veluvolu, K., & Defoort, M. (2015). Simultaneous estimation of road profile and tyre road friction for automotive vehicle. *IEEE Transactions on Vehicular Technology, 64*(10), 4461–4471.
Ren, W., & Beard, R. W. (2005). Consensus seeking in multiagent systems under dynamically changing interaction topologies. *IEEE Transactions on Automatic Control, 50*(5), 655–661.

Shtessel, Y., Taleb, M., & Plestan, F. (2012). A novel adaptive-gain supertwisting sliding mode controller: Methodology and application. *Automatica*, *48*(5), 759–769.

Shtessel, Y. B., Moreno, J. A., Plestan, F., Fridman, L. M., and Poznyak, A. S. (2010). Super-twisting adaptive sliding mode control: A Lyapunov design. In *49th IEEE Conference on Decision and Control, Atlanta, USA*, (pp. 5109–5113).

Weigang, L., de Souza, B. B., Crespo, A. M. F., & Alves, D. P. (2008). Decision support system in tactical air traffic flow management for air traffic flow controllers. *Journal of Air Transport Management*, *14*, 329–336.

Yoon, S., & Qiao, C. (2011). Cooperative search and survey using autonomous underwater vehicles (AUVs). *IEEE Transactions on Parallel and Distributed Systems*, *22*(3), 364–379.

Yu, S., & Long, X. (2015). Finite-time consensus for second-order multi-agent systems with disturbances by integral sliding mode. *Automatica*, *54*, 158–165.

Zhang, W. & Wang, Z. (2015). Adaptive output consensus tracking of uncertain multiagent systems. *International Journal of Systems Science*, *46*(13), 2367–2379.

Zhao, D., Zou, T., Li, S., & Zhu, Q. (2011). Adaptive backstepping sliding mode control for leader-follower multi-agent systems. *IET Control Theory and Applications*, *6*(8), 1109–1117.

Zuo, Z. (2015). Nonsingular fixed-time consensus tracking for second-order multi-agent networks. *Automatica*, *54*, 305–309.

Chapter 12
MPPT Controllers Based on Sliding-Mode Control Theory and Fuzzy Logic in Photovoltaic Power Systems: A Comparative Study

Radhia Garraoui, Mouna Ben Hamed and Lassaad Sbita

Abstract In this chapter we will deal with a comparative study between two control methods for maximum power point tracking (MPPT) algorithms in photovoltaic (PV) systems. The two MPPT controllers considered in this chapter are: the Fuzzy Logic Controller (FLC) and the Sliding Mode Controller (SMC). The MPPT controller based on the fuzzy-logic-algorithm uses directly the DC-DC converter duty cycle as a control variable and it provides a fast response and good performances against the climatic and load changes. The SMC exhibits a very fast response for tracking the maximum power point (MPP) for photovoltaic systems. The input parameters are the voltage and the current, the duty cycle is used to generate the optimal MPP under different operating conditions. Simulation results show that both algorithms can effectively perform the MPPT hence improving the efficiency of PV systems.

Keywords Photovoltaic systems · MPPT · Fuzzy logic controller · Sliding mode controller

R. Garraoui (✉) · M. Ben Hamed · L. Sbita
Research Unit on Photovoltaic, Wind and Geothermal Systems, National Engineering School of Gabes, University of Gabes, Omar Ibn Elkhattab Street 6029, Gabes, Tunisia
e-mail: radhiagarraoui@gmail.com

M. Ben Hamed
e-mail: mouna.benhamed@enig.rnu.tn

L. Sbita
e-mail: lassaad.sbita@enig.rnu.tn

N. Derbel et al. (eds.), *Applications of Sliding Mode Control*,
Studies in Systems, Decision and Control 79, DOI 10.1007/978-981-10-2374-3_12

Nomenclature

G	: Global insulation (W/m^2)	α	: Duty cycle
G_n	: Reference insulation (W/m^2)	K_b	: Boltzmann constant
T	: Cell Junction temperature (°C)		$K_b = 1.3806 \times 10^{-23}$
T_r	: Reference cell temperature (°C)	E_g	: Band gap energy (eV)
i_{pv}	: Output PV current (A)	q	: Charge of an electron (C)
I_L	: Inductance current (A)	C	: Capacitors (F)
I_{ph}	: Light-generated current (A)	L	: Inductance (H)
I_{rr}	: Saturation current (A) at T_{ref}	K_i	: a I_{sc}/T Coefficient (A/K)
V_{pv}	: PV output voltage (V)	r_{ci}	: Parasitic resistance (Ω)
A	: Ideality factor	$r_{l,pv}$	: Parasitic resistance (Ω)

12.1 Introduction

Photovoltaic (PV) is a manner of converting solar energy into direct current electricity using semi conducting materials that manifest the photovoltaic effect, a phenomenon commonly studied in physics. A photovoltaic system employs solar panels composed of a number of solar cells to supply usable solar power.

Generating power from solar PV has long been seen as a clean, sustainable energy technology, which draws upon the planet's most plentiful and widely distributed renewable energy source which is the sun. The direct conversion of sunlight to electricity occurs without any moving parts or environmental emissions during operation. Conventional energy resources in the world are threatened. Moreover, they emit large amounts of greenhouse gas emissions responsible for climate change.

Many pollution problems can be avoided if electrical power is generated from renewable energy sources rather than traditional fossil fuels (Bose 2010). The use of some types of renewable energy increased notably in recent years, with its advantages of being pollution-free. In particular, wind, solar, and hydroelectric systems can generate electricity without air pollution emissions (Ran and Boyce 1996; Young 1964) and can help to reduce energy prices in the future. For these reasons, renewable energy resources are receiving increasing interest in recent years being photovoltaic (PV) energy one of the most attractive sources due to its abundance.

PV generators have a nonlinear current-voltage or power-voltage characteristic with a maximum power point (MPP) depending on the atmospheric conditions, namely the irradiance and the temperature. Therefore, a maximum power point tracking (MPPT) system is required to ensure that the PV always operates at its MPP regardless of the temperature, insulation and load changes. The output of this controller is fed-back to a DC-DC switching converter so that the array delivers a maximum of energy.

Generally speaking, a PV system is composed of three key elements: The first one consists of the PV array as a source of energy, the second one is the DC-DC converter

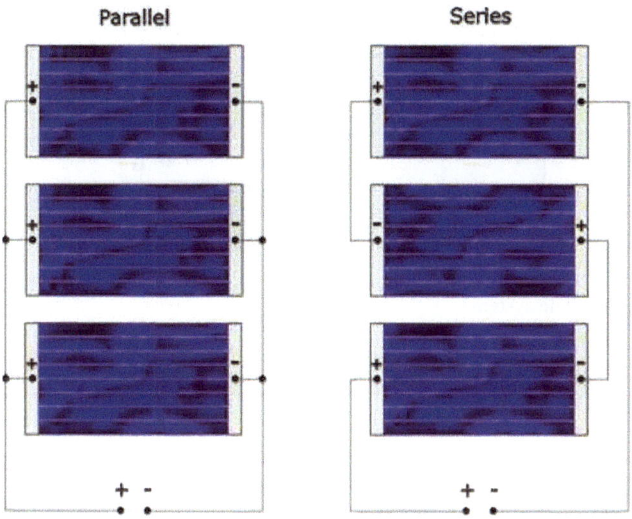

Fig. 12.1 PV cell connecting

used as an interface which allows the adaptation of the energy flow between the PV array and the load. The third block is the control system carrying the regulation of some variables of interest with the aim to extract the MPP from the PV array. For an appropriate operation of PV systems at maximum power, there are a variety of methods that can be used (Koutroulis and Blaabjerg 2013).

12.2 Description of the PV System

A photovoltaic array called also a solar array consists of multiple photovoltaic modules, casually referred to solar panels, to convert solar radiation (sunlight) into usable direct current (DC) electricity. In order to investigate the reliability of MPPT controllers, a PV power system with a boost converter is considered. Solar PV modules can be grouped together as an array of series and parallel connected modules to provide any level of power requirements (Agorreta et al. 2013), from watts (W) to kilowatt (KW) and megawatt (MW) size as shown in Fig. 12.1.

12.2.1 The Photovoltaic Effect

The photovoltaic effect is the basic process in which a solar cell converts sunlight into electricity. The light is composed of tiny particles of electromagnetic energy called photons. When photons are absorbed by a photovoltaic cell, which contains a

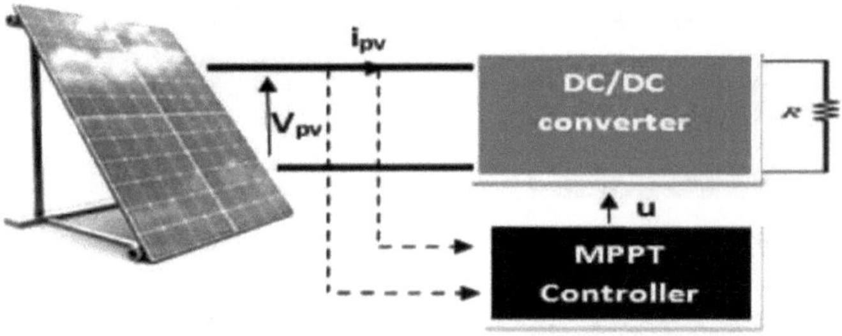

Fig. 12.2 Photovoltaic system

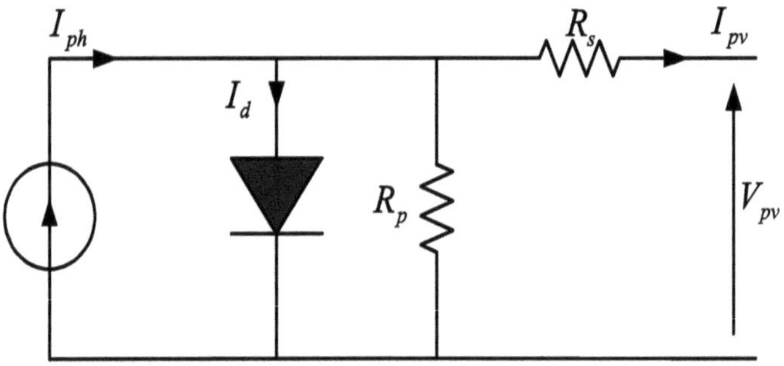

Fig. 12.3 Model of a solar cell

semiconductor material such as silicon or platinum, the energy of the photon is transferred to an electron in an atom of the solar cell. The energized electron is then able to escape its bond with the atom and generates an electric current. This leaves behind a hole. Combined with a P-N junction, which is a layer within the photovoltaic cell that is formed by the intimate contact of P-type and N-type semiconductors that create an electric field, the holes move in the opposite direction from electrons, thereby producing an electric current. The system under study is presented in Fig. 12.2.

We present also in Fig. 12.3 a proposition of an equivalent circuit model of a PV cell which is composed of a light generator source, a parallel resistor expressing a leakage current, a diode and a series resistor describing an internal resistance to the current flow. The equivalent model of a PV cell can be described by the mathematical expressions (12.1–12.3).

The parallel resistor R_p is neglected because of its large resistance and the series resistor R_s is also neglected due to its very small resistance value. Neglecting these two parameters simplifies significantly the numerical simulation without affecting the accuracy. The first equation describes the output current of the cell:

$$i_{\mathrm{pv}} = I_{\mathrm{ph}} - I_{\mathrm{d}} \left[\exp(\frac{q V_{\mathrm{pv}}}{k_{\mathrm{b}} T A}) - 1 \right] \tag{12.1}$$

Table 12.1 PV cell parameters

q	$1.6022.10^{-19}$ C	K_b	$1.3806.10^{-23}$	T_r	298 K
E_g	1.1557 (eV)	G_n	1000 (W/m^2)	I_{scr}	3.45 A
K_i	0.60095 (A/K)	I_{rr}	$5.98.10^{-8}$ A	A	1.2

The equation of the PV current as a function of temperature and irradiance can be written as follows:

$$I_{\text{ph}} = \frac{G}{G_n}\left[I_{\text{scr}} + k_i(T - T_r)\right] \tag{12.2}$$

The saturation current equation is given by:

$$I_{\text{d}} = I_{rr}\left[\frac{T}{T_r}\right]^3 \exp\left(\frac{qE_g}{k_bA}\left[\frac{1}{T_r} - \frac{1}{T}\right]\right) \tag{12.3}$$

The PV cell parameters used in this study are shown in Table 12.1.

12.2.2 *The Effect of Climatic Conditions on the PV Panel*

Figures 12.4 and 12.5 show the effect of varying weather conditions at MPP locations at P-V curves. Figure 12.4 shows that the open circuit voltage and the power increase in the important solar irradiance value. It is thus observed that the reduction in the intensity of irradiation involves a reduction in the power. According to the results shown in the preceding Fig. 12.5, we note that, for the constant illuminations, while the temperature increases, the open circuit voltage (V_{oc}) will decrease. This is the result of the reduction in the energy gap. The increase in the temperature causes also a weak increase in the short circuit current. The fact that the reduction in the open circuit voltage (V_{oc}) is more significant with respect to the increase in the short circuit current, the power production also will decrease.

The power production of the cell decreases gradually with the temperature and it is proportional to the intensity of the incidental light. Thus we note that the effect of the increase in solar cell's temperature results a degradation of their performances.

12.3 Modeling the Boost Converter

Boost converters are appropriate to be used for performing MPPT in PV systems because of their capability of delivering an output voltage larger than the low input voltage of the PV panel. The DC/DC converters (Garraoui et al. 2013) is widely used in regulating switch mode DC power supplies. The input of these converters

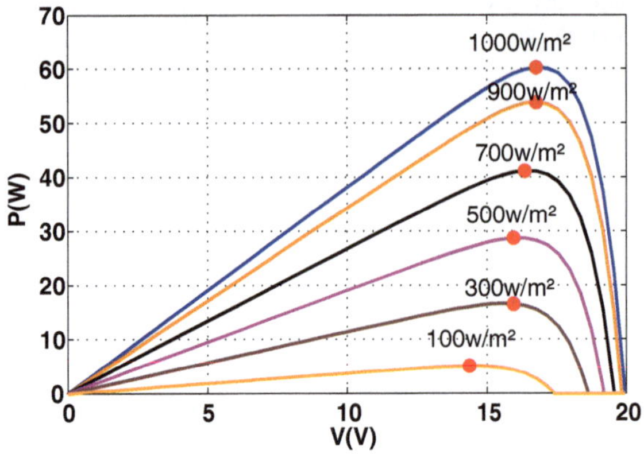

Fig. 12.4 P-V characteristics under different irradiance and T = 25 °C

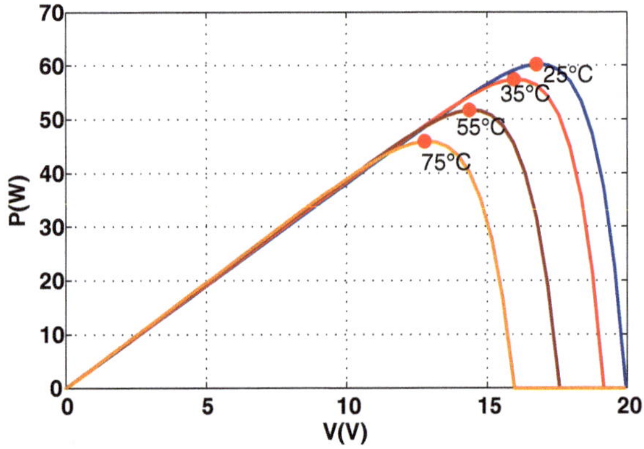

Fig. 12.5 P-V characteristics under different temperature and G = 1000 W/m^2

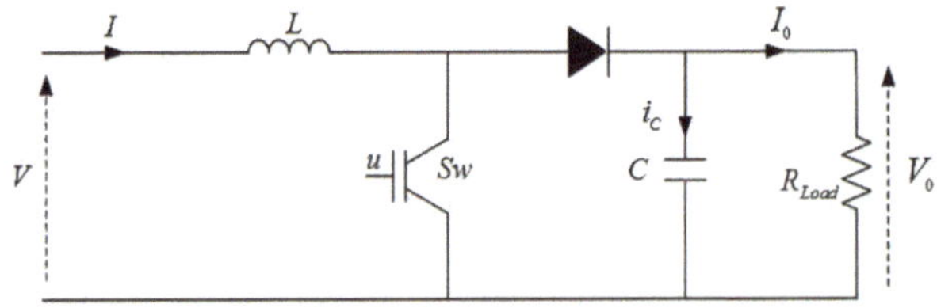

Fig. 12.6 Model of a boost converter

is an unregulated DC voltage, which is obtained by PV array and therefore it will be fluctuated due to changes in radiation and temperature. In these converters the average DC output voltage must be controlled to be equated to the desired value, although the input voltage is changing. The conventional DC/DC boost converter circuit is shown in Fig. 12.6.

From the energy point of view, output voltage regulation in the DC/DC converter is achieved by constantly adjusting the amount of energy absorbed from the source

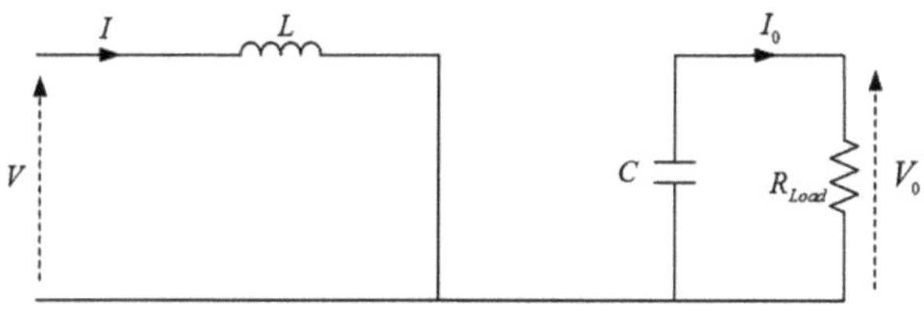

Fig. 12.7 Boost converter operating principle with the switch closed

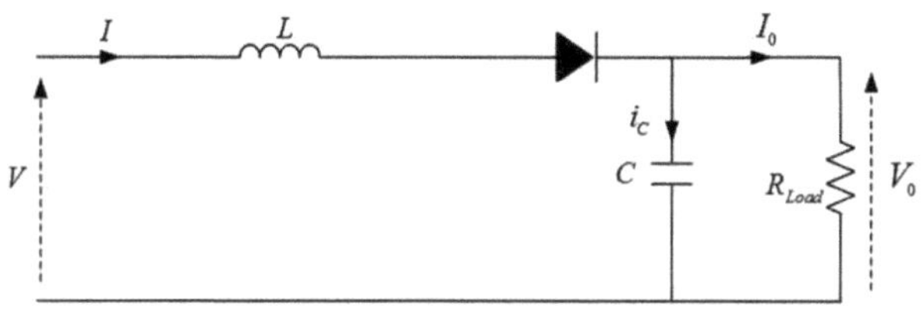

Fig. 12.8 Boost converter operating principle with the switch open

and that injected into the load, which is in turn controlled by the relative durations of the absorption and injection intervals.

These two basic processes of energy absorption and injection constitute a switching cycle. The conventional DC-DC boost converter circuit is shown in Fig. 12.6. Depending on the position of the switch Sw, the system has two operating topologies as shown in Figs. 12.7 and 12.8:

By applying Kirchhoff's voltage law, we get the mathematical model of the system. When the switch is **OFF**, the dynamics of the circuit are governed by the following equation:

$$\begin{cases} I = \frac{1}{L}\int V dt \\ V_0 = -\frac{1}{C}\int \left(\frac{V_0}{R_{Load}}\right) dt \end{cases} \tag{12.4}$$

When the switch is **ON**, the circuit is described by:

$$\begin{cases} I = \frac{1}{L}\int (V - V_0) dt \\ V_0 = \frac{1}{C}\int \left(I - \frac{V_0}{R_{Load}}\right) dt \end{cases} \tag{12.5}$$

The average model of the system is given by:

$$\begin{cases} I = \int \frac{1}{L}[V + (1-u)V_0] dt \\ V_0 = \int \frac{1}{C}\left[I(1-u) - \frac{V_0}{R_{Load}}\right] dt \end{cases} \tag{12.6}$$

12.4 MPPT Controller

The extraction of the maximum of power from the PV generator is then indispensable. Therefore, maximum power point tracking (MPPT) controller accuracy becomes a key control in the device operation for successful PV applications. The MPPT control is a challenging, because the sunshine condition that determines the amount of sun energy into the PVG may change at any time.

Therefore, the PV system can be considered as a non-linear complex system. Numerous MPPT methods have been developed and implemented in previous studies, for instance, perturb and observe (PδO) (Femia et al. 2005), fractional open-circuit voltage approaches. These techniques have high tracking accuracy under stable internal and external condition, but still reveal some tradeoff between tracks speed and tracking reliability when load values or weather conditions rapidly changes. In this chapter, we choose to work with two MPPT controllers based on fuzzy logic and sliding mode control.

12.4.1 Fuzzy Logic MPPT Controller

The conventional control theory is based on the use of mathematical models of the process to be controlled and specification about the expected closed loop behavior. Under parametric changes, these performances can be degraded. Fuzzy logic (Rekioua et al. 2009; Patcharaprakiti 2005; Ammasai et al. 2009) based control does not require a mathematical model of the system.

It uses general information on the process to be controlled and the required performances are achieved by establishing some rules in the form of if-then sentences according to which the controller takes the appropriate decision. The process of Fuzzy Logic Controller (FLC) can be subdivided into three steps: first, fuzzification, then rule evaluation, and finally defuzzification.

FLC is applied in designing the MPPT controller in this section. The input variables of the FLC in this paper are the error and the change in error. The error is calculated as the change in the PV power to the change in the PV voltage.

The Fuzzy inference is determined by using Mandamni's method, and the defuzzification uses the centre of gravity to compute the output of this FLC which is the duty cycle u, the error e, the change in error Δe The control output variable (u) is defined as the output for fuzzy controller, they are considered as a Fuzzy linguistic variable. Each input and output variable is described by seven fuzzy sets: NB, NM, NS, Z, PS, PM and PB (Table 12.2).

The fuzzy logic controller is a rule-based controller and the rules are in the if-then format, e.g. "*If error is* αi *, and change in error is* βi *then output is* ψi". The process of Fuzzy Logic Controller (FLC) can be subdivided into three steps, first, fuzzification, and then rule evaluation, and finally defuzzification. The general architecture of an FLC is shown in Fig. 12.9.

Table 12.2 Fuzzy set partition

NB	Negative big
NM	Negative medium
NS	Negative small
Z	Zero
PS	Positive small
PM	Positive medium
PB	Positive big

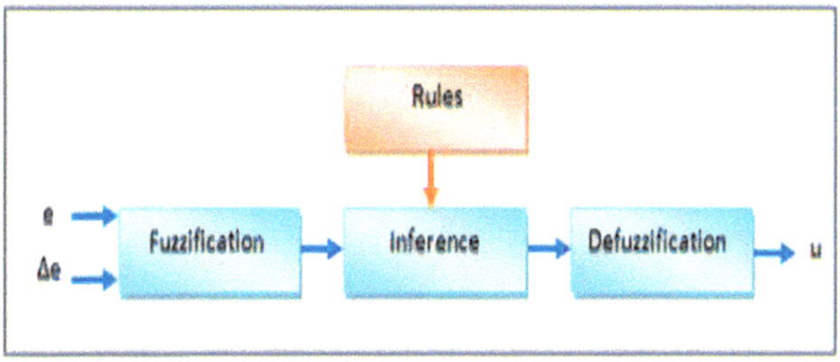

Fig. 12.9 The FLC controller

Fuzzy logic was applied in designing the MPPT controller. The error e is calculated as the change in the PV power to the change in the PV voltage. The change in the duty cycle is the output from FLC which is obtained by applying a set of linguistic rules, the error e, the change in error Δe and the duty cycle u are represented as follows:

$$e(k) = \frac{P(k) - P(k-1)}{V(k) - V(k-1)} \quad (12.7)$$

$$\Delta e = e(k) - e(k-1) \quad (12.8)$$

$$u = f(e, \Delta e) \quad (12.9)$$

and the defuzzification uses the centre of gravity to compute the output of this FLC which is the duty cycle:

$$u = \frac{\sum_{j=1}^{n} \mu(u_j) - u_j}{\sum_{j=1}^{n} \mu(u_j)} \quad (12.10)$$

Table. I. The fuzzy system rules.

	NB	NM	NS	Z	PS	PM	PB
NB	NB	NB	NB	NB	NM	NS	Z
NM	NB	NB	NB	NM	NS	Z	PS
NS	NB	NB	NM	NS	Z	PS	PM
Z	NB	NM	NS	Z	PS	PM	PB
PS	NM	NS	Z	PS	PM	PB	PB
PM	NS	Z	PS	PM	PB	PB	PB
PB	Z	PS	PM	PB	PB	PB	PB

Fig. 12.10 The Fuzzy stem rules

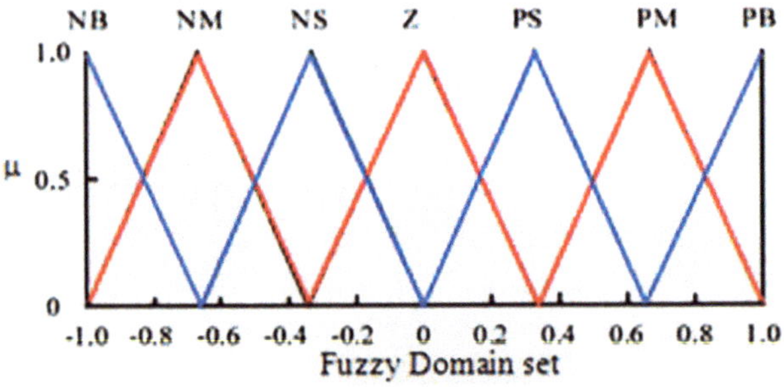

Fig. 12.11 Fuzzy logic control membership function for input and output

The duty cycles from 49 rules must be computed and combined for a specified value. The fuzzy system rules can be designed as shown in Fig. 12.10. The control rules are indicated in Fig. 12.10 with e and Δe as inputs and u as the output.

The fuzzy inference of the proposed FLC is based on the Mamdani's method. The Fuzzy logic control memberships function for input and output take the following form in Fig. 12.11.

12.4.2 MPPT Based on Sliding Mode Control Theory

Sliding Mode Control is known to be a robust control method appropriate for controlling switched systems. High robustness are maintained against various kinds of uncertainties such as external disturbances and measurement error (Il-Song 2007). In traditional sliding mode control, or First Order Sliding Mode controller (SMC) design, the sliding variable is selected such that it has relative degree one with respect to the control. For the photovoltaic system the design procedure for a state based sliding mode controller can be divided into two steps: Finding the switching function, such that the internal dynamics in the sliding mode are stable. The sliding surface will be selected by imposing the maximum power point identifier factor $\partial P/\partial I$ equal to zero as shown in (12.11).

$$\frac{\partial P}{\partial I} = I\left(\frac{\partial V}{\partial I} + \frac{V}{I}\right) = 0 \tag{12.11}$$

The non trivial solution of (12.11) is: $\left(\frac{\partial V}{\partial I} + \frac{V}{I}\right) = 0$. Thus, a proper sliding manifold in the state space can be defined as:

$$\sum = \left\{x \,/\, S(t,x) = \frac{\partial V}{\partial I} + \frac{V}{I} = 0\right\} \tag{12.12}$$

The surface Σ is called a switching surface. The most important task is to design a control that will drive the plant state to the switching surface and maintain it there. The switching control of the SMC is presented as follows:

$$u_d = \left| \begin{array}{ll} u^+ & \text{if } S(z) > 0 \\ u^- & \text{if } S(z) < 0 \end{array} \right. \tag{12.13}$$

A simple sliding mode control design can be written as:

$$u = u_{eq} + u_d \tag{12.14}$$

u_{eq} is called equivalent control. Note that, under the action of the equivalent control u_{eq} any trajectory starting from the manifold $S(z) = 0$ remains on it, since $\dot{S}(z) = 0$, u_{eq} is given by:

$$S(t,z) = \dot{S}(t,z) = 0 \tag{12.15}$$

This ensures the invariance of the sliding surface. The proposed discontinuous control, u_d guarantees a convergence in finite time on the surface; it is defined as:

$$u_d = M\ \text{sign(S)} \tag{12.16}$$

Equation (12.15) determines the equivalent control of our PV system. It takes the following forms:

$$\dot{S} = \left[\frac{\partial S}{\partial Z}\right]^{\top} \dot{Z} = \left[\frac{\partial S}{\partial Z}\right]^{\top} \left(\phi(Z) + \psi(Z)u_{eq}\right) = 0 \tag{12.17}$$

Now, clearly the equivalent control will be defined as:

$$u_{eq} = -\frac{\left[\frac{\partial S}{\partial Z}\right]^{\top} \phi(Z)}{\left[\frac{\partial S}{\partial Z}\right]^{\top} \psi(Z)} = 1 - \frac{V_{pv}}{V_{out}} \tag{12.18}$$

The real control signal is proposed as:

$$u = \begin{vmatrix} 1 & \text{if} & u_{eq} + M\ \text{sign}(S) > 1 \\ u_{eq} + M\ \text{sign}(S) & \text{if} & 0 < u_{eq} + M\ \text{sign}(S) < 1 \\ 0 & \text{if} & u_{eq} + M\ \text{sign}(S) < 0 \end{vmatrix} \tag{12.19}$$

12.4.3 Stability Analysis

To conduct stability analysis for the corresponding control, it is common to use Lyapunov approach (Feng et al. 2009), which is adopted to identify this mission. This Lyapunov function must be defined from the sliding surface already selected, for sliding mode; the function V is generally defined by:

$$V = \frac{1}{2} S^2 \tag{12.20}$$

In this case, the sufficient condition is given by:

$$\dot{V} = S\dot{S} < 0 \tag{12.21}$$

In this way, an asymptotic convergence to the surface S = 0 is guaranteed. The derivative of the surface is calculated as:

$$\dot{S} = \left[\frac{\partial S}{\partial Z}\right]^{\top} \dot{Z} = \left(3\,\frac{\partial R_{pv}}{\partial I_{pv}} + I_{pv}\,\frac{\partial^2 R_{pv}}{\partial {I_{pv}}^2}\right)\left(-\frac{V_{out}}{L}(1-u) + \frac{V_{pv}}{L}\right) \tag{12.22}$$

where:

$$\frac{\partial R_{pv}}{\partial I_{pv}} = \frac{\partial}{\partial I_{pv}}\left[\frac{V_{pv}}{I_{pv}}\right] = \frac{1}{I_{pv}}\frac{\partial V_{pv}}{\partial I_{pv}} - \frac{V_{pv}}{{I_{pv}}^2} \tag{12.23}$$

and:

$$\frac{\partial^2 R_{pv}}{\partial^2 I_{pv}} = \frac{1}{I_{pv}}\frac{\partial^2 V_{pv}}{{I_{pv}}^2} - \frac{2}{{I_{pv}}^2}\frac{\partial V_{pv}}{\partial I_{pv}} + \frac{2V_{pv}}{{I_{pv}}^3} \tag{12.24}$$

On the other hand, the PV voltage is needed to be expressed in terms of mathematical functions using the PV panel parameters as follows:

$$V_{pv} = \frac{K_b T A}{q} \ln\left(\frac{I_{ph} + I_d - I_{pv}}{I_d}\right) \tag{12.25}$$

The first derivative of the voltage can be expressed by:

$$\frac{\partial V_{pv}}{\partial I_{pv}} = -\frac{K_b T A}{q}\frac{I_d}{I_{ph} + I_d - I_{pv}} < 0 \tag{12.26}$$

Clearly the first derivative of the voltage has a negative sign. Turning now to the expression of the second derivative:

$$\frac{\partial^2 V_{pv}}{\partial^2 I_{pv}} = -\frac{K_b T A}{q} \frac{I_d}{(I_{ph} + I_d - I_{pv})^2} < 0 \tag{12.27}$$

It is obvious from (12.26) that the second voltage derivative also has a negative sign. Moreover, the first derivative of the sliding surface can be presented as:

$$\dot{S} = \left[\frac{\partial S}{\partial Z}\right]^{\mathrm{T}} \dot{Z} = \left(3\frac{\partial R_{pv}}{\partial I_{pv}} + I_{pv}\frac{\partial^2 R_{pv}}{\partial {I_{pv}}^2}\right)\left(-\frac{V_{out}}{L}(1-u) + \frac{V_{pv}}{L}\right) \tag{12.28}$$

Denoting by "A" the following term:

$$A = \frac{1}{I_{pv}}\frac{\partial V_{pv}}{\partial I_{pv}} - \frac{V_{pv}}{{I_{pv}}^2} + \frac{\partial^2 V_{pv}}{\partial {I_{pv}}^2} \tag{12.29}$$

So:

$$\dot{S} = A\left(-\frac{V_{out}}{L}(1-u) + \frac{V_{pv}}{L}\right) \tag{12.30}$$

This term (12.29) has always a negative sign. In order to verify the stability, we should ensure the inequality (12.21). We consider $0 < u < 1$.

So:

$$\begin{aligned}\dot{Z} &= -\frac{V_{out}}{L}(1-u) + \frac{V_{pv}}{L} = -\frac{V_{out}}{L}\left(1 - u_{eq} - u_d\right) + \frac{V_{pv}}{L}\\ &= \frac{V_{out}}{L}\left(1 - \left[1 - \frac{V_{pv}}{V_{out}}\right] - u_d\right) + \frac{V_{pv}}{L} = \frac{V_{out}}{L}u_d\end{aligned} \tag{12.31}$$

So, it is obvious that the derivative of the surface has been always opposite in sign to the surface S. Moreover:

$$\dot{S} = A\,\dot{Z} = A\,u_d = A\,M\,\mathrm{sign}(S) = N\,\mathrm{sign}(S) \tag{12.32}$$

And $N = A.M$ is obviously considered as a negative quantity $N < 0$ so $S\dot{S} < 0$. So the system is stable.

12.5 Simulation of the Fuzzy Logic and the Sliding MPPT Controllers and Results

Figure 12.12 shows the functional diagram of the simulated photovoltaic system using the two MPPT controllers under standard conditions: temperature (25 °C) and

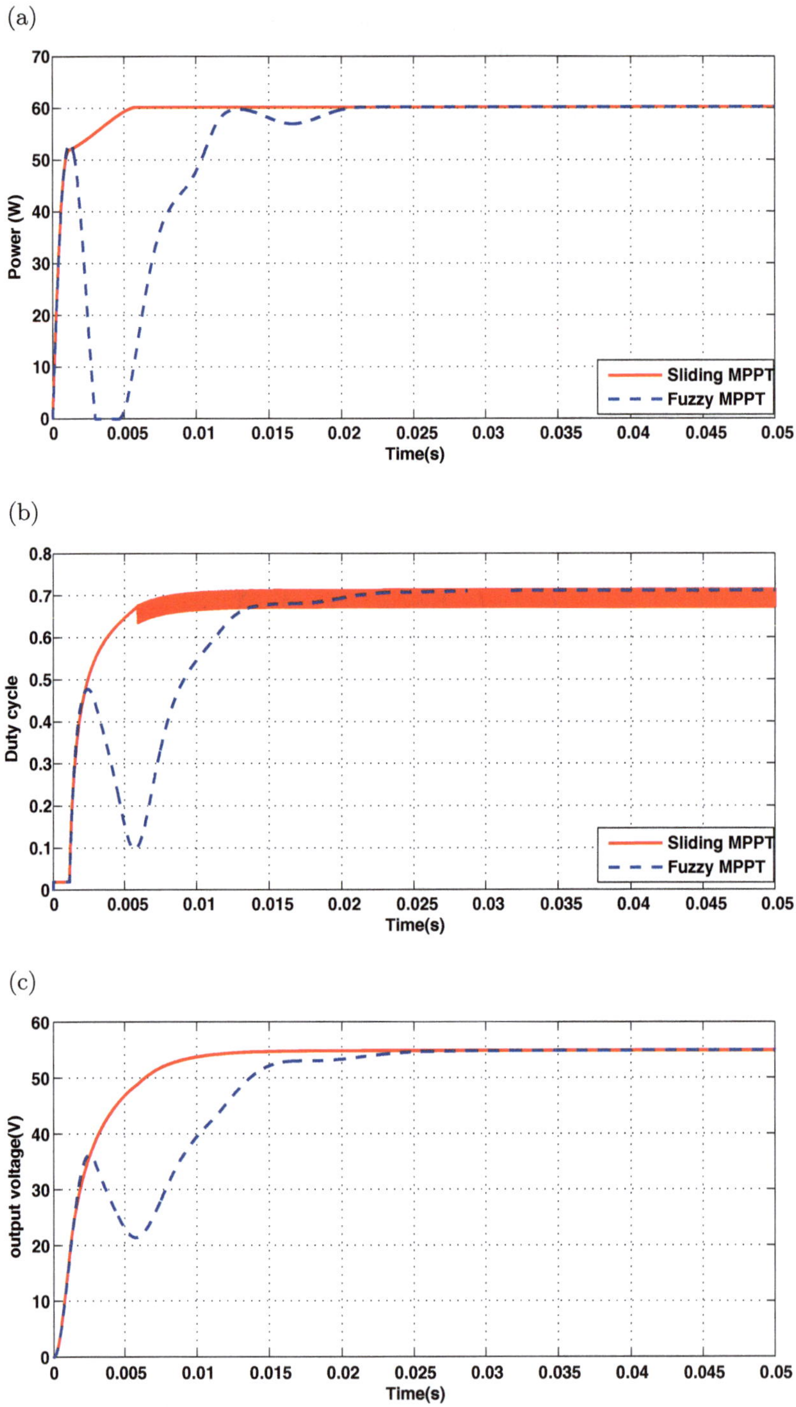

Fig. 12.12 The PV power (**a**), the duty cycle (**b**) and the output voltage (**c**) under standard climatic conditions

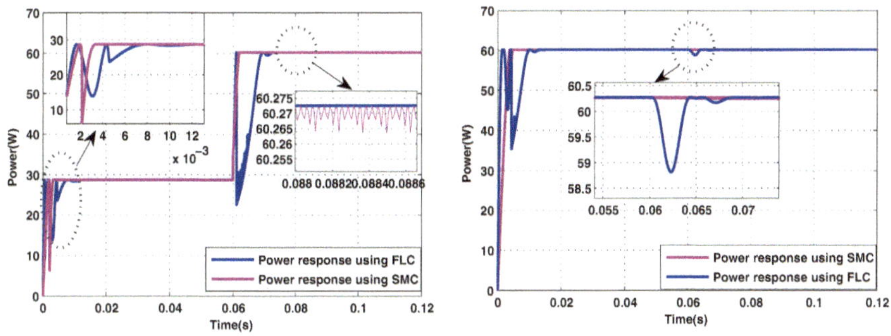

Fig. 12.13 The power under variable irradiance (**a**) and load change (**b**)

(a) (b)

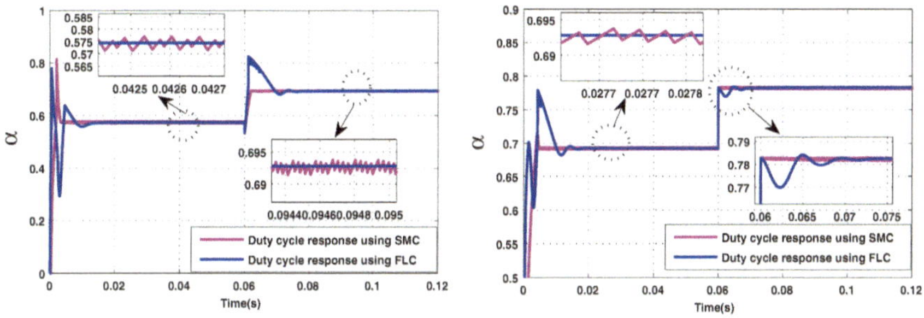

Fig. 12.14 The duty cycle under variable irradiance (**a**) and load change (**b**)

(a) (b)

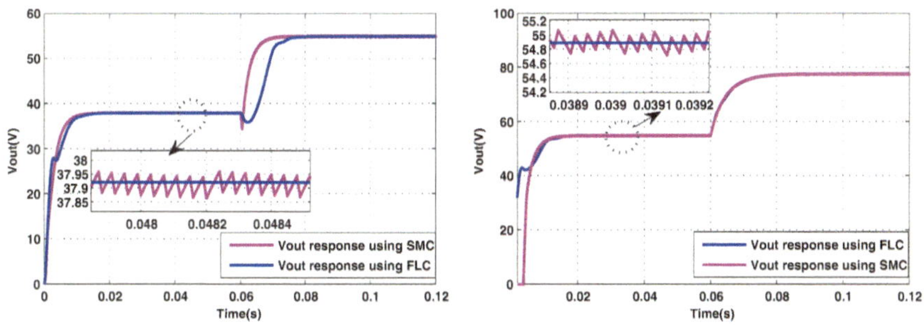

Fig. 12.15 The output voltage under variable irradiance (**a**) and load change (**b**)

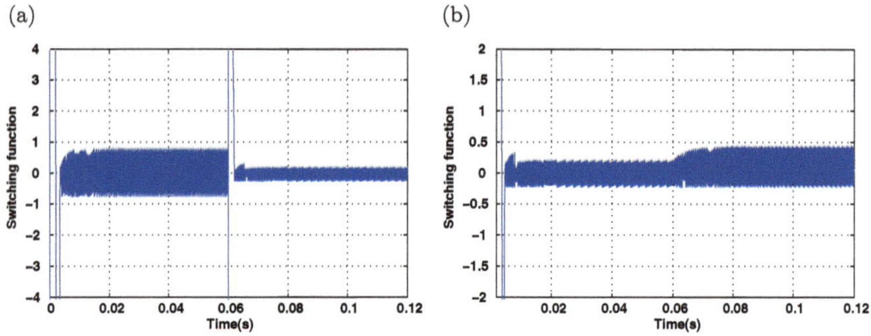

Fig. 12.16 The switching function under variable irradiance (**a**) and load change (**b**)

solar insolation (1000 W/m^2). Obviously, it can be deduced that the Sliding controller is faster than the Fuzzy controller in the transitional state, but the Fuzzy presents a much smoother signal with less fluctuations in steady state.

The irradiance profile used in this work takes the following form: it increases from 500 to 1000 W/m^2 in steps of 500 units every 0.06 s. We note the following answers provided by the FLC and the SMC on the power, which is shown in Fig. 12.13a. The photovoltaic system has responded correctly, the MPPT based on SMC provides the fastest and most efficient response against trade lighting, the two MPPT algorithms resisted with success the abrupt change of irradiance.

This is clear if we examine the nature of the generated duty cycle while observing the simulation result in Fig. 12.14a. The influence of the good nature of command generated by the FLC is clear at the output voltage V_{out} in Fig. 12.15a. It is also confirmed by these plots that the MPPT based on FLC has a very small error due to its special structure that is why it has the most precise response.

In addition, the MPPT based on SMC ensures a good dynamics around his switching surface which is stable and attractive, as it is depicted in Fig. 12.16a. Now, assume that R is suddenly changed from 50 to 100 Ω at 0.06 s, for such variation of R, the MPPT system is simulated to obtain the result of power as shown in Fig. 12.13b. Obviously, the MPPT system still holds on the power value of 60.27 W during the interval of R variation.

So, by using our two MPPT controllers, the system still holds on the MPP in spite of the variation of load resistance, and according to the responses on duty cycle and output voltage presented respectively in Figs. 12.14b and 12.15b we could insure this. But, if we pay attention to the difference between the natures of the responses occurred by the FLC and the robust command based on the SMC, we can say that the first order SMC is more robust against the load change because of its very low sensitivity in comparison with FLC. But it provides a problem which is characterized by the high oscillation frequency. This is clear when observing the plot of the sliding function in Fig. 12.16b. However, the FLC seems to be the most precise one due to its special structure.

12.6 Conclusion

In this chapter, we have presented two MPPT controllers which are the Sliding Mode Controller (SMC) and the Fuzzy Logic Controller (FLC). The FLC makes the output PV system value close to the theoretical maximum value. The simulation results show that in the case of irradiance and load abrupt change, the SMC can quickly find a new maximum power point, and has a high power output efficiency. The FLC has also performing results. The experimental validation of these two algorithms will be the topic of future work.

References

Agorreta J. L., Reinaldos L., Gonzalez, R., Borrega M., Balda, J., & Marroyo, L. (2013). Review of maximum-power-point tracking techniques for solar-photovoltaic systems. *Energy Technology, 1*, 438–448.

Ammasai, G. N., Nallandula, H. A., & Krithiga, S. (2009). Fuzzy logic controller with mppt using line commutated inverter for three phase grid connected photovoltaic systems. *Renewable Energy, 34*, 909–915.

Bose, B. K. (2010). Global warming: Energy, environmental pollution, and the impact of power electronics. *IEEE Industrial Electronics Magazine*, *4*, 6–17.

Femia, N., Petrone, G., Spagnuolo, G., & Vitelli, V. (2005). Optimization of perturb and observe maximum power point tracking method. *IEEE Transactions on Power Electronics*, *20*, 963–973.

Feng, Y., Zheng, J., Yu, X., & Vu. T. N. (2009). Hybrid terminal sliding mode observer design method for a permanent magnet synchronous motor control system. *IEEE Transaction on Industrial Electronics*, *56*, 3424–3431.

Garraoui, R., Hamed, M. B., & Sbita, L. (2013). MPPT controller for a photovoltaic power system based on fuzzy logic. *10th IEEE International Conference on International Multi-Conference on Systems, Signals and Devices (SSD), Hammamet, Tunisia.*

Il-Song, K. (2007). Robust maximum power point tracker using sliding mode controller for the three-phase grid-connected photovoltaic system. *Solar Energy*, *81*, 405–414.

Koutroulis, E., & Blaabjerg, F. (2013). Design optimization of transformer less grid-connected pv inverters including reliability. *IEEE Transactions on Power Electronics*, *28*, 325–335.

Ran, B., & Boyce, D. E. (1996). *Modeling Dynamic Transportation Network*. Germany: Springer.

Rekioua, D., Lalouni, S., Rekioua, T., & Matagne, E. (2009). Fuzzy logic control of stand-alone photovoltaic system with battery storage. *Journal of Power Sources*, *193*, 899–907.

Patcharaprakiti, N., Premrudeepreechacharn, S., & Sriuthaisiriwong, Y. (2005). Maximum power point tracking using adaptive fuzzy logic control for grid-connected photovoltaic system. *Renewable Energy*, *30*, 1771–1788.

Young, G. O. (1964). Synthetic structure of industrial plastics. *Polymers of Hexadromicon* 2nd ed., J. Peters, Ed., New York: McGraw-Hill, *3*, 15–64.

12.6 Conclusions

[illegible]

Chapter 13
Sliding Bifurcations and Sliding Mode Controller for a Two-Cell DC/DC Buck Converter

Karama Koubaâ

Abstract In this chapter, we analyze the sliding bifurcations that occur in the two-cell DC/DC buck converter controlled using a dynamic feedback controller, then we apply the sliding mode controller to the converter in order to inhibit bifurcations and chaotic behavior. We use a simplified discrete model to analyze the bifurcations in the two-cell converter, which can be regarded as a piecewise smooth nonlinear system with discontinuous iterated maps. Then, we give theoretical conditions of stability according to the parameters values of the dynamic feedback controller. The presence of discontinuities in the converter leads to several types of non-smooth bifurcations namely border collision bifurcation, degenerate flip bifurcation and sliding bifurcations such as switching-sliding, grazing-sliding and adding-sliding also called multi-sliding. Non-smooth bifurcations, and more particularly, sliding bifurcations are caused by structural changes in the system dynamics, then we apply the sliding mode controller which is a variable structure control system (VSCS) to avoid sliding modes in the DC/DC buck converter. Numerical simulations confirm the analytical results and explain the bifurcations and the strange phenomena encountered in the two-cell converter.

Keywords Sliding bifurcations · Two-cell DC/DC buck converter · Dynamic feedback controller · Sliding mode controller

13.1 Introduction

Piecewise-smooth systems are characterized by periods of smooth evolutions interrupted by instantaneous events. Traditional analysis of dynamical systems has restricted its attention to smooth problems, thus preventing the investigation of non-smooth processes such as impact, switching, sliding and other discrete state transitions. These phenomena arise, for example, in any application involving friction, collision, intermittently constrained systems or processes with switching compo-

K. Koubaâ (✉)
CEMLab, National Engineering School of Sfax, Sfax, Tunisia
e-mail: karama1484@gmail.com

N. Derbel et al. (eds.), *Applications of Sliding Mode Control*,
Studies in Systems, Decision and Control 79, DOI 10.1007/978-981-10-2374-3_13

nents (di Bernardo et al. 2008). In fact, power converters are known to exhibit border collision bifurcation (di Bernardo and Tse 2002; Tse 2002; Yuan et al. 1998).

In addition, piecewise-smooth systems can undergo sliding bifurcations due to interactions between a periodic orbit and the boundary of the sliding region. Basically, four distinct cases of such bifurcations can be identified:

- Crossing-sliding: under parameter variation, the trajectory intersects the boundary of the sliding region and then switches from a subspace to the other with no sliding portion.
- Grazing-sliding: a section of trajectory lying in one region grazes the boundary of the sliding region. This causes the formation of a section of sliding motion.
- Switching-sliding: under parameter variation, the trajectory initially completely contained in the subspace and in the sliding region, intersects its boundary and then passes from the subspace to the other, and from this to the sliding region.
- Adding-sliding: it differs from the scenarios presented above because the segment of the trajectory that undergoes the bifurcation lies entirely within the sliding region. Further variation of the parameter causes the formation of an additional segment of trajectory above (or below) the sliding region (Santos 2006).

Sliding bifurcations may be encountered in several dynamical systems. In fact, the two-cell DC/DC buck converter controlled with a dynamic PI controller exhibits grazing-sliding also called spiralling bifurcation with multi-sliding orbits (Koubaâ and Feki 2014a). A previous work (di Bernardo et al. 1998) provides an analytical explanation for the grazing-sliding scenario observed in the one-cell DC/DC buck converter in a double spiralling form. The existence of sliding orbits has been pointed out in the context of relay feedback systems (di Bernardo et al. 2001, 1999), dry-friction oscillator (Guardia et al. 2010) and superconducting resonator (Jeffrey et al. 2010).

We notice that sliding bifurcations are caused by structural changes in the system dynamics, then we apply the sliding mode controller which is a variable structure control system (VSCS) to avoid sliding modes in the DC/DC buck converter. Sliding mode control (SMC) is well known for efficiently dealing with uncertain dynamical systems even when no full information of their models is available (Edwards 1998; Utkin 1992). The purpose of the SMC law is to force the tracking error to approach the sliding surface and then move along the sliding surface to the origin. However, the drawbacks of the conventional SMC such as chattering phenomenon, which reveals itself as high-frequency dangerous vibrations of the whole system (Levant 2010), and *a priori* knowledge of the bounds of uncertainties, can be destructive (Amer et al. 2012).

The multi-cell converters introduced more than ten years ago make it possible to distribute the voltage constraints among series-connected switches and to improve the output waveforms. The balance of the constraints requires an appropriate distribution of the flying voltages (Gateau et al. 2002). The use of a discrete time model (El-Aroudi et al. 2008, 2006; Robert and El-Aroudi 2006) in this type of power converters combined with a digital pulse width modulation (DPWM) controller leads to discontinuous iterated maps. Due to the existence of various operating modes and control saturations, the overall operation is compared to a piecewise smooth nonlinear dynamical system (Robert et al. 2006), which can undergo sliding bifurcations.

In this chapter, we analyze the sliding bifurcations in the behavior of a two-cell DC/DC buck converter controlled using a generalized dynamic feedback controller. This converter is generally characterized by a switching-sliding bifurcation, and especially a grazing-sliding bifurcation for a particular choice of the control parameters, and adding-sliding also called multi-sliding, which results in the appearance of an additional sliding section. Furthermore, we notice the presence of limit cycles and undesirable saturating regimes caused by the saturation in the dynamic control law. Sliding bifurcations are caused by structural changes in the system dynamics, then we apply a discrete sliding mode controller which is a variable structure control system (VSCS) to avoid sliding modes in the two-cell DC/DC buck converter.

The rest of the chapter is outlined as follows. In Sect. 13.2, we present the simplified discrete model for the two-cell DC/DC buck converter. Section 13.3 deals with the stability analysis and the bifurcations that occur in the generalized dynamic feedback controller. In Sect. 13.4, we adopt a discrete sliding mode controller to improve the response of the two-cell DC/DC buck converter and to eliminate the sliding bifurcations. Finally, concluding remarks relating the overall study and future work will be drawn in the last section.

13.2 Discrete Model of the Two-Cell DC/DC Buck Converter

The structure of the two-cell DC/DC buck converter that we deal with in this chapter is depicted in Fig. 13.1.

It consists of two cells separated by a flying capacitor C to balance the switch voltages. Each cell contains a switch S_j and a diode D_j $(1 \leq j \leq 2)$ that behave in a complementary manner such that when the switch is open, the diode is closed and vice versa. U_1 and U_2 are the control signals delivered by a Digital Pulse Width Modulator (DPWM) controller and can take two boolean states 0 and 1. These control

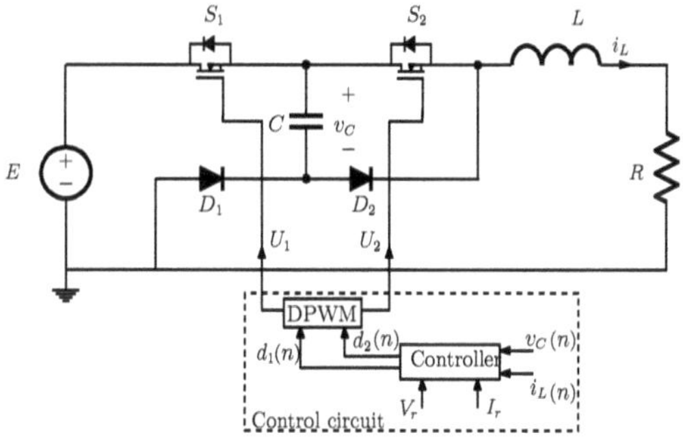

Fig. 13.1 Structure of the two-cell DC/DC buck converter

Table 13.1 Topologies in the two-cell DC/DC converter

Topologies	State of S_1	State of S_2
Topology 1 (T_1)	OFF	ON
Topology 2 (T_2)	ON	ON
Topology 3 (T_3)	ON	OFF
Topology 4 (T_4)	OFF	OFF

signals represent the evolution of the duty cycles d_1 and d_2 of switches S_1 and S_2 with respect to their OFF state, which should vary in the interval [0 1].

The two-cell DC/DC converter is used to feed an *RL* load. i_L is the inductor current and v_C is the voltage across the capacitor. We will suppose that U_1 and U_2 are phase shifted by π in order to obtain optimum waveforms of the inductor current. According to the states of the switches, we can define four different topologies in Table 13.1.

The current and the voltage are normalized by their maximum values $i_{L_{max}} = \frac{E}{R}$ and $v_{C_{max}} = E$ and time is also normalized by the switching period T, then the new state variables are given by:

$$x_i = \frac{R}{E} i_L \tag{13.1}$$

$$x_v = \frac{1}{E} v_C \tag{13.2}$$

By applying Kirchhoff laws, the system can be described by a linear continuous model for the kth topology:

$$\dot{x} = A_k x + B_k \quad \text{for} \quad 1 \le k \le 4 \tag{13.3}$$

where x is the state vector:

$$x = \begin{bmatrix} x_i \\ x_v \end{bmatrix} \tag{13.4}$$

and A_k and B_k are the matrices defined by:

$$A_1 = \begin{bmatrix} -\delta_L & \delta_L \\ -\delta_C & 0 \end{bmatrix}, \ A_2 = \begin{bmatrix} -\delta_L & 0 \\ 0 & 0 \end{bmatrix}, \ A_3 = \begin{bmatrix} -\delta_L & -\delta_L \\ \delta_C & 0 \end{bmatrix}, \ A_4 = \begin{bmatrix} -\delta_L & 0 \\ 0 & 0 \end{bmatrix}$$

$$B_1 = B_4 = \begin{bmatrix} 0 \\ 0 \end{bmatrix}, \ B_2 = B_3 = \begin{bmatrix} \delta_L \\ 0 \end{bmatrix}$$

δ_L and δ_C are the time constants given by:

$$\delta_L = \frac{RT}{L}, \ \delta_C = \frac{T}{RC} \tag{13.5}$$

In order to reduce the ripple current through the load and the ripple voltage across the capacitor, the circuit parameters should satisfy the following condition:

$$\delta_L \ll 1 \text{ and } \delta_C \ll 1 \tag{13.6}$$

Toggling between different topologies occurs according to the values of the duty cycles d_1 and d_2 at the beginning of the period. As a matter of fact, we can define six configurations, to each of which we can assign a recurrent system to express the values of the states at the beginning of the $(n+1)^{th}$ period in terms of its value at the nth period. Then, we obtain the simplified discrete model (Feki et al. 2007):

$$x(n+1) = \mathrm{A}\, x(n) + \mathrm{B} \tag{13.7}$$

where:

$$\mathrm{A} = \begin{bmatrix} 1-\delta_L & \delta_L(d_1-d_2) \\ \delta_C(d_2-d_1) & 1 \end{bmatrix}, \ \mathrm{B} = \begin{bmatrix} \delta_L(1-d_1) \\ 0 \end{bmatrix} \tag{13.8}$$

The simplified discrete model for the two-cell DC/DC converter can be written as follows:

$$x_i(n+1) = (1-\delta_L)x_i(n) + \delta_L\big(d_1(n)-d_2(n)\big)x_v(n) + \delta_L\big(1-d_1(n)\big) \tag{13.9}$$

$$x_v(n+1) = \delta_C\big(d_2(n)-d_1(n)\big)x_i(n) + x_v(n) \tag{13.10}$$

where the duty cycles have the following forms:

$$d_1(n) = \mathrm{sat}[k_i e_i(n) + k_v e_v(n) + u(n)] \tag{13.11}$$

$$d_2(n) = \mathrm{sat}[k_i e_i(n) + u(n)] \tag{13.12}$$

k_i and k_v denote respectively the current gain and the voltage gain, $e_i(n)$ and $e_v(n)$ are the errors defined with respect to the reference values I_r and V_r as:

$$e_i(n) = x_i(n) - I_r \tag{13.13}$$

$$e_v(n) = x_v(n) - V_r \tag{13.14}$$

$u(n)$ is the control law that will be defined in the following sections, and the saturation function is also defined by:

$$\mathrm{sat}(d) = \begin{cases} 0 \text{ if } d < 0 \\ d \text{ if } 0 \le d \le 1 \\ 1 \text{ if } d > 1 \end{cases} \tag{13.15}$$

In the sequel, we choose the following values for numerical simulations:

$$\delta_L = \delta_C = 0.1, \ I_r = 0.6\,A, \ V_r = 0.5\ V, \ E = 1\ V.$$

13.3 Sliding Bifurcations in the Two-Cell DC/DC Buck Converter Controlled Using a Dynamic Feedback Controller

13.3.1 Dynamic Feedback Controller

One of the most popular methods in chaos control research is the delayed feedback control (DFC) method proposed by Pyragas (Pyragas 1992). The main idea relies on the fact that a chaotic attractor has typically embedded in it a dense set of unstable periodic orbits (UPOs) that can be stabilized by a small feedback perturbation. The DFC is based on applying a feedback proportional to the deviation of the current state of the system from its state one period in the past so that the control signal vanishes when the stabilization of the desired orbit is attained. This method has the advantage of not requiring prior knowledge of anything but the period of the desired orbit (Pyragas 2002), and the delayed feedback controller is given by:

$$u(n) = K\left[\, x(n) - x(n-\tau) \,\right] \tag{13.16}$$

where K is the feedback gain, τ is the time delay determined as the period of the unstable periodic orbit to be stabilized, $x(n)$ is the current state and $x(n-\tau)$ is the delayed state. Successful real implementations of this method include mechanical pendulums (Landry et al. 2005), power electronics (Koubaâ et al. 2009; Koubaâ and Feki 2014b; Robert et al. 2006), chemical systems (Xu and Wu 2014), lasers (Liu and Ohtsubo 1994; Schikora et al. 2006), the human respiratory control system (Batzel and Tran 2000a, b), cardiac systems (Hall et al. 1997), financial systems (Chen 2008), fractional-order chaotic systems (Gjurchinovski et al. 2010). Recently, Socolar et al. (1994) have proposed an extended delayed-feedback control (EDFC) method which uses many previous states of the system and permits to widen the stability region as follows:

$$u(n) = K\sum_{j=0}^{+\infty} r^{j}\left[x(n-j) - x(n-j-1)\right] \tag{13.17}$$

which can be written in a simplified form:

$$u(n) = K\left[x(n) - (1-r)\sum_{j=1}^{+\infty} r^{j-1}x(n-j)\right] \tag{13.18}$$

Although the first form of the extended delayed feedback controller (13.17) can be seen as an analogy of the Pyragas control, this form is not feasible for numerical implementation because it requires to store information of all states in the past. However, there is also the equivalent recursive form given in (13.19) which involves next to the time-delayed control signal $x(n-1)$ the delayed version of the control law $u(n-1)$ itself. This form becomes more suitable for an experiment.

$$u(n) = K\big(x(n) - x(n-1)\big) + ru(n-1) \tag{13.19}$$

The absolute value of the real constant r is smaller than unity, i.e., $|r| < 1$, such that it can be interpreted as a memory parameter that weights information of states further in the past. Note that the case $r = 0$ recovers the original Pyragas control scheme (Hövel 2011). In (Robert et al. 2006), the EDFC has been used to stabilize a current-programmed PWM single phase inverter and to widen the stability region. Other applications of the EDFC include lasers (Gauthier 1998) and mechanical pendulums (de Paula and Savi 2009). However, both of the DFC and the EDFC methods have a limitation such that they can not stabilize any systems with an odd number of real eigenvalues greater than one, also called odd number limitation (Konishi et al. 1999; Ushio 1996). To overcome this limitation, Ahlborn and Parlitz (2004) have suggested Multiple Delay Feedback Control (MDFC) given by:

$$u(n) = \sum_{j=1}^{N} k_j\big(x(n) - x(n-\tau_j)\big) \tag{13.20}$$

where two or more delayed feedback signals with different delay times τ_j are used, and these delay times are not integer multiples of each other and may enter independent control terms. In comparison to other delay based chaos control methods, Multiple Delay Feedback Control (MDFC) is superior for controlling steady states and works also for relatively large delay times, which are sometimes unavoidable in experiments due to system dead times (Ahlborn and Parlitz 2004). Many researchers have demonstrated the effectiveness of this approach through numerical simulations of the Chua's circuit (Ahlborn and Parlitz 2004) and a successful experimental application for stabilizing a chaotic frequency doubled Nd:YAG laser (Ahlborn and Parlitz 2004). In addition, (Chen 2008) indicates that chaotic behavior in economic systems can be controlled under appropriate feedback strengths and delay times.

Other schemes using modulation of the control parameters with a period different from the periodic target state are used to overcome the so-called odd number limitation from which time-delayed feedback control suffers. It has been reported recently that time-dependent modulation improves the control performance in autonomous systems by several orders of magnitude. Whereas the phase of a periodic target state does not play a significant role, a phase selection mechanism may take place when a time-dependent control loop is applied. As time-dependent modulation breaks the time translation invariance, certain phase lags between the periodic orbit and the controller are selected due to enhanced stability properties. Consider the logistic map subjected to time-delayed feedback control for stabilizing a periodic orbit of period p:

$$\mathbf{x}_{n+1} = 1 - \mu\mathbf{x}_n^2 + K_n\big(\mathbf{x}_n - \mathbf{x}_{n-p}\big) \tag{13.21}$$

with μ is a real parameter, K_n is a control parameter, $\mathbf{x}_{n+1}$ and $\mathbf{x}_n$ are the magnitudes of the population respectively in generation $n+1$ and in the preceding generation n. Whereas the original Pyragas scheme corresponds to a fixed value of the control amplitude $K_n = K$, the authors consider in (Fichtner et al. 2004) a time-dependent parameter having the same period as the target state $K_n = K_{n+p}$. The modulated feedback controller has been successfully applied to the logistic map (Fichtner et al. 2004) and the Rössler model (Just et al. 2003).

Moreover, a Full Delayed Feedback Controller (FDFC) has been proposed to stabilize the unstable fixed points (UFPs) for continuous chaotic systems, where the two parameters of a FDFC, i.e., the controller gain and the upper bound of controller time delay, can be simultaneously obtained by solving some linear matrix inequalities (Guan et al. 2006). Illustrative examples of chaotic systems such as the chua and the Rössler systems are presented to show the effectiveness of the proposed method in (Guan et al. 2007, 2006). Another recent method, called the Double Delayed Feedback Control (DDFC) with two mutually prime delays and the relation between the feedback gain matrices and the controller delays, can be implicitly represented by some stabilization criteria using the Lyapunov theory and matrix inequality technique (Lu et al. 2009).

In (Pyragas and Novičenko 2013), Pyragas et al. have presented an algorithm for a delayed feedback control design to stabilize periodic orbits with an odd number of positive Floquet exponents in autonomous systems. Due to the so-called odd number theorem, such orbits have been considered as uncontrollable by time-delayed feedback methods. However, this theorem has been refuted by a counterexample and recently a corrected version of the theorem has been proved, where the control matrix is designed using a relationship between Floquet multipliers of the systems controlled by time-delayed and proportional feedbacks. Pyragas et al. have adopted this algorithm (Pyragas and Novičenko 2013) to stabilize the periodic orbits in the Lorenz and Chua systems, which have been considered as classical examples unaccessible for the conventional delayed feedback controller.

To overcome the odd number limitation, Yamamoto et al. (2001) have supplemented DFC by dynamic DFC with a function which stores the past differences as the state of the controller, and can be described in term of a difference equation of the controller's state. The proposed controller is given in the following form:

$$x_d(n+1) = A\,x_d(n) + B\,y(n) \tag{13.22}$$

$$u(n) = C\,x_d(n) + D\,y(n) \tag{13.23}$$

where:

$$y(n) = x(n) - x(n-1) \tag{13.24}$$

$A \in \mathbb{R}^{n_c \times n_c}$, $B \in \mathbb{R}^{n_c \times n}$, $C \in \mathbb{R}^{l \times n_c}$, $D \in \mathbb{R}^{l \times n}$, $x_d \in \mathbb{R}^{n_c}$ is the state of the controller, $u(n) \in \mathbb{R}^l$ is the dynamic control law and $x \in \mathbb{R}^n$ is the state (Yamamoto et al. 2001).

This controller has been used successfully in (Koubaâ et al. 2010) to analyze the stability of the two-cell DC/DC buck converter according to the control parameters values.

13.3.1.1 Controller Design

In this section, we apply the generalized dynamic DFC without a delay at the levels of the duty cycles d_1 and d_2 defined by Eqs. (13.11) and (13.12) to the two-cell DC/DC buck converter. In order to guarantee a zero static error, we modify the input of the dynamic delayed controller in a feedback current error. This choice is well justified in a previous work (Koubaâ et al. 2012), where it has been proved that we can achieve a zero static voltage error without inserting an extra state variable in the voltage loop, since its transfer function is a pure integrator. Then, the dynamic controller can be defined in the following form:

$$x_d(n+1) = a\, x_d(n) + b\, e_i(n) \tag{13.25}$$

$$u(n) = c\, x_d(n) + d\, e_i(n) \tag{13.26}$$

where a, b, c and d are scalar parameters.

The state vector is given by:

$$x(n) = \begin{bmatrix} x_i(n) \\ x_v(n) \\ x_d(n) \end{bmatrix} \tag{13.27}$$

By using Eqs. (13.9)–(13.12), (13.25)–(13.26), the closed-loop system becomes as follows:

$$\begin{aligned} x_i(n+1) &= [1 - \delta_L(1 + d + k_i)]x_i(n) - \delta_L k_v(1 + V_r)x_v(n) - \delta_L c x_d(n) \\ &\quad + \delta_L k_v x_v^2(n) + \delta_L[1 + (d + k_i)I_r + k_v V_r] \end{aligned} \tag{13.28}$$

$$x_v(n+1) = \delta_C k_v V_r x_i(n) + x_v(n) - \delta_C k_v x_v(n) x_i(n) \tag{13.29}$$

$$x_d(n+1) = b x_i(n) + a x_d(n) - b I_r \tag{13.30}$$

The fixed point denoted by (x_i^*, x_v^*, x_d^*) is obtained by solving the equation:

$$x^*(n+1) = x^*(n) \tag{13.31}$$

According to the values of the parameters, we can distinguish two different cases:

- Case 1: $0 \le a < 1$:

$$x_i^* = \frac{1 - a + [(1-a)(d + k_i) + bc]I_r}{(1-a)(1 + d + k_i) + bc} \tag{13.32}$$

$$x_v^* = V_r \tag{13.33}$$

$$x_d^* = \frac{b(1 - I_r)}{(1-a)(1 + d + k_i) + bc} \tag{13.34}$$

- Case 2: $a = 1$:

$$x_i^* = I_r \tag{13.35}$$

$$x_v^* = V_r \tag{13.36}$$

$$x_d^* = \frac{1 - I_r}{c} \tag{13.37}$$

We notice that in both two cases, we obtain a zero static error for the voltage. It is more interesting to choose $a = 1$, which corresponds to case 2 to ensure zero static errors for the current and the voltage.

13.3.1.2 Stability Analysis

In the sequel, we carry out a stability analysis for the error system in case 2, therefore, we consider the error vector:

$$e(n) = x(n) - x^* \tag{13.38}$$

Then, the error system is defined by:

$$\begin{aligned} e_i(n+1) &= [1 - \delta_L(1 + d + k_i)]e_i(n) + \delta_L k_v (V_r - 1)e_v(n) \\ &\quad -\delta_L c e_d(n) + \delta_L k_v e_v^2(n) \end{aligned} \tag{13.39}$$

$$e_v(n+1) = (1 - \delta_C k_v I_r)e_v(n) - \delta_C k_v e_i(n) e_v(n) \tag{13.40}$$

$$e_d(n+1) = b e_i(n) + e_d(n) \tag{13.41}$$

with the characteristic equation:

$$[\lambda - (1 - \delta_C k_v I_r)]P(\lambda) = 0 \tag{13.42}$$

where:

$$P(\lambda) = \lambda^2 + a_1 \lambda + a_0 \tag{13.43}$$

and the parameters:

$$a_1 = \delta_L(1 + d + k_i) - 2 \tag{13.44}$$

$$a_0 = 1 - \delta_L(1 + d + k_i) + \delta_L bc \tag{13.45}$$

The stability condition for discrete systems is that the norm of each eigenvalue should be less than one.

By applying this condition to the real eigenvalue $\lambda_0 = 1 - \delta_C k_v I_r$, we get:

$$0 < k_v < \frac{2}{\delta_C I_r} \tag{13.46}$$

According to the Jury criterion for the other eigenvalues, we obtain the following conditions:

$$|a_0| < 1 \tag{13.47}$$
$$1 + a_1 + a_0 > 0 \tag{13.48}$$
$$1 - a_1 + a_0 > 0 \tag{13.49}$$

which leads to:

$$-k_i + bc - 1 < d < -k_i + bc + \frac{2}{\delta_L} - 1 \tag{13.50}$$
$$bc > 0 \tag{13.51}$$
$$d < -k_i + \frac{bc}{2} + \frac{2}{\delta_L} - 1 \tag{13.52}$$

From these conditions, we depict in Fig. 13.2 the stability zone in the plane $k_i - d$, where the lines are given by:

$$L_1 : d(k_i) = -k_i + bc - 1 \tag{13.53}$$
$$L_2 : d(k_i) = -k_i + bc + \frac{2}{\delta_L} - 1 \tag{13.54}$$
$$L_3 : d(k_i) = -k_i + \frac{bc}{2} + \frac{2}{\delta_L} - 1 \tag{13.55}$$

The stability conditions (13.50)–(13.52) can be reformulated in the plane $b - c$ as follows:

$$\frac{1 + d + k_i - \frac{2}{\delta_L}}{b} < c < \frac{1 + d + k_i}{b} \tag{13.56}$$
$$bc > 0 \tag{13.57}$$
$$c > \frac{2(1 + d + k_i) - \frac{4}{\delta_L}}{b} \tag{13.58}$$

and graphically presented in Fig. 13.3, where C_1, C_2 and C_3 are the curves given by:

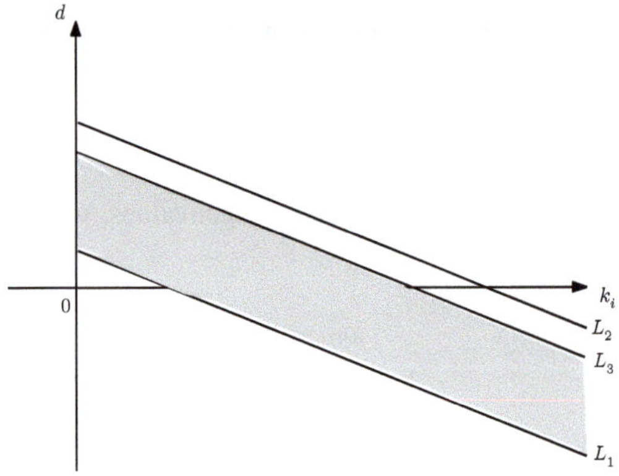

Fig. 13.2 Stability zone in the plane $k_i - d$

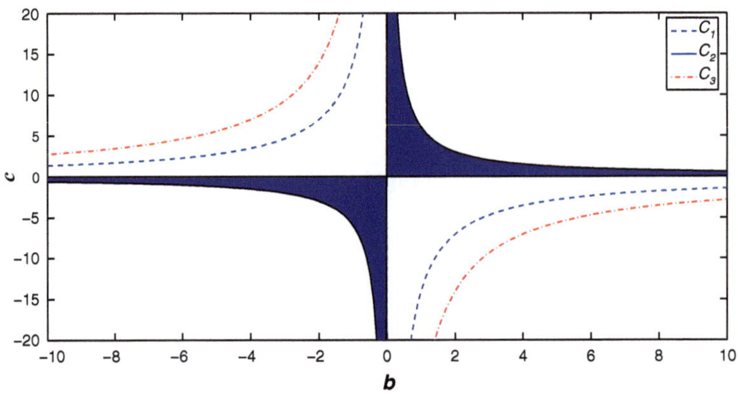

Fig. 13.3 Stability zone in the plane $b - c$

$$C_1 : c(b) = \frac{1 + d + k_i - \dfrac{2}{\delta_L}}{b} \tag{13.59}$$

$$C_2 : c(b) = \frac{1 + d + k_i}{b} \tag{13.60}$$

$$C_3 : c(b) = \frac{2(1 + d + k_i) - \dfrac{4}{\delta_L}}{b}. \tag{13.61}$$

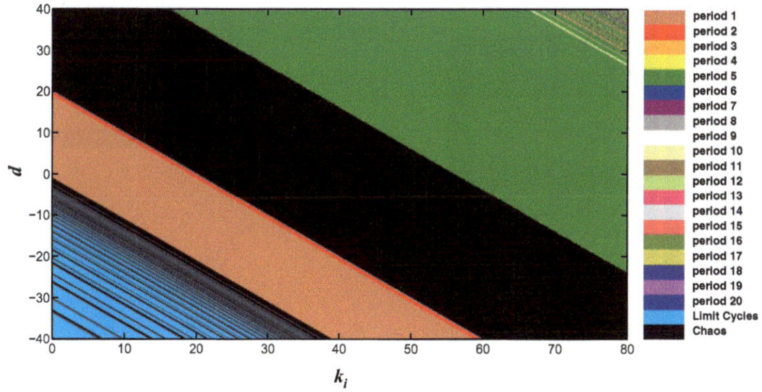

Fig. 13.4 2D bifurcation diagram of the current behavior in the plane $k_i - d$ for the parameters $a = 1, b = 1$ and $c = 0.1$

13.3.2 Sliding Bifurcations

In this section, we study the complex dynamics and the strange phenomena observed in the two-cell DC/DC buck converter controlled using a dynamic feedback controller. We can optimize the choice of the voltage gain k_v to obtain the fastest response by choosing the eigenvalue λ_0 at the origin, and the optimal value of k_v is:

$$k_{v_{opt}} = \frac{1}{\delta_C I_r} = 16.6667 \tag{13.62}$$

Figure 13.4 presents the 2D bifurcation diagram of the current behavior in the plane $k_i - d$ for the parameters $a = 1, b = 1$ and $c = 0.1$. In this figure, we use different colors to indicate the periodic behaviors from period 1 to period 20 and the limit cycles. Brown refers to the one periodic behavior, black is used for chaos and the limit cycles are presented in light blue. The 2D bifurcation diagram confirms the stability zone obtained theoretically in Fig. 13.2. We remark the existence of a two periodic behavior plotted in red and a zone of a five periodic behavior presented in green.

In Fig. 13.5, we present the 1D bifurcation diagram of the current behavior versus the current gain k_i for parameters chosen from the 2D bifurcation diagram and $d = -30$.

We can distinguish four main parts in this diagram. Indeed, for low values of the current gain $k_i < 29.1$, we obtain phases of limit cycles, and a one periodic behavior with a zero static error after a subcritical Neimark-Sacker bifurcation based on two complex conjugate eigenvalues crossing the unit circle. Then, we obtain a two periodic behavior for $49 \leq k_i < 50$, then a four periodic behavior at one bifurcation point for $k_i = 50$ followed by a chaotic behavior in a degenerate flip bifurcation, where cycles of double period are degenerated.

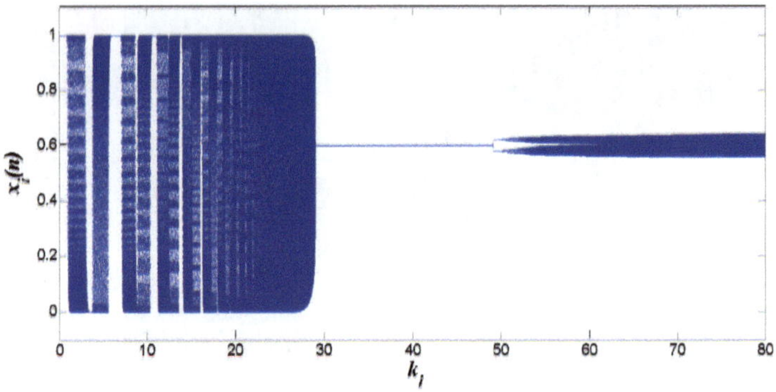

Fig. 13.5 1D bifurcation diagram of the current behavior versus the current gain k_i for the parameters $a = 1, b = 1, c = 0.1$ and $d = -30$

In Fig. 13.6, we plot the evolution of the current $x_i(n)$ and the duty cycles $d_1(n)$ and $d_2(n)$ for the parameters $a = 1, b = 1, c = 0.1, k_i = 20$ and $d = -30$ at the steady regime (i.e. the initial condition $x_v(1) = 0.5$), that leads to a limit cycle.

The current response shows the existence of saturating regimes $x_i(n) = 0$ and $x_i(n) = 1$ in the limit cycle. We remark the occurrence of repeated saturations of the duty cycles at 0 and 1 alternated with phases of non saturation $\big(0 < d_j(n) < 1,\ j \in \{1, 2\}\big)$. At the beginning, the duty cycles saturate at 1, then their values decrease and remain always between 0 and 1 until the saturation at 0. In a similar way, we obtain a phase of non saturation, then the values of the duty cycles increase up to 1. We recall that the two-cell DC/DC buck converter is described by the following discrete model:

$$x_i(n+1) = \big(1 - \delta_L\big)x_i(n) + \delta_L\big(d_1(n) - d_2(n)\big)x_v(n) + \delta_L\big(1 - d_1(n)\big) \quad (13.63)$$
$$x_v(n+1) = x_v(n) + \delta_C\big(d_2(n) - d_1(n)\big)x_i(n) \quad (13.64)$$

At the steady regime, we have equal duty cycles $d_1(n) = d_2(n)$, and system (13.63) and (13.64) becomes:

$$x_i(n+1) = \big(1 - \delta_L\big)x_i(n) + \delta_L\big(1 - d_1(n)\big) \quad (13.65)$$
$$x_v(n+1) = x_v(n) \;=\; x_v(1) \;=\; 0.5 \quad (13.66)$$

In the sequel, we are interested in studying the current behavior only since the voltage is stabilized to its reference value, and the following functioning is repeated in four phases to ensure the limit cycle:

1. Phase 1: Saturation at 1 ($d_1(n) = d_2(n) = 1$):

   In this case, the current is given in a simplified form:

$$x_i(n+1) = \big(1 - \delta_L\big)x_i(n) \quad (13.67)$$

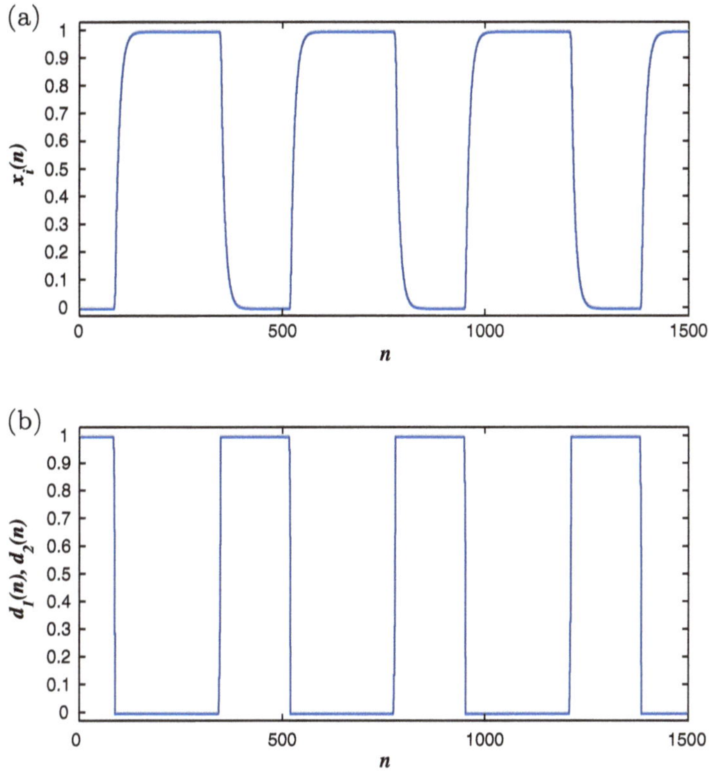

Fig. 13.6 Evolution of the response curves for $a = 1, b = 1, c = 0.1, k_i = 20$ and $d = -30$: **a** The current $x_i(n)$, **b** The duty cycles $d_1(n)$ and $d_2(n)$

During the phase of saturation at 1, the evolution of the current is:

$$x_i(n) = \left(1 - \delta_L\right)^{n-1} x_i(1) \tag{13.68}$$

The initial condition $x_i(1) = 0$, which explains that the current $x_i(n) = 0$.

2. Phase of non saturation:

The current is described by Eq. (13.65), and when the values of the duty cycles decrease, the current increases to reach the reference current $I_r = 0.6$.

3. Phase of saturation at 0 ($d_1(n) = d_2(n) = 0$):

The evolution of the current becomes as follows:

$$x_i(n+1) = \left(1 - \delta_L\right) x_i(n) + \delta_L \tag{13.69}$$

We notice that the phase of saturation at 1 is long, then the current tends to the fixed point $x_i^* = 1$.

4. Phase of non saturation:

In a similar way, the evolution of the current is obtained from Eq. (13.65). The duty cycles values increase and the current decreases simultaneously until reaching the fixed point $x_i^* = 0$.

Figure 13.7 depicts the phase space $x_d - x_i$ for the same parameters that shows the existence of a limit cycle.

We plot the lower and upper boundaries of saturation of the duty cycles at 0 and 1 respectively in light blue and red and their expressions are given by:

$$d_1(n) = d_2(n) = 0 \Longrightarrow x_i(n) = -\frac{c}{k_i + d} x_d(n) + I_r \tag{13.70}$$

$$d_1(n) = d_2(n) = 1 \Longrightarrow x_i(n) = -\frac{c}{k_i + d} x_d(n) + \frac{1}{k_i + d} + I_r \tag{13.71}$$

The shape of the limit cycle changes slightly when the system hits the saturation boundaries. In Fig. 13.8, we present the 1D bifurcation diagram of the current behavior versus the parameter d for $a = 1, b = 1, c = 0.1$ and $k_i = 40$.

There are four main parts in the 1D bifurcation diagram. For low values of the parameter $d < -21$, we obtain a one periodic behavior followed by a two periodic one for $d \in [-21, -19.9]$ in a degenerate flip bifurcation leading to a four periodic behavior at one bifurcation point $d = -19.9$, then a chaotic behavior in cyclical chaotic sets. In addition, for high values of the parameter d, we observe an abrupt transition from the chaotic behavior to a five periodic behavior.

To explain the type of bifurcation that leads to the appearance of the five periodic behavior, we plot in Fig. 13.9 the phase spaces $x_d - x_i$ for $d = 16$ before the bifurcation, which indicates the presence of a chaotic behavior, and for $d = 30$ after the bifurcation that shows a five periodic behavior.

In Fig. 13.10, we carry out a zoom in the previous figure for $d = 16$ and $d = 30$ that confirms the existence of a chaotic attractor and a five periodic behavior respectively.

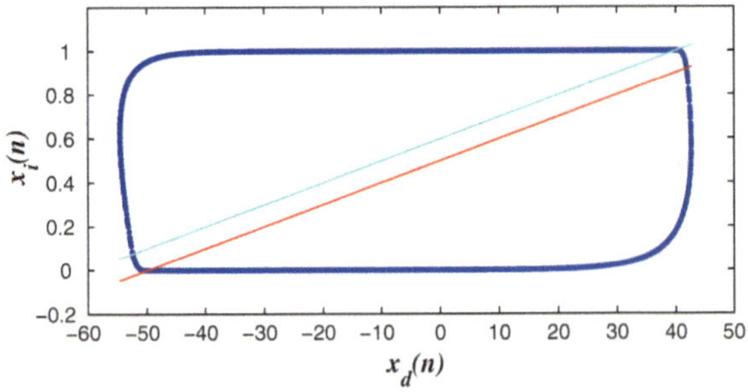

Fig. 13.7 Phase space $x_d - x_i$ for the parameters $a = 1, b = 1, c = 0.1, k_i = 20$ and $d = -30$: Limit cycle

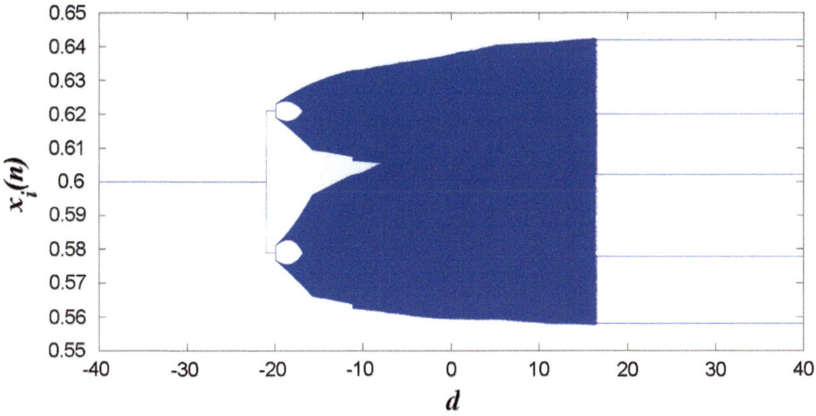

Fig. 13.8 1D bifurcation diagram of the current behavior versus the parameter d for $a = 1, b = 1, c = 0.1$ and $k_i = 40$

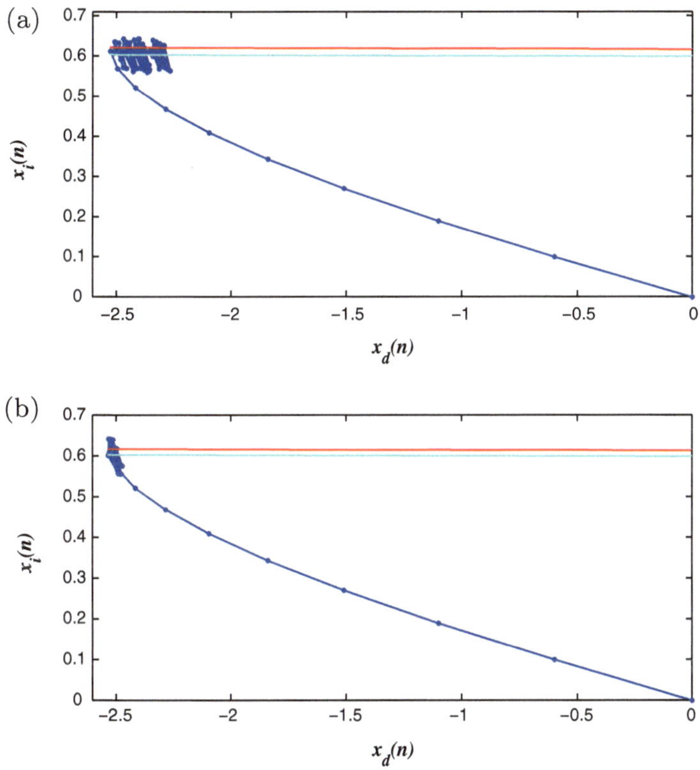

Fig. 13.9 Phase spaces $x_d - x_i$ for the parameters $a = 1, b = 1, c = 0.1, k_i = 40$: **a** $d = 16$: Chaotic behavior, **b** $d = 30$: Five periodic behavior

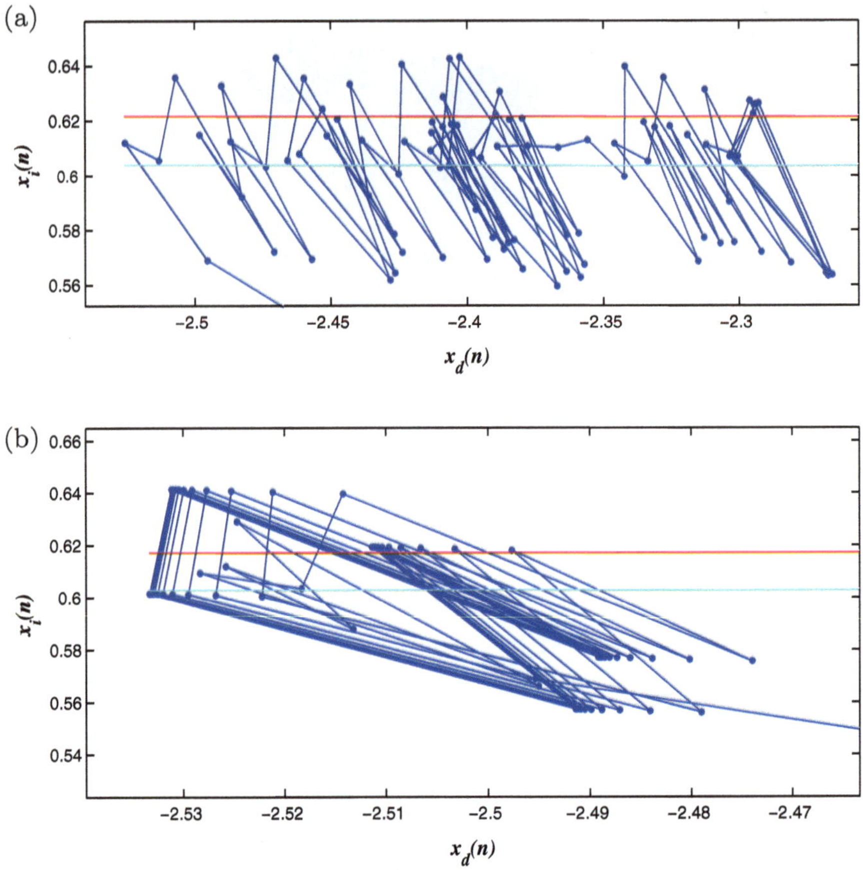

Fig. 13.10 Zoom in the phase spaces $x_d - x_i$ for $a = 1$, $b = 1$, $c = 0.1$, $k_i = 40$: **a** $d = 16$: Chaotic behavior, **b** $d = 30$: Five periodic behavior

For more explanation, we plot the evolution of the current and the duty cycles for the same parameters in Fig. 13.11. The current response presents a five periodic behavior, which is due to the saturation of the duty cycles at 0 and 1.

A sliding bifurcation is shown in the steady regime in Table 13.2, where we give the corresponding duty cycles values in the transient and the steady regimes.

In order to have sliding motion in the discrete case, the necessary and sufficient condition is:

$$|S(n+1)| < |S(n)| \tag{13.72}$$

where $S(n)$ is the switching function (Sarpturk et al. 1987).

The sliding orbit is obtained for the non saturated value of $d_1(14) = d_2(14) = 0.4495$ and the next value $d_1(15) = d_2(15) = 0.0338$ which verifies condition (13.72):

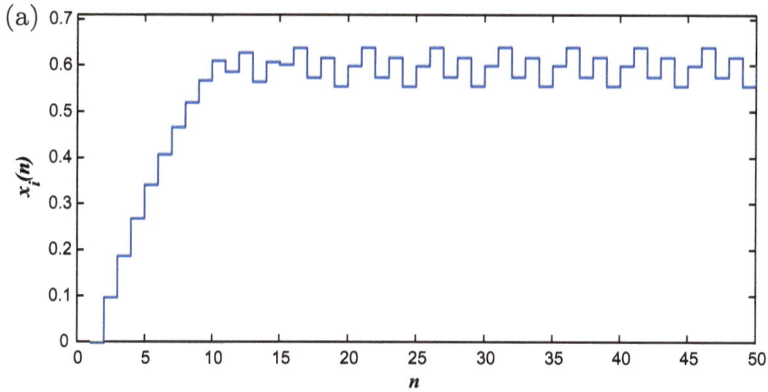

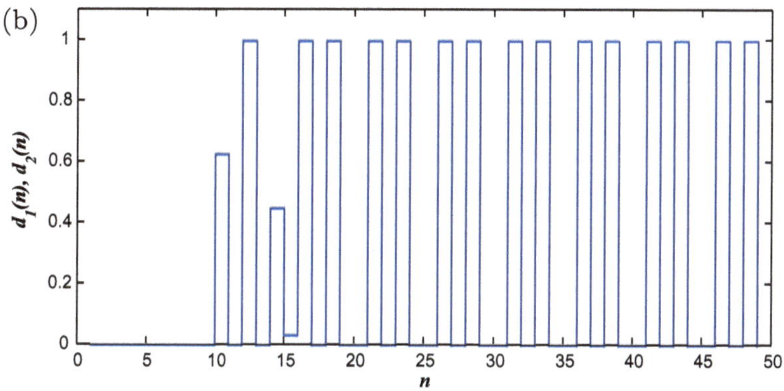

Fig. 13.11 Evolution of the response curves for $a = 1$, $b = 1$, $c = 0.1$, $k_i = 40$ and $d = 30$: **a** The current $x_i(n)$, **b** The duty cycles $d_1(n)$ and $d_2(n)$

$$|d_1(15)| < |d_1(14)| \tag{13.73}$$

The current is characterized by successive switchings between the three regions: region of saturation at 0, at 1 and the region of non saturation, then the sliding bifurcation obtained is switching-sliding. In Fig. 13.12, we plot the 2D bifurcation diagram of the current behavior in the plane $b - c$ for the parameters $a = 1$, $k_i = 40$ and $d = -30$.

The stable zone for the one periodic behavior is delineated in brown. In addition, we represent periodic behaviors from period 1 to period 20 and two undesirable saturating regimes that correspond to the current $x_i(n) = 0$ and $x_i(n) = 1$ respectively in light blue and red.

Figure 13.13 shows the 1D bifurcation diagram of the current $x_i(n)$ versus b for the parameters $a = 1$, $c = 6$, $k_i = 40$ and $d = -30$. For negative values of the parameter b, we obtain a saturating regime ($x_i(n) = 0$), which is confirmed in the 2D bifurcation diagram. In addition, we remark a peak in the current until 1 for negative

Table 13.2 Duty cycles values for the parameters $a = 1, b = 1, c = 0.1, k_i = 40$ and $d = 30$: Five periodic behavior

Regime	Period n	Duty cycles $d_1(n) = d_2(n)$
Transient regime	$1 \leq n \leq 9$	0
	$n = 10$	0.628
	$n = 11$	0
	$n = 12$	1
	$n = 13$	0
	$n = 14$	0.4495
	$n = 15$	0.0338
Steady regime	$n \geq 16$	Switching cycle: 1 0 1 0 0

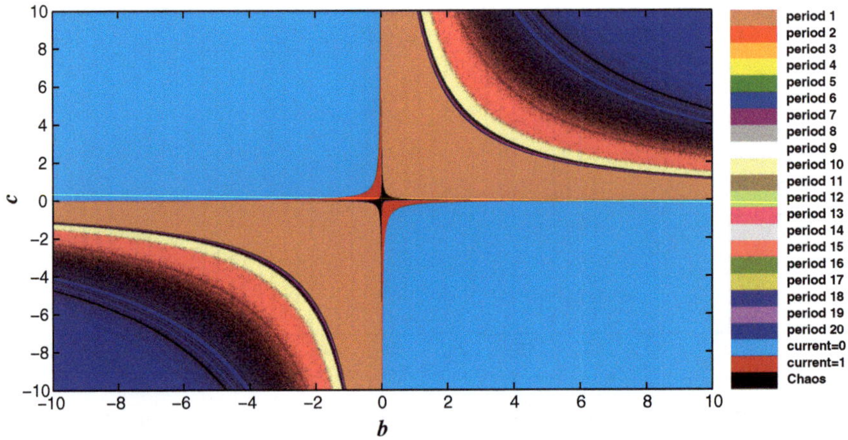

Fig. 13.12 2D bifurcation diagram of the current behavior in the plane $b - c$ for the parameters $a = 1, k_i = 40$ and $d = -30$

ones close to 0. The case $b = 0$ corresponds to the proportional controller, which is a particular case of the dynamic feedback controller according to Eqs. (13.25) and (13.26). For positive values of the parameter b, three behaviors can be shown:

- one periodic behavior for $0 < b < 1.82$.
- phase-locking phenomenon characterized by high periodic windows alternated with quasi-periodic strips for $1.82 \leq b \leq 2.04$.
- high periodic orbits for $b > 2.04$.

It is worth noting that the saturating regime $x_i(n) = 0$, obtained for negative values of the parameter b, is due to the saturation of the duty cycles at 1 (see Fig. 13.14).

Figure 13.15 depicts the evolution of the current, the voltage and the duty cycles for a negative value of the parameter b but close to 0, that shows a saturating regime $x_i(n) = 1$ caused by the saturation of the duty cycles at 0 with a long settling time.

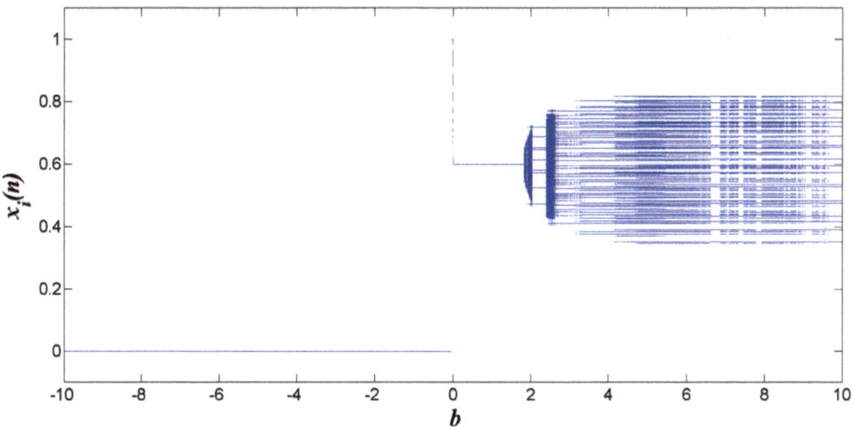

Fig. 13.13 1D bifurcation diagram of the current behavior versus b for the parameters $a = 1$, $c = 6$, $k_i = 40$ and $d = -30$

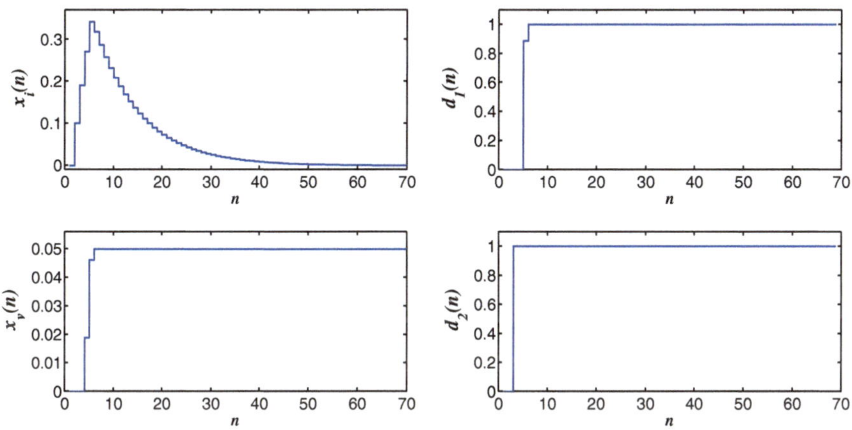

Fig. 13.14 Evolution of the current, the voltage and the duty cycles for the parameters $a = 1$, $b = -1$, $c = 6$, $k_i = 40$ and $d = -30$

In Fig. 13.16, we present the current evolution for $b = 0$ and we remark the existence of a static error in the current response, which characterizes the traditional proportional controller.

The high periodic orbits obtained for a positive value of the parameter b, for instance $b = 3$, are treated in Fig. 13.17. We notice the presence of repeated saturations of the duty cycles at 0 then at 1. Figure 13.18 shows high periodic orbits in the phase space $x_d - x_i$ for the parameters $a = 1$, $b = 3$, $c = 6$, $k_i = 40$ and $d = -30$. A peculiar route to sliding is detected following a spiralling bifurcation also called grazing-sliding bifurcation. The global behavior consists of a transient regime and a steady regime. The transient regime shows grazing bifurcation points at the corners

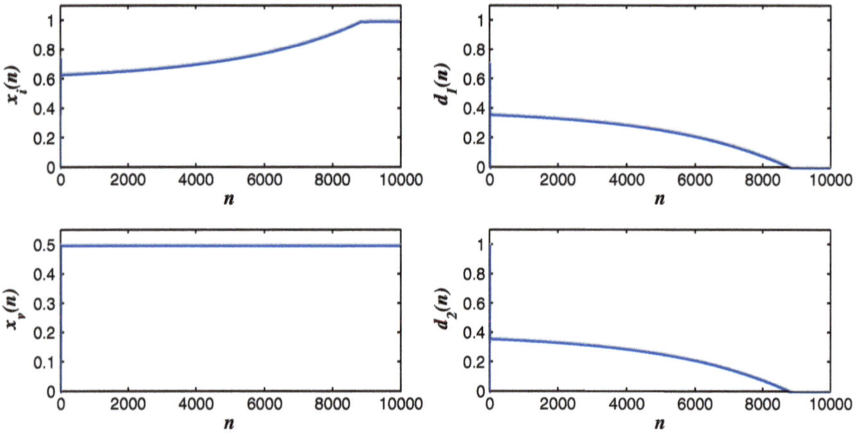

Fig. 13.15 Evolution of the current, the voltage and the duty cycles for the parameters $a = 1$, $b = -5 \times 10^{-4}$, $c = 6$, $k_i = 40$ and $d = -30$

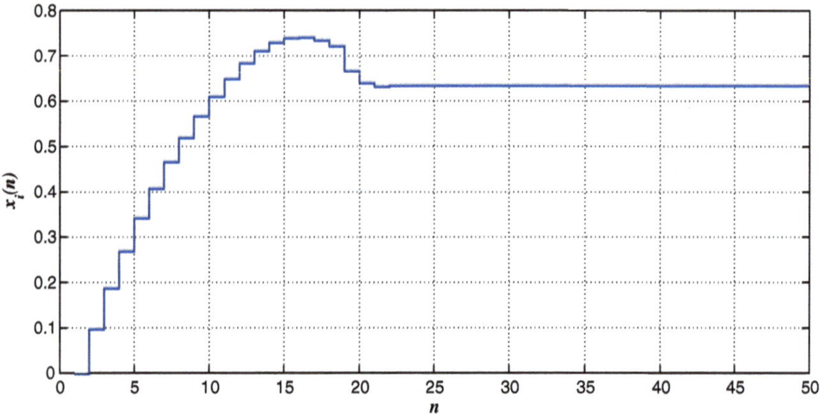

Fig. 13.16 Evolution of the current for the parameters $a = 1$, $b = 0$, $c = 6$, $k_i = 40$ and $d = -30$

of the spiral after successive saturation phases of the duty cycles at 0 then at 1 (see Table 13.3), and the steady regime is a sliding motion of a high periodic orbit in the middle of the spiral (see Table 13.4).

Condition (13.72) for discrete sliding bifurcation is verified:

$$|d_1(95)| < |d_1(94)| \text{ and } |d_2(95)| < |d_2(94)| \tag{13.74}$$

with the non saturated values:

$$d_1(94) = 3.0611,\ d_2(94) = 3.5075 \text{ and } d_2(95) = 1.1942 \tag{13.75}$$

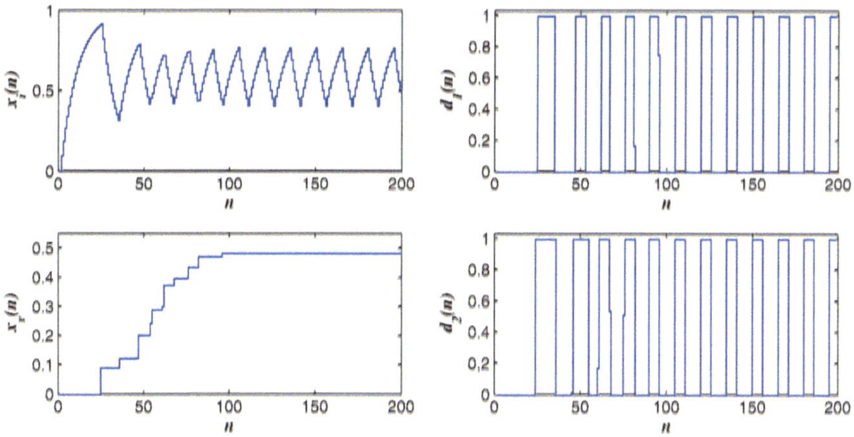

Fig. 13.17 Evolution of the current, the voltage and the duty cycles for the parameters $a = 1$, $b = 3$, $c = 6$, $k_i = 40$ and $d = -30$

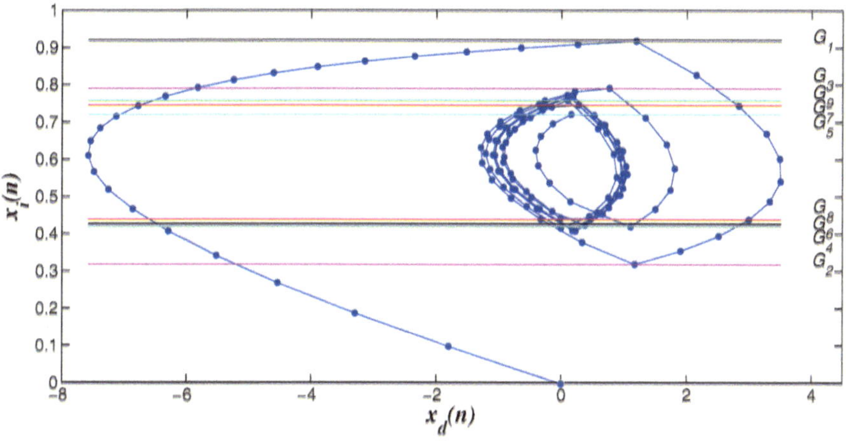

Fig. 13.18 Phase space $x_d - x_i$ for the parameters $a = 1$, $b = 3$, $c = 6$, $k_i = 40$ and $d = -30$: High periodic orbits

At the steady regime, the duty cycles follow a switching cycle composed of 9 saturations at 0 then 6 saturations at 1, which confirms the fifteen periodic behavior according to the 2D bifurcation diagram depicted in Fig. 13.12. The existence of different sliding orbits as it is shown in Fig. 13.13 proves the existence of a multi-sliding bifurcation with more than one sliding section. Grazing-sliding and multi-sliding bifurcations are found in the two-cell DC/DC buck converter controlled using a dynamic PI controller in a previous work (Koubaâ and Feki 2014a), which is a particular case of the dynamic feedback controller described by Eqs. (13.25) and (13.26) such that $a = c = 1$, $b = d = \dfrac{k_i}{\tau_i}$, where k_i is a proportional gain and τ_i is an integral constant.

Table 13.3 The Transient regime for $a = 1, b = 3, c = 6, k_i = 40$ and $d = -30$: A Grazing bifurcation

Period n	$d_1(n)$	$d_2(n)$	Boundaries
$1 \leq n \leq 23$	0	0	
$n = 24$	0	1	Grazing 1: $G_1 : x_i(25) = 0.9202$
$25 \leq n \leq 34$	1	1	Grazing 2: $G_2 : x_i(35) = 0.3209$
$n = 35$	0	1	
$36 \leq n \leq 44$	0	0	
$n = 45$	0	0.0157	
$n = 46$	0	1	Grazing 3: $G_3 : x_i(47) = 0.7927$
$47 \leq n \leq 52$	1	1	Grazing 4: $G_4 : x_i(53) = 0.4213$
$53 \leq n \leq 54$	0	1	
$55 \leq n \leq 59$	0	0	
$n = 60$	0	0.1734	Grazing 5: $G_5 : x_i(61) = 0.7231$
$n = 61$	0	1	
$62 \leq n \leq 66$	1	1	Grazing 6: $G_6 : x_i(67) = 0.4255$
$n = 67$	0	0.5392	
$68 \leq n \leq 74$	0	0	
$n = 75$	0	0.5156	Grazing 7: $G_7 : x_i(76) = 0.7482$
$76 \leq n \leq 80$	1	1	Grazing 8: $G_8 : x_i(81) = 0.4418$
$n = 81$	0.1677	1	
$82 \leq n \leq 89$	0	0	Grazing 9: $G_9 : x_i(90) = 0.7609$

Table 13.4 The steady regime for $a = 1, b = 3, c = 6, k_i = 40$ and $d = -30$: A sliding bifurcation

Period n	$d_1(n)$	$d_2(n)$
$90 \leq n \leq 94$	1	1
$n = 95$	0.7478	1
$n \geq 96$	Switching cycle: 9 saturations at 0, 6 saturations at 1	

13.4 Sliding Mode Controller

13.4.1 Previous Work

Recently, a particular attention has been paid to this type of controllers. Indeed, the authors in (Márquez et al. 2014), have applied a second-order sliding using reduced order of derivatives to the inverted pendulum and the stability conditions are derived with LMI. In the work presented in (Jovanović and Bučevac 2015), fuzzy logic was combined with sliding mode to control the DC servo motor.

In (Amer et al. 2012), the authors have suggested a robust adaptive control strategy for robot manipulators based on a decoupled fuzzy proportional integral sliding mode control approach. Moreover, a multi-surface sliding control scheme for multi-agent networks has been applied in (Khoo et al. 2014), where the agent is described by a class of high-order uncertain nonlinear systems in chain form, and we add a power integrator approach to drive the sliding variables to the sliding surfaces in fast finite time.

As a matter of fact, Levant and Michael (2009) have proposed a high-order sliding-mode (HOSM) regularization procedure to diminish chattering and solve long-lasting problems of HOSM design. Three main HOSM algorithms have been studied in the literature:

- The Twisting Algorithm (TA) (Levant 1993) is one of the simplest and most popular among the second order sliding mode algorithms. It is used for the control of systems with relative degree two or one with introduction of an integrator in the loop (twisting-as-a-filter) (Boiko et al. 2004). It has been recently applied to control a robot manipulator (Guendouzi et al. 2013) and an inverted pendulum (Mahjoub et al. 2015).
- The Super-twisting Sliding Mode algorithm (STA) (Levant 1993) is a second order sliding mode control algorithm, which ensures all the main properties of first order sliding mode control for the systems with Lipschitz matched uncertainties and bounded gradients. Super-twisting algorithm does not require the knowledge of the values of the derivatives and the perturbation (Kareem and Azeem 2012).

A particular attention was paid to the implementation of this algorithm on examples of real systems. In fact, Kareem et al. have presented in (Kareem and Azeem 2012) a novel fuzzy logic based adaptive STA for the DC-DC buck converter. In (Lin and Chiang 2013), the authors use the STA controller for a synchronous reluctance motor. Another version of this algorithm with variable gains was applied by Evangelista et al. (Evangelista et al. 2013) to control the variable-speed wind energy conversion system (WECS) with slip power recovery to maximize the energy extracted from the wind. In addition, the authors in (Derafa et al. 2012) deal with the design and implementation of the STA for the attitude tracking of a four-rotor helicopter known as quadrotor. A modified version of the STA has been applied to a class of linear uncertain and multivariable fractional-order systems in (Pisano et al. 2010).

Moreover, the authors in (Saadaoui et al. 2006) have constructed STA observers for switched chaotic systems. A novel Lyapunov based design of a generalized super-twisting observer for a class of 2-dimensional nonlinear systems has been tested in a mathematical model regarding to the reduced Glucose-Insulin process in (Salgado et al. 2011). An adaptive STA derived using Lyapunov function technique has been applied to position control of an electropneumatic actuator in (Shtessel et al. 2011).

- The sub-optimal algorithm has been proposed by Bartolini et al. (Bartolini et al. 1997b), where the solution of a tracking problem for a second order nonlinear system with uncertain dynamics and incomplete state measurement is obtained by means of a procedure directly inspired by the solution of the classical minimum-time optimal control problem (Bartolini et al. 1997b), and has been successfully applied for a two arms constrained planar manipulator even when the accelerations are not available (Bartolini et al. 1997a). In (Liu and Li 2014), the authors have proposed an nth order sub-optimal integral sliding mode controller for a class of nonlinear affine systems, and to verify the effectiveness of this method, an application example of an overhead crane system is provided.

It should be noted that the previous three algorithms do not require knowledge of the system model. Nevertheless, it is beneficial to use all available information from it to improve the performances of the system controlled using the sliding mode. Therefore, instead of taking a classic sliding surface, many researchers have adopted a dynamic sliding one. In fact, dynamic sliding mode (DSM) control methods for MIMO systems in a proper differential Input-Output form have been proposed in (Lu and Spurgeon 1999) using dynamic feedback controllers resulting from both the direct and indirect sliding mode method based on an approach known as the equivalent control method, which assumes the invariance conditions and is combined with a semi-high gain observer. The DSM has been applied also to non-minimum phase tracking control for boost and buck-boost power converters (Shtessel et al. 2003). Another scheme has been presented in (Raoufi et al. 2010) to design robust sliding mode observers (SMO) with H_∞ performance for uncertain nonlinear Lipschitz systems where both faults and disturbances are considered and a numerical example of Matsumoto-Chua-Kobayashi (MCK) (Matsumoto et al. 1986) hyperchaotic circuit demonstrates the high performance of the results compared with a pure SMO.

The different methods cited previously deal with the continuous systems. For the discrete case, Sarpturk et al. (1987) have proposed a version of the sliding mode controller. Indeed, many applications of the discrete sliding mode controller have been reported in literature: A design synthesis procedure for a discrete time output feedback sliding-mode controller, which incorporates integral action, has been presented in (Lai et al. 2006), and the controller has been implemented on a small DC motor test rig in real time. In (Mihoub et al. 2009), the discrete sliding mode controller has been successfully applied to the temperature control of a chemical reactor, and as shown by experimental results, this control law resolves the chattering problem while ensuring good robustness of the closed loop system behavior. Recently, a Chebyshev Neural Network based sliding mode control has been proposed for a class of unknown nonlinear discrete-time systems in the presence of fixed time delay in (Goyal et al. 2015).

Some innovative approaches in the sliding mode controller are intended to regulate the behavior of the one-cell DC/DC converters. In (Utkin 2013), a direct voltage control and harmonic cancellation principle along with switching frequency control was applied for a DC/DC buck converter. A discrete-time sliding mode control scheme has been designed in (Rivera et al. 2014) for the tracking of a DC-biased sinusoidal signal in a boost power converter, where the discrete-time model was obtained by means of a variational integrator scheme based on the discrete Lagrangian formulation of the boost power converter that uses the midpoint rule integration method.

13.4.2 Application of the Sliding Mode Controller to the Two-Cell DC/DC Buck Converter

In this section, we apply a discrete sliding mode controller to the two-cell DC/DC buck converter in order to suppress the sliding bifurcations and the chaotic behavior. The dynamics of the converter are described by the following simplified model:

$$x_i(n+1) = (1-\delta_L)x_i(n) + \delta_L\big(d_1(n) - d_2(n)\big)x_v(n) + \delta_L\big(1 - d_1(n)\big) \quad (13.76)$$

$$x_v(n+1) = \delta_C\big(d_2(n) - d_1(n)\big)x_i(n) + x_v(n) \quad (13.77)$$

where the duty cycles are given by:

$$d_1(n) = k_i e_i(n) + k_v e_v(n) + u(n) \quad (13.78)$$

$$d_2(n) = k_i e_i(n) + u(n) \quad (13.79)$$

We define the sliding surface as:

$$S(n) = Ce(n) \quad (13.80)$$

where:

$$C = \begin{bmatrix} c_i & c_v \end{bmatrix} \text{ and } e(n) = \begin{bmatrix} e_i(n) \\ e_v(n) \end{bmatrix} \quad (13.81)$$

The discrete sliding mode controller verifies:

$$S(n) = S(n+1) = 0 \quad (13.82)$$

The control law $u(n)$ is deduced from the relation $S(n+1)=0$ and is expressed as:

$$u(n)=\left(\frac{1}{\delta_L}-1-k_i\right)e_i(n)+\left[k_v(V_r-1)+\frac{c_v}{\delta_L c_i}\left(1-\delta_C k_v I_r\right)\right]e_v(n)$$
$$+k_v e_v(n)^2-\frac{c_v\delta_C k_v}{c_i\delta_L}e_v(n)e_i(n)+1-I_r \tag{13.83}$$

By inserting the control law $u(n)$ in the expressions of the duty cycles given by Eqs. (13.78) and (13.79), the closed-loop error system becomes as follows:

$$e_i(n+1)=-\frac{c_v}{c_i}\left(1-\delta_C k_v I_r\right)e_v(n)+\frac{c_v}{c_i}\delta_C k_v e_v(n)e_i(n) \tag{13.84}$$
$$e_v(n+1)=\left(1-\delta_C k_v I_r\right)e_v(n)-\delta_C k_v e_v(n)e_i(n) \tag{13.85}$$

which guarantees zero error fixed point:

$$e_i^*=e_v^*=0 \tag{13.86}$$

and consequently the system converges to the reference values:

$$x_i^*=I_r \tag{13.87}$$
$$x_v^*=V_r \tag{13.88}$$

The characteristic equation of the linearized error system (13.84)–(13.85) around its fixed point is:

$$\lambda\left[\lambda-\left(1-\delta_C k_v I_r\right)\right]=0 \tag{13.89}$$

and has two real eigenvalues:

$$\lambda_1=0 \tag{13.90}$$
$$\lambda_2=1-\delta_C k_v I_r \tag{13.91}$$

The stability conditions are obtained when the norm of each eigenvalue is less than 1, then the voltage gain k_v verifies:

$$0<k_v<\frac{2}{\delta_C I_r}=33.3333 \tag{13.92}$$

The coefficients c_i and c_v don't intervene in the stability conditions of the discrete sliding mode controller depending only on the choice of k_v.

The equation $S(n)=Ce(n)=0$ enables the reduction of system order, and the desired system dynamics in sliding mode can be designed by an appropriate choice of matrix C (Rivera et al. 2014). The shaded area in Fig. 13.19 depicts the stability zone in the plane k_v-k_i that confirms condition (13.92).

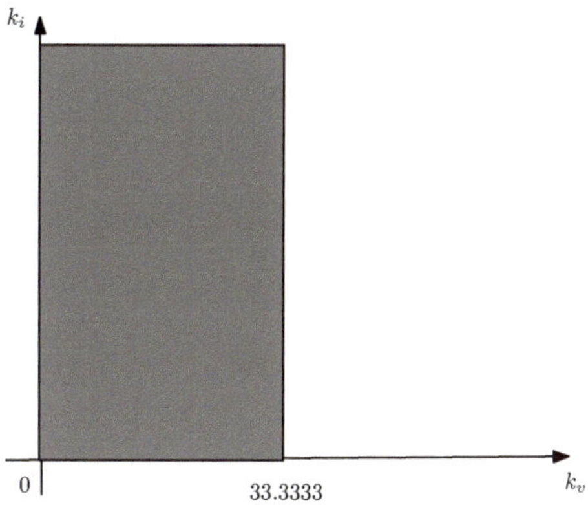

Fig. 13.19 Stability zone in the plane $k_v - k_i$

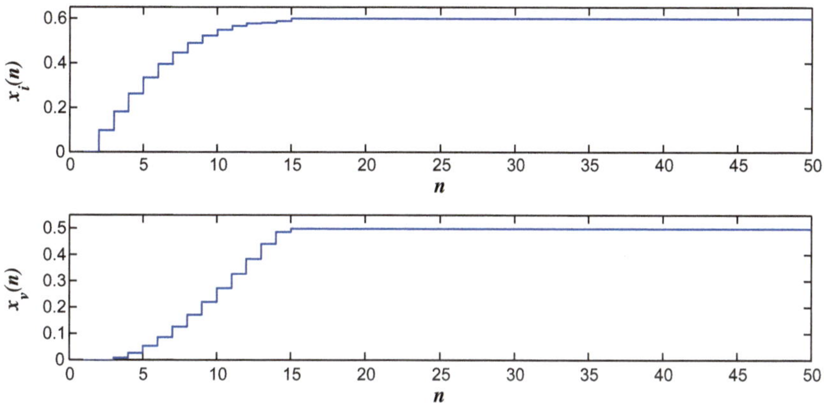

Fig. 13.20 Evolution of the current $x_i(n)$ and the voltage $x_v(n)$ for the parameters $c_i = 2$, $c_v = -2$ and $k_i = 50$

To ensure the rapidity of the closed-loop system, we choose the optimal value of the voltage gain k_v obtained for the eigenvalue $\lambda_2 = 0$, given by:

$$k_{v_{opt}} = \frac{1}{\delta_C I_r} = 16.6667 \tag{13.93}$$

Figure 13.20 shows the evolution of the current $x_i(n)$ and the voltage $x_v(n)$ for the parameters $c_i = 2$, $c_v = -2$ and $k_i = 50$, which guarantees zero overshoot and static errors for the current and the voltage.

In Fig. 13.21, we plot the curves of the duty cycles for the same parameters. At the transient regime, d_1 saturates at 0 and d_2 saturates at 1, which improves the response of the closed-loop system.

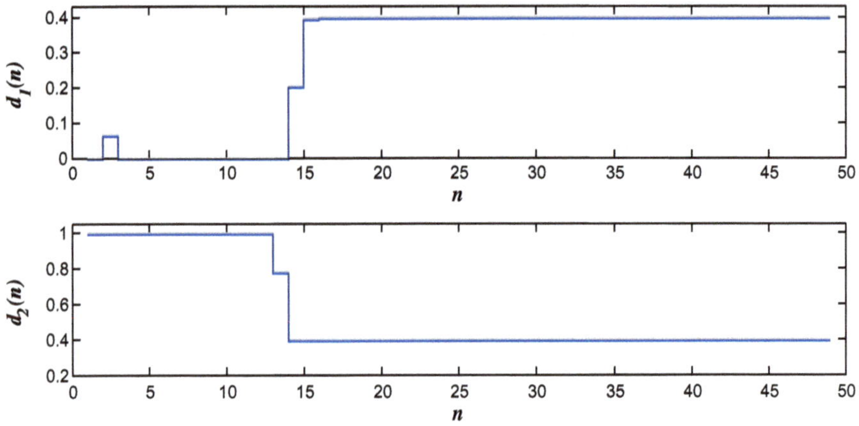

Fig. 13.21 Evolution of the duty cycles $d_1(n)$ and $d_2(n)$ for the parameters $c_i = 2$, $c_v = -2$ and $k_i = 50$

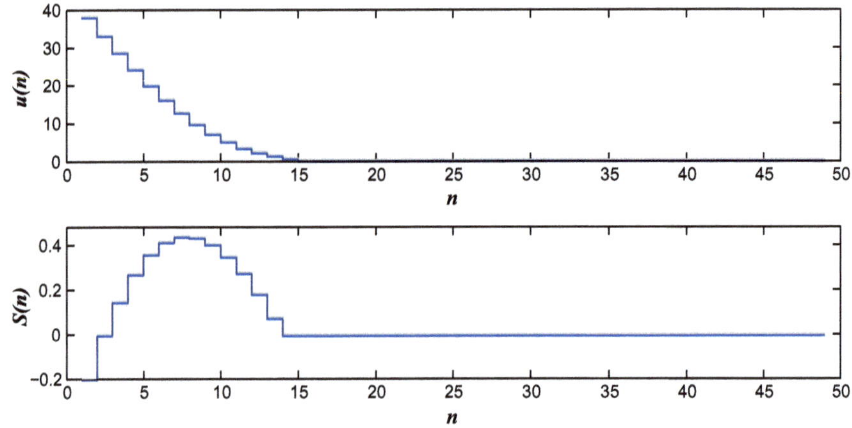

Fig. 13.22 Evolution of the control law $u(n)$ and the sliding surface $S(n)$ for the parameters $c_i = 2$, $c_v = -2$ and $k_i = 50$

Figure 13.22 presents the evolution of the control law $u(n)$ and the sliding surface $S(n)$ that verifies condition (13.82).

Figure 13.23 shows the 2D bifurcation diagram of the current behavior for the parameters $c_i = 2$ and $c_v = -2$. We can distinguish three behaviors in this diagram: the one periodic behavior is presented in brown for $k_v < 33.3333$ (see Fig. 13.19), a red strip indicates the existence of a two periodic behavior followed by a chaotic behavior for high values of the voltage gain k_v.

In Fig. 13.24, we present the 1D bifurcation diagram of the current behavior versus the voltage gain k_v for the parameters $c_i = 2$, $c_v = -2$ and $k_i = 50$.

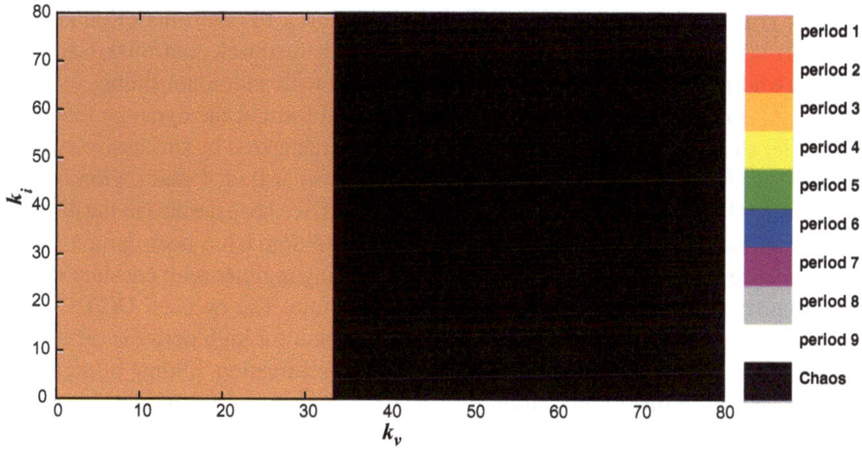

Fig. 13.23 2D bifurcation diagram of the current behavior in the plane $k_v - k_i$ for the parameters $c_i = 2$ and $c_v = -2$

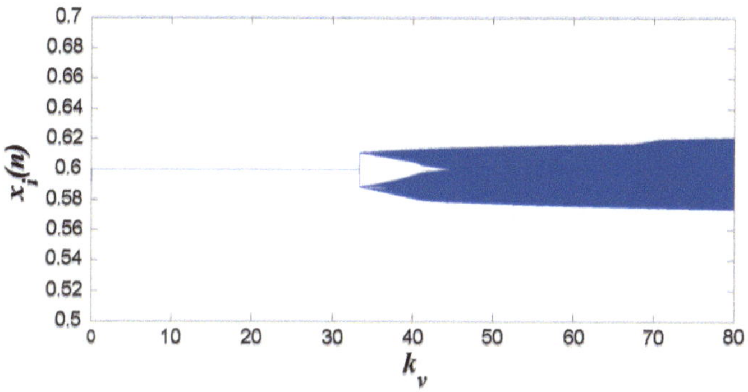

Fig. 13.24 1D bifurcation diagram of the current behavior versus the voltage gain k_v for the parameters $c_i = 2$, $c_v = -2$ and $k_i = 50$

This diagram proves the degenerate flip bifurcation scenario found in the previous 2D bifurcation diagram and the two periodic behavior is obtained at one bifurcation point.

13.5 Conclusion

Sliding bifurcations are non-smooth bifurcations characterized by sections of sliding motion giving rise to the formation of the so-called sliding orbits. Different routes to sliding have been pointed out in the current behavior of the two-cell DC/DC buck converter controlled using a generalized dynamic feedback controller: switching-sliding, grazing-sliding and adding-sliding. Adopting a PI controller (Koubaâ and Feki 2014a), which is a particular case of the dynamic feedback controller, leads to a grazing-sliding bifurcation in a one spiralling form, with additional sliding sections also called multi-sliding orbits. The use of a general form of the dynamic feedback controller shows a switching-sliding bifurcation characterized by successive switchings of the duty cycles between regions of saturation at 0, at 1 and region of non saturation. The conditions to obtain a sliding motion have been derived in the discrete case according to the values of the duty cycles. In addition, it has been demonstrated the existence of limit cycles and undesirable saturating regimes with constant values of the current at 0 and 1 when the duty cycles saturate. The two-cell DC/DC buck converter exhibits also the phase-locking phenomenon for high periodic orbits, the Neimark-Sacker bifurcation and the degenerate flip bifurcation. Sliding bifurcations are non-smooth bifurcations caused by structural changes in the system dynamics, then we apply the sliding mode controller to avoid sliding modes in the DC/DC buck converter. The analysis of the bifurcations that occur in other types of power converters such as multi-level and stacked multi-cell converters and the use of a dynamic sliding mode controller to improve the performances of the controlled system remain directions for future research.

References

Ahlborn, A., & Parlitz, U. (2004). Stabilizing unstable steady states using multiple delay feedback control. *Physical Review Letters*, *93*, 264101.

Amer, A. F., Sallam, E. A., & Elawady, W. M. (2012). Quasi sliding mode-based single input fuzzy self-tuning decoupled fuzzy PI control for robot manipulators with uncertainty. *International Journal of Robust and Nonlinear Control*, *22*, 2026–2054.

Bartolini, G., Ferrara, A., & Usai, E. (1997a). Applications of a sub-optimal discontinuous control algorithm for uncertain second order systems. *International Journal of Robust and Nonlinear Control*, *7*, 299–319.

Bartolini, G., Ferrara, A., & Usai, E. (1997b). Output tracking control of uncertain nonlinear second-order systems. *Automatica*, *33*, 2203–2212.

Batzel, J. J., & Tran, H. T. (2000a). Stability of the human respiratory control system I. analysis of a two-dimensional delay state-space model. *Journal of Mathematical Biology*, *41*, 45–79.

Batzel, J. J., & Tran, H. T. (2000b). Stability of the human respiratory control system II. analysis of a three-dimensional delay state-space model. *Journal of Mathematical Biology*, *41*, 80–102.

Boiko, I., Castellanos, M. I., & Fridman, L. (2004). Analysis of second order sliding mode algorithms in the frequency domain. *IEEE Transactions on Automatic Control*, *49*, 946–950.

Chen, W.-C. (2008). Dynamics and control of a financial system with time-delayed feedbacks. *Chaos, Solitons & Fractals*, *37*, 1198–1207.

de Paula, A. S., & Savi, M. A. (2009). Controlling chaos in a nonlinear pendulum using an extended time-delayed feedback control method. *Chaos, Solitons & Fractals*, *42*, 2981–2988.

Derafa, L., Benallegue, A., & Fridman, L. (2012). Super twisting control algorithm for the attitude tracking of a four rotors UAV. *Journal of the Franklin Institute*, *349*, 685–699.

di Bernardo, M., Budd, C., & Champneys, A. (1998). Grazing, skipping and sliding: analysis of the non-smooth dynamics of the DC/DC buck converter. *Nonlinearity*, *11*, 858–890.

di Bernardo, M., Budd, C. J., Champneys, A. R., & Kowalczyk, P. (2008). *Piecewise-smooth dynamical systems, theory and applications* (Vol. 163)., Applied mathematical sciences. London: Springer.

di Bernardo, M., Johansson, K., & Vasca, F. (1999). Sliding orbits and their bifurcations in relay feedback systems. In *Proceedings of the 38th IEEE Conference on Decision & Control* (pp. 708–713).

di Bernardo, M., Johansson, K. H., & Vasca, F. (2001). Self-oscillations and sliding in relay feedback systems: symmetry and bifurcations. *International Journal of Bifurcation and Chaos*, *11*, 1121–1140.

di Bernardo, M., & Tse, C. K. (2002). *Chaos in circuits and systems*, chapter Chaos in power electronics: an overview (pp. 317–340). New York: World Scientific.

Edwards, C. (1998). *Sliding mode control: theory and applications*. London: Taylor and Francis.

El-Aroudi, A., Robert, B., & Martínez-Salamero, L. (2006). Modelling and analysis of multi-cell converters using discrete time models. In *Proceedings of the IEEE International Symposium on Circuits and Systems*, (pp. 2161–2164).

El-Aroudi, A., Robert, B. G. M., Cid-Pastor, A., & Martínez-Salamero, L. (2008). Modeling and design rules of a two-cell buck converter under a digital PWM controller. *IEEE Transactions on Power Electronics*, *23*, 859–870.

Evangelista, C., Puleston, P., Valenciaga, F., & Fridman, L. (2013). Lyapunov designed super-twisting sliding mode control for wind energy conversion optimization. *IEEE Transactions on Industrial Electronics*, *60*, 538–545.

Feki, M., El-Aroudi, A., & Robert, B. G. M. (2007). *Multicell dc/dc converter: modeling, analysis and control*. Technical report, National Engineering School of Sfax.

Fichtner, A., Just, W., & Radons, G. (2004). Analytical investigation of modulated time-delayed feedback control. *Journal of Physics A: Mathematical and General*, *37*, 3385–3391.

Gateau, G., Fadel, M., Maussion, P., Bensaid, R., & Meynard, T. A. (2002). Multicell converters: active control and observation of flying-capacitor voltages. *IEEE Transactions on Industrial Electronics*, *49*, 998–1008.

Gauthier, D. J. (1998). Controlling lasers by use of extended time-delay autosynchronization. *Optics Letters*, *23*, 703–705.

Gjurchinovski, A., Sandev, T., & Urumov, V. (2010). Delayed feedback control of fractional-order chaotic systems. *Journal of Physics A: Mathematical and Theoretical*, *43*, 445102.

Goyal, V., Deolia, V. K., & Sharma, T. N. (2015). Robust sliding mode control for nonlinear discrete-time delayed systems based on neural network. *Intelligence Control and Automation*, *6*, 75–83.

Guan, X., Feng, G., & Chen, C. (2006). A stabilization method of chaotic systems based on full delayed feedback controller design. *Physics Letters A*, *348*, 210–221.

Guan, X., Feng, G., Chen, C., & Chen, G. (2007). A full delayed feedback controller design method for time-delay chaotic systems. *Physica D: Nonlinear Phenomena*, *227*, 36–42.

Guardia, M., Hogan, S. J., & Seara, T. M. (2010). An analytical approach to codimension-2 sliding bifurcations in the dry-friction oscillator. *SIAM Journal of Applied Dynamical Systems*, *9*, 769–798.

Guendouzi, A., Boubakir, A., & Hamerlain, M. (2013). Higher order sliding mode control of robot manipulator. In *Proceedings of the 9th International Conference on Autonomic and Autonomous Systems*.

Hall, K., Christini, D. J., Tremblay, M., Collins, J. J., Glass, L., & Billette, J. (1997). Dynamic control of cardiac alternans. *Physical Review Letters*, *78*, 4518–4521.

Hövel, P. (2011). *Control of complex nonlinear systems with delay*, chapter Time-delayed feedback control, (pp. 7–36). Berlin: Springer.

Jeffrey, M. R., Champneys, A. R., di Bernardo, M., & Shaw, S. W. (2010). Catastrophic sliding bifurcations and onset of oscillations in a superconducting resonator. *Physical Review E*, *81*, 016213.

Jovanović, R., & Bučevac, Z. M. (2015). Discrete-time exponentially stabilizing fuzzy sliding mode control via Lyapunov's method. *Advances in Fuzzy Systems*.

Just, W., Popovich, S., Amann, A., Baba, N., & Schöll, E. (2003). Improvement of time-delayed feedback control by periodic modulation: analytical theory of floquet mode control scheme. *Physical Review E*, *67*, 026222.
Kareem, A., & Azeem, M. F. (2012). A novel fuzzy logic based adaptive super-twisting sliding mode control algorithm for dynamic uncertain systems. *International Journal of Artificial Intelligence & Applications*, *3*, 21–34.
Khoo, S., Xie, L., Zhao, S., & Man, Z. (2014). Multi-surface sliding control for fast finite-time leader-follower consensus with high order SISO uncertain nonlinear agents. *International Journal of Robust and Nonlinear Control*, *24*, 2388–2404.
Konishi, K., Ishii, M., & Kokame, H. (1999). Stability of extended delayed-feedback control for discrete-time chaotic systems. *IEEE Transactions on Circuits and Systems I: Fundamental Theory and Applications*, *46*, 1285–1288.
Koubaâ, K., & Feki, M. (2014a). Bifurcation analysis and anti-windup approach for a dynamic PI controller in a switched two-cell DC/DC buck converter. In *Proceedings of the 21st Annual Seminar on Automation, Industrial Electronics and Instrumentation.*
Koubaâ, K., & Feki, M. (2014b). Quasi-periodicity, chaos and coexistence in the time delay controlled two-cell DC-DC buck converter. *International Journal of Bifurcation and Chaos*, *24*, 1450124.
Koubaâ, K., Feki, M., El-Aroudi, A., & Robert, B. G. M. (2009). Coexistence of regular and chaotic behavior in the time-delayed feedback controlled two-cell DC/DC converter. In *Proceedings of the 6th International Multi-Conference on Systems, Signals and Devices.*
Koubaâ, K., Feki, M., El-Aroudi, A., Robert, B. G. M., & Derbel, N. (2010). Stability analysis of a two-cell DC/DC converter using a dynamic time delayed feedback controller. In *Proceedings of the 7th International Multi-Conference on Systems, Signals and Devices.*
Koubaâ, K., Pelaez-Restrepo, J., Feki, M., Robert, B. G. M., & El-Aroudi, A. (2012). Improved static and dynamic performances of a two-cell DC-DC buck converter using a digital dynamic time-delayed control. *International Journal of Circuits Theory and Applications*, *40*, 395–407.
Lai, N., Edwards, C., & Spurgeon, S. K. (2006). Discrete output feedback sliding-mode control with integral action. *International Journal of Robust and Nonlinear Control*, *16*, 21–43.
Landry, M., Campbell, S. A., Morris, K., & Aguilar, C. O. (2005). Dynamics of an inverted pendulum with delayed feedback control. *SIAM Journal of Applied Dynamical Systems*, *4*, 333–351.
Levant, A. (1993). Sliding order and sliding accuracy in sliding mode control. *International Journal of Control*, *58*, 1247–1263.
Levant, A. (2010). Chattering analysis. *IEEE Transactions on Automatic Control*, *55*, 1380–1389.
Levant, A., & Michael, A. (2009). Adjustment of high-order sliding-mode controllers. *International Journal of Robust and Nonlinear Control*, *19*, 1657–1672.
Lin, W.-B., & Chiang, H.-K. (2013). *Super-twisting algorithm second-order sliding mode control for a synchronous reluctance motor speed drive* (p. 632061). Mathematical Problems in Engineering.
Liu, R., & Li, S. (2014). Suboptimal integral sliding mode controller design for a class of affine systems. *Journal of Optimization Theory and Applications*, *161*, 877–904.
Liu, Y., & Ohtsubo, J. (1994). Experimental control of chaos in a laser-diode interferometer with delayed feedback. *Optics Letters*, *19*, 448–450.
Lu, J., Ma, Z., & Li, L. (2009). Double delayed feedback control for the stabilization of unstable steady states in chaotic systems. *Communications in Nonlinear Science and Numerical Simulation*, *14*, 3037–3045.
Lu, X.-Y., & Spurgeon, S. K. (1999). Output feedback stabilization of MIMO non-linear systems via dynamic sliding mode. *International Journal of Robust and Nonlinear Control*, *9*, 275–305.
Mahjoub, S., Mnif, F., & Derbel, N. (2015). Second-order sliding mode approaches for the control of a class of underactuated systems. *International Journal of Automation and Computing*, *12*, 134–141.
Márquez, R., Tapia, A., Bernal, M., & Fridman, L. (2014). LMI-based second-order sliding set design using reduced order of derivatives. *International Journal of Robust and Nonlinear Control.* doi:10.1002/rnc.3295. In Press.

Matsumoto, T., Chua, L. O., & Kobayashi, K. (1986). Hyperchaos: laboratory experiment and numerical confirmation. *IEEE Transactions on Circuits Systems*, *33*, 1143–1147.

Mihoub, M., Nouri, A. S., & Ben-Abdennour, R. (2009). Real-time application of discrete second order sliding mode control to a chemical reactor. *Control Engineering Practice*, *17*, 1089–1095.

Pisano, A., Rapaić, M. R., Jeličiá, Z. D., & Usai, E. (2010). Sliding mode control approaches to the robust regulation of linear multivariable fractional-order dynamics. *International Journal of Robust and Nonlinear Control*, *20*, 2045–2056.

Pyragas, K. (1992). Continuous control of chaos by self-controlling feedback. *Physics Letters A*, *170*, 421–428.

Pyragas, K. (2002). Analytical properties and optimization of time-delayed feedback control. *Physical Review E*, *66*, 026207.

Pyragas, K., & Novičenko, V. (2013). Time-delayed feedback control design beyond the odd-number limitation. *Physical Review E*, *88*, 012903.

Raoufi, R., Marquez, H. J., & Zinober, A. S. I. (2010). H_∞ sliding mode observers for uncertain nonlinear lipschitz systems with fault estimation synthesis. *International Journal of Robust and Nonlinear Control*, *20*, 1785–1801.

Rivera, J., Chavira, F., Loukianov, A., Ortega, S., & Raygoza, J. J. (2014). Discrete-time modeling and control of a boost converter by means of a variational integrator and sliding modes. *Journal of the Franklin Institute*, *351*, 315–339.

Robert, B., & El-Aroudi, A. (2006). Discrete time model of a multi-cell dc/dc converter: non linear approach. *Mathematics and Computers in Simulation*, *71*, 310–319.

Robert, B., Feki, M., & Iu, H. H. C. (2006). Control of a PWM inverter using proportional plus extended time-delayed feedback. *International Journal of Bifurcation and Chaos*, *16*, 113–128.

Saadaoui, H., Djemai, M., Manamanni, N., & Benmansour, K. (2006). Super twisting algorithm observer for a class of switched chaotic systems. In *Proceedings of the 2nd International Symposium on Communications, Control and Signal Processing*.

Salgado, I., Chairez, I., Moreno, J., Fridman, L., & Poznyak, A. (2011). Generalized super-twisting observer for nonlinear systems. In *Proceedings of the 18th IFAC World Congress* (vol. 18).

Santos, I. M. (2006). *Modeling and numerical study of nonsmooth dynamical systems. applications to mechanical and power electronic systems*. Ph.D. thesis, Spain: Technical University of Catalonia.

Sarpturk, S. Z., Istefanopulos, Y., & Kaynak, O. (1987). On the stability of discrete-time sliding mode control systems. *IEEE Transactions on Automatic Control*, *32*, 930–932.

Schikora, S., Hövel, P., Wünsche, H.-J., Schöll, E., & Henneberger, F. (2006). All-optical noninvasive control of unstable steady states in a semiconductor laser. *Physical Review Letters*, *97*, 213902.

Shtessel, Y., Plestan, F., and Taleb, M. (2011). Super-twisting adaptive sliding mode control with not-overestimated gains: application to an electropneumatic actuator. In *Proceedings of the 18th IFAC World Congress*, (vol. 18).

Shtessel, Y., Zinober, A. S. I., & Shkolnikov, I. A. (2003). Sliding mode control of boost and buck-boost power converters using the dynamic sliding manifold. *International Journal of Robust and Nonlinear Control*, *13*, 1285–1298.

Socolar, J. E. S., Sukow, D. W., & Gauthier, D. J. (1994). Stabilizing unstable periodic orbits in fast dynamical systems. *Physical Review E*, *50*, 3245–3248.

Tse, C. K. (2002). *Chaos in circuits and systems, chapter experimental techniques for investigating chaos in electronics*. New York: World Scientific.

Ushio, T. (1996). Limitation of delayed feedback control in nonlinear discrete-time systems. *IEEE Transactions on Circuits and Systems I: Fundamental and Theory Applications*, *43*, 815–816.

Utkin, U. (1992). *Sliding modes in control and optimization*. Berlin: Springer.

Utkin, V. (2013). Sliding mode control of DC/DC converters. *Journal of the Franklin Institute*, *350*, 2146–2165.

Xu, C.-J., & Wu, Y.-S. (2014). Chaos control of a chemical chaotic system via time-delayed feedback control method. *International Journal of the Automation and Computing*, *11*, 392–398.

Yamamoto, S., Hino, T., & Ushio, T. (2001). Dynamic delayed feedback controllers for chaotic discrete-time systems. *IEEE Transactions on Circuits and Systems I: Fundamental Theory and Applications*, *48*, 785–789.

Yuan, G., Banerjee, S., Ott, E., & Yorke, J. A. (1998). Border-collision bifurcations in the buck converter. *IEEE Transactions on Circuits and Systems I: Fundamental Theory and Applications*, *45*, 707–716.

Chapter 14
DTC-SVM-Based Sliding Mode Controllers with Load Torque Estimators for Induction Motor Drives

Fatma Ben Salem and Nabil Derbel

Abstract In order to improve performances of the induction motor (IM) speed control under sliding mode (SM) DTC-SVM, the paper proposes a SM DTC-SVM based adaptive load torque. The aim of the suggested approach consists to discard load disturbances effects on the IM operating. For this reason, an adaptive estimator of the load torque has been developed to overcome this drawback. The load torque has been considered as a combination of the following three types: a constant load torque, a linear load torque and a quadratic load torque. Simulation results clearly show that SM DTC-SVM strategy with load estimator offers best performances and load disturbances effects can be completely discarded. Thus, the improvement of the proposed adaptive approach has been observed.

Keywords Induction motor · DTC-SVM · Sliding mode control · Load torque variations · Adaptive estimator

14.1 Introduction

Since 1980, the Direct Torque Control (DTC) has become popular in industrial drives for the speed control of induction motor drives. In fact, DTC of IM drives has been introduced by *Takahashi* (Takahashi and Noguchi 1986). Moreover, DTC and field-oriented control (FOC) are most popular for electric machine vector control methods. Compared with FOC (Sadoui and Meroufel 2012), it has been shown that DTC has a very simple control scheme (Ben Salem et al. 2005; Chlebis et al. 2010; Chaikhy et al. 2011; Allirani and Jagannathan 2014). Nevertheless, the switching frequency of the conventional DTC, which is uncontrolled, can induce vibrations and noises, and also only one voltage space vector is applied for the entire sampling period (Ben Salem

F. Ben Salem (✉) · N. Derbel
Control and Energy Management Laboratory (CEMLab), Sfax Engineering School, University of Sfax, Sfax, Tunisia
e-mail: fatma_bs@yahoo.fr; fatma.bensalem@enis.rnu.tn

N. Derbel
e-mail: n.derbel@enis.rnu.tn

N. Derbel et al. (eds.), *Applications of Sliding Mode Control*,
Studies in Systems, Decision and Control 79, DOI 10.1007/978-981-10-2374-3_14

et al. 2005; Ben Salem and Masmoudi 2007). In order to overcome this drawback, several DTC-SVM (direct torque control-space vector modulation) strategies have been developed so far (Habetler et al. 1996; Bounadja et al. 2009; Joseline Metilda et al. 2011; Rashag et al. 2013; Ahammad et al. 2014; Rashag et al. 2014). Using the space vector modulation (SVM) technique in DTC, the switching frequency can be maintained constant. This could produce accurate control of the stator flux linkage, and the torque (Habetler et al. 1996). In order to improve the DTC-SVM performances, hysteresis comparators of electromagnetic torque and stator flux have been replaced by PI controllers (Chen 2009).

Despite these performance improvements, DTC-SVM using PI controllers is sensitive to the system-parameter variations (Ben Salem and Derbel 2014). Moreover, PI controllers are inadequate to reject external disturbances and load variations. Moreover, since DTC-SVM along with induction motor (IM) is mostly nonlinear, sliding mode controllers become more suitable. Sliding mode controllers perform well in nonlinear systems (Veera and Brahmananda Reddy 2012; Ben Salem and Derbel 2014; Carmelia and Maurib 2011; Venkateswarlu et al. 2013). Indeed, The sliding mode control is a variable structure system characterized by a high robustness against to parameters variations and external disturbances. In fact, in SM regime, the dynamic of the system is insensitive to model uncertainties and external disturbances. Several schemes of sliding mode control dedicated to the control of induction motor have been studied in the literature (Carmelia and Maurib 2011; Ben Salem and Derbel 2007; Venkateswarlu et al. 2013; Zhang Yan et al. 2000; Boucheta et al. 2012). Despite the robustness of SM controllers, the major limitation of the variable structure control is the high-frequency switching (Ben Salem and Derbel 2007; Zhang Yan et al. 2000). DTC-SVM using sliding mode controllers seems as an effective solution offering (i) high performances in terms of the robustness to parameter variations, (ii) the torque ripple reduction, (iii) the simplicity of design and implementation, etc.

Within this approach, this paper proposes a sliding mode DTC-SVM approach to control the speed of an induction motor and to discard load disturbances effects on the induction machine operating, by considering load torque containing (i) a constant load, (ii) a linear load, and (iii) a quadratic load. For this, an adaptive estimator of the load torque has been added to the speed loop in order to increase its robustness and to overcome the problem of torque disturbances acting on the motor shaft which can affect motor performances. In order to study performances of the drive system under the steady state and dynamic conditions during starting, various speed range and load disturbances have been considered. Simulation results show that the proposed adaptive SM controller achieves robust and satisfactory performances.

14.2 Mathematical Model of Induction Machines

In the literature, there are several mathematical models representing the dynamical behavior of induction machines. In the following, a state space model related to α and β axes, for electrical variables, is considered:

$$\begin{cases} \dfrac{d\phi_{\alpha s}}{dt} = v_{\alpha s} - R_s i_{\alpha s} \\ \dfrac{d\phi_{\beta s}}{dt} = v_{\beta s} - R_s i_{\beta s} \\ \dfrac{d\phi_{\alpha r}}{dt} = -R_r i_{\alpha r} - \omega_m \phi_{\beta r} \\ \dfrac{d\phi_{\beta r}}{dt} = -R_r i_{\beta r} + \omega_m \phi_{\alpha r} \end{cases} \tag{14.1}$$

where subscripts s and r refer to stator and rotor, subscripts α and β refer to components in (α, β) frame, v, i and ϕ refer to voltage, current and flux, R_s and R_r refer to stator and rotor resistances, and ω_m refers to the machine speed ($\omega_m = N_p \Omega_m = \omega_s - \omega_r$ and N_p is the pole pair number).

Currents and flux relationships are:

$$\begin{bmatrix} \phi_{\alpha s} \\ \phi_{\alpha r} \end{bmatrix} = \begin{bmatrix} L_s & M \\ M & L_r \end{bmatrix} \begin{bmatrix} i_{\alpha s} \\ i_{\alpha r} \end{bmatrix}, \quad \begin{bmatrix} \phi_{\beta s} \\ \phi_{\beta r} \end{bmatrix} = \begin{bmatrix} L_s & M \\ M & L_r \end{bmatrix} \begin{bmatrix} i_{\beta s} \\ i_{\beta r} \end{bmatrix} \tag{14.2}$$

where L and M refer to the inductance and the mutual one.

The mechanical part of the machine is described by:

$$\frac{d\Omega_m}{dt} = \frac{T_{em} - T_L}{J} \tag{14.3}$$

where J is the motor inertia, T_{em} is the electromagnetic torque and T_L is the load torque.

14.2.1 Voltage Source Inverter

The made constant DC voltage by the rectifier is delivered to the inverter input, which thanks to controlled transistor switches, converts this voltage to three-phase AC voltage signal with wide range variable voltage amplitude and frequency.

The inverter one leg consists of two transistor switches. A simple transistor switch consists of a feedback diode connected in anti-parallel with the transistor. Feedback diode conducts current when the load current direction is opposite to the voltage direction.

Assuming that the power devices are ideal: when they are conducting, the voltage across them is zero and they present an open circuit in their blocking mode. Therefore, each inverter leg can be represented as an ideal switch. It gives the possibility to connect each of the three motor phase coils to a positive or negative voltage of the dc link (E_0).

Considering a 2-level inverter, presented by Fig. 14.1, the voltage vector of the three-phase voltage inverter is expressed as follows:

Fig. 14.1 2-level inverter

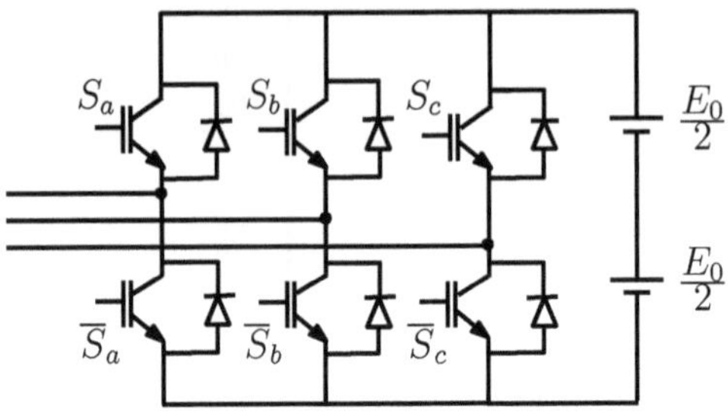

$$\vec{V}_s = \sqrt{\frac{2}{3}}\left[S_a + S_b e^{j\frac{2\pi}{3}} + S_c e^{j\frac{4\pi}{3}}\right] \tag{14.4}$$

where S_a, S_b and S_c are three-phase inverter switching functions, which can take a logical value of either 0 or 1.

14.2.2 Basic Principle of the Conventional DTC

The basic DTC strategy, to which is referred most if not all the literature in this area, is the one developed by *Takahashi* (Takahashi and Noguchi 1986) consisting of a two of hysteresis comparators, a torque and a flux estimators, a voltage vector selected by a look-up table and three-phase voltage inverter. In fact, The principle of DTC strategies consists of describing the way in which the stator flux and the torque are directly controlled by selecting a suitable inverter's voltage vector.

14.2.3 DTC-SVM Principle

The space vector modulation (SVM) technique has been widely used in industrial applications of PWM (pulse width modulation) inverters because of lower current harmonics, higher modulation index, fast transient response and simple digital implementation.

In the DTC system, the same active voltage vectors are applied during the whole sample period, and possibly several consecutive samples which give rise to relatively high ripple levels in stator current, flux linkage and torque. One of proposals to minimize these problems is to introduce Space Vector Modulation (SVM), which is a pulse width modulation technique able to involve the determination of the power switch conduction times in each modulation period, leading to control the switching frequency DTC strategy.

14.3 DTC-SVM-Based PI Controllers

The basic idea of this strategy is the decoupling between the amplitude and the argument of the stator flux vector. Indeed, the amplitude of this vector will be imposed, but the argument will be calculated according to desired performances, namely the reduction of electromagnetic torque ripples. The control of the electromagnetic torque is provided by a predictive controller. This approach differs from the basic DTC approach by using a predictive controller and a space vector modulation (SVM) which ensures the working with a constant commutation frequency. The vector selection table and hysteresis controllers are eliminated. The generation of control pulses applied to the switches of the inverter is provided by a predictive controller which receives signals of the torque error, the mechanical speed, the amplitude of the stator flux reference vector, the estimated stator flux and stator current vector coordinates. Then the controller determines the coordinates of the reference voltage vector in the (α, β) frame (Ben Salem and Derbel 2014; Rashag et al. 2013).

14.3.1 Torque Controller

In these works (Ben Salem and Derbel 2014; Rashag et al. 2013), it has been shown that for a constant stator flux $|\Phi_s| = |\Phi_s^*|$, the electromagnetic torque T_{em} and the rotor angular speed ω_r are related by a linear transmittance expressed as: (Ben Salem and Derbel 2014; Srirattanawichaikul et al. 2010):

$$G_1(p) = \frac{T_{em}}{\omega_r} = \frac{a}{1+\tau p} \tag{14.5}$$

where:

$$\tau = \sigma\frac{L_r}{R_r}, \quad \sigma = 1 - \frac{M^2}{L_s L_r}, \quad a = N_p\frac{M^2}{R_r L_s^2}|\Phi_s^*|^2$$

For a linear load described by the load torque $T_L = K_1 \Omega_m$, the transfer function between the speed Ω_m and the torque T_{em} is given by:

$$G_2(p) = \frac{\Omega_m}{T_{em}} = \frac{1}{K_1 + Jp} \tag{14.6}$$

14.3.2 Flux Reference Coordinates Computing

The slip angular reference speed ω_r^*, which is the output of the PI controller, will be used to calculate the argument of the stator flux reference. In the reference frame (α, β), coordinates of the reference stator flux $\phi_{\alpha s}^*$ and $\phi_{\beta s}^*$ are calculated from the polar coordinates according to the following expressions:

$$\begin{cases} \phi_{\alpha s}^{*} = |\Phi_s^{*}| \cos \theta_s^{*} \\ \phi_{\beta s}^{*} = |\Phi_s^{*}| \sin \theta_s^{*} \end{cases} \tag{14.7}$$

14.3.3 Voltage Reference Coordinates Computing

The coordinates of references of voltage vectors $v_{\alpha s}^{*}$ and $v_{\beta s}^{*}$ in (α, β) frame are determined by the following equations:

$$\begin{cases} v_{\alpha s}^{*} = \dfrac{\phi_{\alpha s}^{*} - \phi_{\alpha s}}{T_{em}} + R_s i_{\alpha s} \\ v_{\beta s}^{*} = \dfrac{\phi_{\beta s}^{*} - \phi_{\beta s}}{T_{em}} + R_s i_{\beta s} \end{cases} \tag{14.8}$$

These vectors are introduced to the SVM block, which uses them to control the inverter switches (S_a, S_b, S_c).

14.3.4 PI-DTC-SVM Scheme

The implementation scheme of the DTC-SVM strategy with PI controllers applied to the speed regulation of an induction motor drive is shown in Fig. 14.2. This strategy differs from the basic DTC strategy by using a bloc of voltage reference coordinates computing and a space vector modulation (SVM) which ensures the working with a constant commutation frequency. The vector selection table and hysteresis controllers are eliminated.

14.4 DTC-SVM-Based SM Controllers

The main drawbacks of DTC-SVM using PI controllers are the sensitivity of the performances to the system-parameter variations and the inadequate rejection of external disturbances and load changes (Ben Salem and Derbel 2014, 2016). The variable structure control can offer a good insensitivity to parameter variations, external disturbances rejection and fast dynamics. This approach utilizes discontinuous and robust control laws to drive the system state trajectory onto a sliding surface in the state space.

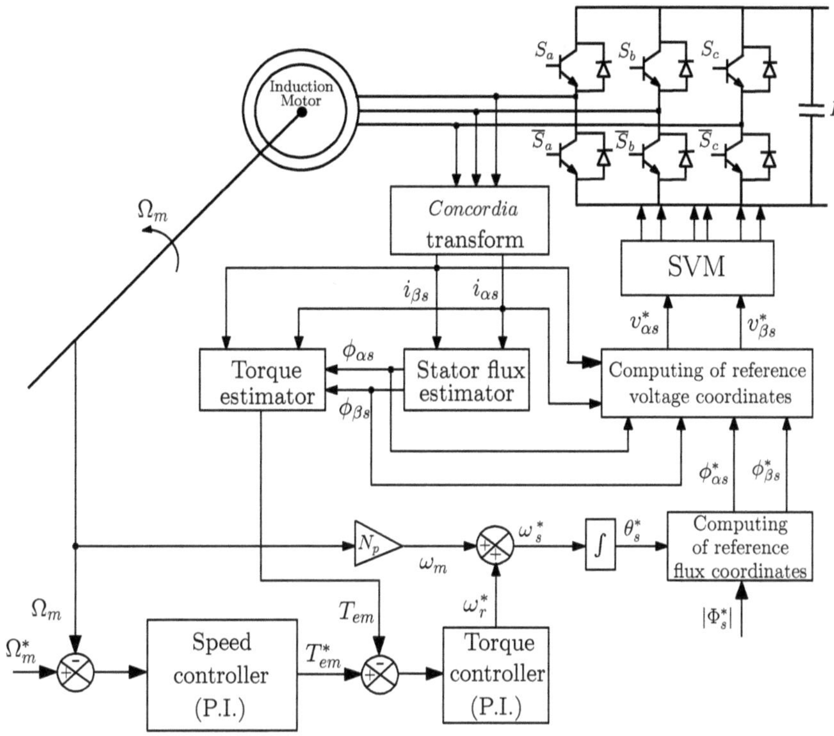

Fig. 14.2 Induction motor speed control based on DTC-SVM with PI controllers

14.4.1 SM Control Principle

Let's consider a system described by the following differential equation:

$$\frac{d^n}{dt^n}Y + a_{n-1}\frac{d^{n-1}}{dt^{n-1}}Y + \cdots + a_1\frac{d}{dt}Y + a_0Y = b_0U \tag{14.9}$$

where $U \in \mathbb{R}^m$ is the input vector and $Y \in \mathbb{R}^m$ is the output vector. b_0 is an $m \times m$ matrix and a_i $(0 \leq i \leq n-1)$ are $m \times m$ matrices.

Differential equations between an input and an output variable of induction machines under DTC-SVM can be represented by this form.

The sliding surface can be expressed as:

$$S = \left(\frac{d}{dt} + \lambda\right)^{n-1} (Y - Y_r) = 0 \tag{14.10}$$

where Y_r is the reference trajectory of the system.

The sliding mode control consists on expressing the control law as (Carmelia and Maurib 2011; Utkin 1993; Young et al. 1999; Mahmoudi et al. 1993):

$$U = U_{eq} + \Delta U \tag{14.11}$$

The so called "equivalent control" U_{eq} can be determined by imposing $\dot{S} = 0$ which can be expressed as:

$$\dot{S} = f(X) + BU \tag{14.12}$$

where X is the state vector and $B = \text{diag}([b_1, b_2, \ldots, b_m]) > 0$ is a positive diagonal matrix (in the case of induction machines under DTC-SVM). This gives:

$$U_{eq} = -B^{-1} f(X) \tag{14.13}$$

and:

$$\Delta U = -U_0 \text{ sign } S \tag{14.14}$$

where $U_0 = \text{diag}([U_{01}, U_{02} \ldots U_{0m}])$ is a diagonal positive definite matrix containing positive terms U_{0i}, $i = 1, 2, \ldots, m$.

The stability of the overall system can be ensured by considering the following Lyapunov candidate function:

$$V = \frac{1}{2} S^T S \tag{14.15}$$

Its differential with respect to time is expressed as:

$$\begin{aligned} \dot{V} &= S^T \dot{S} = S^T [f(X) + BU] = S^T [\underbrace{f(X) + BU_{eq}}_{=0} + B\Delta U] \\ &= S^T B\Delta U = -S^T BU_0 \text{sign} S = -\sum_{i=1}^{m} U_{0i} b_i |S_i| \leq 0 \end{aligned} \tag{14.16}$$

Then, the system is stable.

14.4.2 Speed SM Controller

The subsystem with the input $U_\Omega = \omega_r$ and with the output Ω_m is a second order system. Then, let us consider the sliding surface:

$$S_\Omega = \frac{d}{dt}(\Omega_m - \Omega_r) + \lambda_\Omega (\Omega_m - \Omega_r) \tag{14.17}$$

$$= \frac{1}{J} T_{em} + \left(\lambda_\Omega - \frac{K_1}{J} \right) \Omega_m - \dot{\Omega}_r - \lambda_\Omega \Omega_r \tag{14.18}$$

Ω_r is the speed reference trajectory.

The differential with respect to time of S_Ω gives:

$$\dot{S}_\Omega = \frac{a}{J\tau}U_\Omega + \left(\lambda_\Omega - \frac{1}{\tau} - \frac{K_1}{J}\right)\frac{1}{J}T_{em} - \frac{K_1}{J}\left(\lambda_\Omega - \frac{K_1}{J}\right)\Omega_m - \ddot{\Omega}_r - \lambda_\Omega\dot{\Omega}_r \tag{14.19}$$

For $S_\Omega = \dot{S}_\Omega = 0$, the equivalent control can be expressed as:

$$U_{eq,\Omega} = \frac{J\tau}{a}\left(\ddot{\Omega}_r + \lambda_\Omega\dot{\Omega}_r\right) + \frac{\tau}{a}\left[\left(\frac{1}{\tau} + \frac{K_1}{J} - \lambda_\Omega\right)T_{em} + K_1\left(\lambda_\Omega - \frac{K_1}{J}\right)\Omega_m\right] \tag{14.20}$$

The applied control is:

$$U_\Omega = U_{eq,\Omega} + \Delta U_\Omega \tag{14.21}$$

where:

$$\Delta U_\Omega = -U_{0,\Omega}\ \mathrm{sign}\ S_\Omega \tag{14.22}$$

14.4.3 Flux SM Controller

Let us consider the subsystem with the vector $V_{sref} = U_\phi$ as its input and the vector Φ_s as its output. The sliding surface is chosen as:

$$S_\phi = \left(\Phi_s - \Phi_s^*\right) + \lambda_\phi \int \left(\Phi_s - \Phi_s^*\right) dt \tag{14.23}$$

where: $|\Phi_s^*| = |\Phi_N| = 1\,\mathrm{Wb}$, and:

$$\Phi_s^* = \begin{bmatrix} \phi_{\alpha s}^* \\ \phi_{\beta s}^* \end{bmatrix} = \begin{bmatrix} |\Phi_s^*|\ \cos\theta_s^* \\ |\Phi_s^*|\ \sin\theta_s^* \end{bmatrix} \tag{14.24}$$

Similarly to the last case, and imposing $\dot{S}_\phi = 0$, the expression of the equivalent control is:

$$U_{eq,\phi} = R_s I_s + \dot{\Phi}_s^* - \lambda_\phi\left(\Phi_s - \Phi_s^*\right) \tag{14.25}$$

where:

$$\dot{\Phi}_s^* = \begin{bmatrix} \dot{\phi}_{\alpha s}^* \\ \dot{\phi}_{\beta s}^* \end{bmatrix} = \begin{bmatrix} -|\Phi_s^*|\ \omega_s^*\ \sin\theta_s^* \\ |\Phi_s^*|\ \omega_s^*\ \cos\theta_s^* \end{bmatrix} = \begin{bmatrix} -\omega_s^*\ \phi_{\beta s}^* \\ \omega_s^*\ \phi_{\alpha s}^* \end{bmatrix} \tag{14.26}$$

This leads to the reference stator voltage control:

$$U_\phi = U_{eq,\phi} - U_{0,\phi}\,\mathrm{sign}\left(S_\phi\right) \qquad (14.27)$$

14.4.4 SM-DTC-SVM Scheme

The structure of this control approach is given by the block diagram of Fig. 14.3 when we use only two sliding mode controllers: one for the speed and one for the flux.

14.4.5 Adaptive Load Torque Estimator

The load torque adaptation law will be designed so that the updated gain converges to its desired value. Stability properties are given on the basis of Lyapunov analysis.

It is obvious that the control $U_{eq,\Omega}$ is a function of the load gains K_0, K_1 and K_2 ($T_L = K_0 + K_1\Omega_m + K_2\Omega_m^2$). If the load varies or if it is ill-known, an adaptive estimation law of gains K_0, K_1 and K_2 should be determined.

Let's consider the following notations:

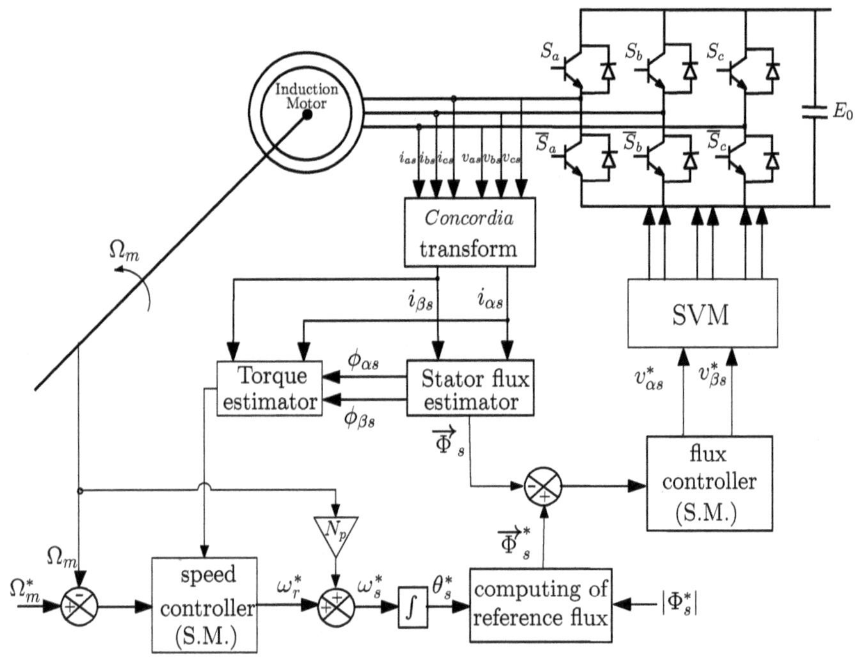

Fig. 14.3 Induction motor speed control based on DTC-SVM with sliding mode controllers

$$U_{eq,\Omega} = U_{eq,\Omega}(K_0, K_1, K_2)$$

and

$$\overline{U}_{eq,\Omega} = U_{eq,\Omega}(\overline{K}_0, \overline{K}_1, \overline{K}_2)$$

where $\overline{K}_0$, $\overline{K}_1$ and $\overline{K}_2$ are estimated values of gains K_0, K_1 and K_2. Then, we can write:

$$U_{eq,\Omega}(K_1) = U_{eq,\Omega}(\overline{K}_1) - \frac{\partial U_{eq,\Omega}}{\partial K_0}\widetilde{K}_0 - \frac{\partial U_{eq,\Omega}}{\partial K_1}\widetilde{K}_1 - \frac{\partial U_{eq,\Omega}}{\partial K_2}\widetilde{K}_2 + o\big(\widetilde{K}_0, \widetilde{K}_1, \widetilde{K}_2\big)^2 \tag{14.28}$$

where:

$$\begin{aligned}
\widetilde{K}_0 &= \overline{K}_0 - K_0 \\
\widetilde{K}_1 &= \overline{K}_1 - K_1 \\
\widetilde{K}_2 &= \overline{K}_2 - K_2
\end{aligned}$$

Moreover, the differential with respect to time of S_Ω can be expressed as:

$$\dot{S}_\Omega = f(X) + bU_\Omega \tag{14.29}$$

where X is the state vector, and b is a positive scalar constant. The equivalent control verifies: $f(X) + bU_{eq,\Omega} = 0$. The applied control is:

$$U_\Omega = \overline{U}_{eq,\Omega} + \Delta U_\Omega \tag{14.30}$$

Thus:

$$\begin{aligned}
\dot{S}_\Omega &= f(X) + b(\overline{U}_{eq,\Omega} + \Delta U_\Omega) = \underbrace{f(X) + bU_{eq,\Omega}}_{=0} + b(\overline{U}_{eq,\Omega} - U_{eq,\Omega}) + b\Delta U_\Omega \\
&= b\widetilde{K}_0\frac{\partial U_{eq,\Omega}}{\partial K_0} + b\widetilde{K}_1\frac{\partial U_{eq,\Omega}}{\partial K_1} + b\widetilde{K}_2\frac{\partial U_{eq,\Omega}}{\partial K_2} + b\Delta U_\Omega + o\big(\widetilde{K}_0, \widetilde{K}_1, \widetilde{K}_2\big)^2
\end{aligned} \tag{14.31}$$

with (for $i = 0, 1, 2$):

$$\frac{\partial U_{eq,\Omega}}{\partial K_i} = \frac{\tau}{a}\frac{\partial T_L}{\partial \Omega_m}\frac{T_e - T_L}{J} - \frac{\tau}{a}\left(\frac{1}{J}\frac{\partial T_L}{\partial \Omega_m} - \lambda\right)\frac{\partial T_L}{\partial K_i} \tag{14.32}$$

Then:

$$\frac{\partial U_{eq,\Omega}}{\partial K_0} = \frac{\tau}{Ja}\left(J\lambda_\Omega - \overline{K}_1 - 2\overline{K}_2\Omega_m\right) \tag{14.33}$$

$$\frac{\partial U_{eq,\Omega}}{\partial K_1} = \frac{\tau}{Ja}\left[T_{em} - (\overline{K}_0 + \overline{K}_1\Omega_m + \overline{K}_2\Omega_m^2) + \left(J\lambda_\Omega - \overline{K}_1 - 2\overline{K}_2\Omega_m\right)\Omega_m\right] \tag{14.34}$$

$$\frac{\partial U_{eq,\Omega}}{\partial K_2} = \frac{\tau\Omega_m}{Ja}\left[2T_{em} - 2(\overline{K}_0 + \overline{K}_1\Omega_m + \overline{K}_2\Omega_m^2) + \left(J\lambda_\Omega - \overline{K}_1 - 2\overline{K}_2\Omega_m\right)\Omega_m\right] \tag{14.35}$$

- Theorem

Control laws (14.22) and (14.30) with the following adaptive laws:

$$\dot{\overline{K}}_0 = -\eta_0\, b\, S_\Omega\, \frac{\partial U_{eq,\Omega}}{\partial K_0} \tag{14.36}$$

$$\dot{\overline{K}}_1 = -\eta_1\, b\, S_\Omega\, \frac{\partial U_{eq,\Omega}}{\partial K_1} \tag{14.37}$$

$$\dot{\overline{K}}_2 = -\eta_2\, b\, S_\Omega\, \frac{\partial U_{eq,\Omega}}{\partial K_2} \tag{14.38}$$

stabilise the speed loop, where η_0, η_1 and η_2 are positive scalars.

- Proof

Let us consider the following Lyapunov function:

$$V_\Omega = \frac{1}{2}S_\Omega^2 + \frac{1}{2\eta_0}\widetilde{K}_0^2 + \frac{1}{2\eta_1}\widetilde{K}_1^2 + \frac{1}{2\eta_2}\widetilde{K}_2^2 \tag{14.39}$$

In the following, it assumed that parameters K_0, K_1 and K_2 are constants or they have slow variations with respect to time, in such away that we can neglect their differentials with respect to time: $\dot{K}_0 \simeq 0$, $\dot{K}_1 \simeq 0$ and $\dot{K}_2 \simeq 0$. Then, we can write: $\dot{\widetilde{K}}_0 \simeq \dot{\overline{K}}_0$, $\dot{\widetilde{K}}_1 \simeq \dot{\overline{K}}_1$ and $\dot{\widetilde{K}}_2 \simeq \dot{\overline{K}}_2$.

The differential with respect to time of function V_Ω is expressed as:

$$\begin{aligned}
\dot{V}_\Omega &= S_\Omega\dot{S}_\Omega + \frac{1}{\eta_0}\widetilde{K}_0\dot{\widetilde{K}}_0 + \frac{1}{\eta_1}\widetilde{K}_1\dot{\widetilde{K}}_1 + \frac{1}{\eta_2}\widetilde{K}_2\dot{\widetilde{K}}_2 \\
&= S_\Omega b\left[\frac{\partial U_{eq,\Omega}}{\partial K_0}\widetilde{K}_0 + \frac{\partial U_{eq,\Omega}}{\partial K_1}\widetilde{K}_1 + \frac{\partial U_{eq,\Omega}}{\partial K_2}\widetilde{K}_2 + \Delta U_\Omega + o(\widetilde{K}_0, \widetilde{K}_1, \widetilde{K}_2)^2\right] \\
&\quad + \frac{1}{\eta_0}\widetilde{K}_0\dot{\widetilde{K}}_0 + \frac{1}{\eta_1}\widetilde{K}_1\dot{\widetilde{K}}_1 + \frac{1}{\eta_2}\widetilde{K}_2\dot{\widetilde{K}}_2 \\
&= S_\Omega b\Delta U_\Omega + o(\widetilde{K}_0, \widetilde{K}_1, \widetilde{K}_2)^2 + \underbrace{\left[S_\Omega b\frac{\partial U_{eq,\Omega}}{\partial K_0}(\overline{K}_0) + \frac{1}{\eta_0}\dot{\widetilde{K}}_0\right]}_{=0}\widetilde{K}_0 \\
&\quad + \underbrace{\left[S_\Omega b\frac{\partial U_{eq,\Omega}}{\partial K_1}(\overline{K}_1) + \frac{1}{\eta_1}\dot{\widetilde{K}}_1\right]}_{=0}\widetilde{K}_1 + \underbrace{\left[S_\Omega b\frac{\partial U_{eq,\Omega}}{\partial K_2}(\overline{K}_2) + \frac{1}{\eta_2}\dot{\widetilde{K}}_2\right]}_{=0}\widetilde{K}_2
\end{aligned}$$

$$= S_{\Omega}\, b\, \Delta U_{\Omega} + o(\widetilde{K}_0, \widetilde{K}_1, \widetilde{K}_2)^2 = -b\, U_{0,\Omega}|S_{\Omega}| + o(\widetilde{K}_0, \widetilde{K}_1, \widetilde{K}_2)^2 \leq 0 \tag{14.40}$$

14.5 Simulation Results and Discussions

Parameters of the induction motor are listed in Table 14.1. It has the following ratings: 220 V, 10 kW and 1470 rpm at 50 Hz.

Simulations have been done for a desired size of the flux $|\Phi_s^*| = 1$ Wb, and a desired speed trajectory defined by: (i) an acceleration from $t = 0$ s to $t = 1$ s varying the speed from 0 rpm to 1200 rpm, (ii) a constant speed equal to 1200 rmp from $t = 1$ s to $t = 2.5$ s, (iii) a deceleration from $t = 2$ s to $t = 2.5$ s varying the speed from 1200 rpm to 600 rpm, and (iv) a constant speed equal to 600 rmp from $t = 2.5$ s to $t = 4$ s.

14.5.1 Case of Known Loads

14.5.1.1 Considering PI Controllers

Figures 14.4, 14.5 and 14.6 present the evolution of the speed Ω_m, the flux $|\Phi_s|$, the electromagnetic torque T_{em}, the stator current i_{as}, one period of the stator current i_{as} for the steady state defined by the speed equal to 1200 rpm, and the control variable ω_r which represents the rotor pulsation.

It is obvious that the flux $|\Phi_s|$ reaches its desired size rapidly, with low ripples, and the speed follows its desired trajectory. Moreover, the electromagnetic torque has a good evolution with low ripples, and the stator current presents sinusoidal variations for constant speeds. Figures 14.4, 14.5 and 14.6 present the response of the system for different loads:

$$\text{Case 1 (Fig. 14.4)}: \; [K_0,\; K_1,\; K_2] = [0,\; K_{1n},\; 0]$$

$$\text{Case 2 (Fig. 14.5)}: \; [K_0,\; K_1,\; K_2] = \frac{1}{2}[0,\; K_{1n},\; K_{2n}]$$

$$\text{Case 3 (Fig. 14.6)}: \; [K_0,\; K_1,\; K_2] = \frac{1}{4}[0,\; K_{1n},\; K_{2n}]$$

Table 14.1 Induction motor parameters

$R_s = 0.29\,\Omega$	$L_s = L_r = 50\,\text{mH}$	$N_p = 2$
$R_r = 0.38\,\Omega$	$M = 47.3\,\text{mH}$	$J = 0.5\,\text{kg} \cdot \text{m}^2$

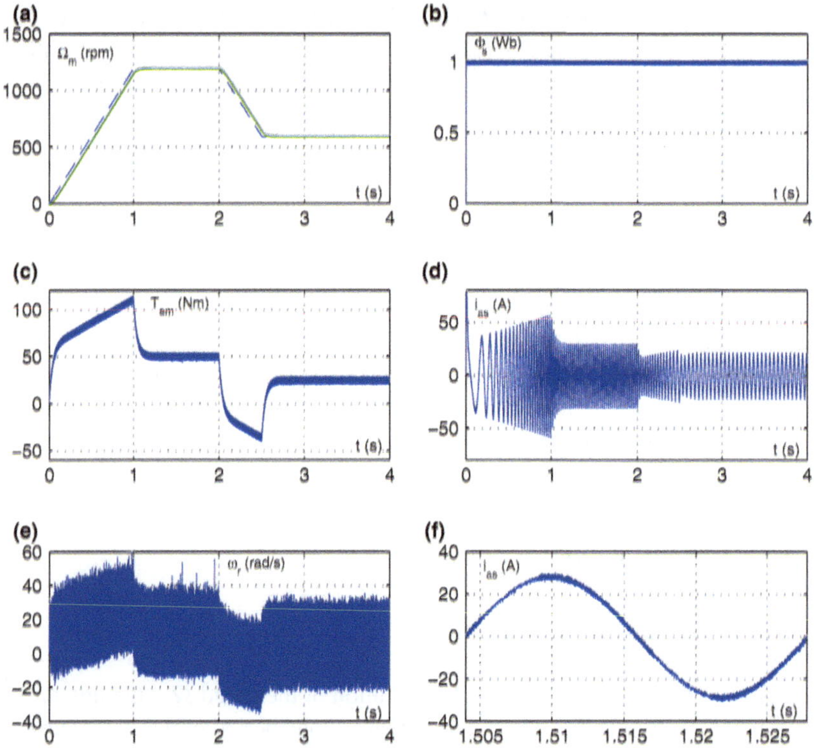

Fig. 14.4 Transient behavior of the IM under DTC-SVM using PI controllers with a linear load torque $T_L = K_1 \Omega_m$, $K_1 = K_{1n}$

The first case needs the highest currents and highest torques. Because, in the first case, the torque requirement is larger than the second and the third case. In fact, the third case presents a half nominal load torque at the nominal speed.

However, the first case and the second case present a nominal load torque at the same speed. Moreover, Using the fact that the nominal torque can be expressed as:

$$T_n = K_{0n} = K_{1n} \Omega_n = K_{2n} \Omega_n^2$$

The expression of the load torque becomes:

$$T_L = K_0 + K_1 \Omega_m + K_2 \Omega_m^2 = T_n \left[\frac{K_0}{K_{0n}} + \frac{K_1}{K_{1n}} \frac{\Omega_m}{\Omega_n} + \frac{K_2}{K_{2n}} \left(\frac{\Omega_m}{\Omega_n} \right)^2 \right]$$

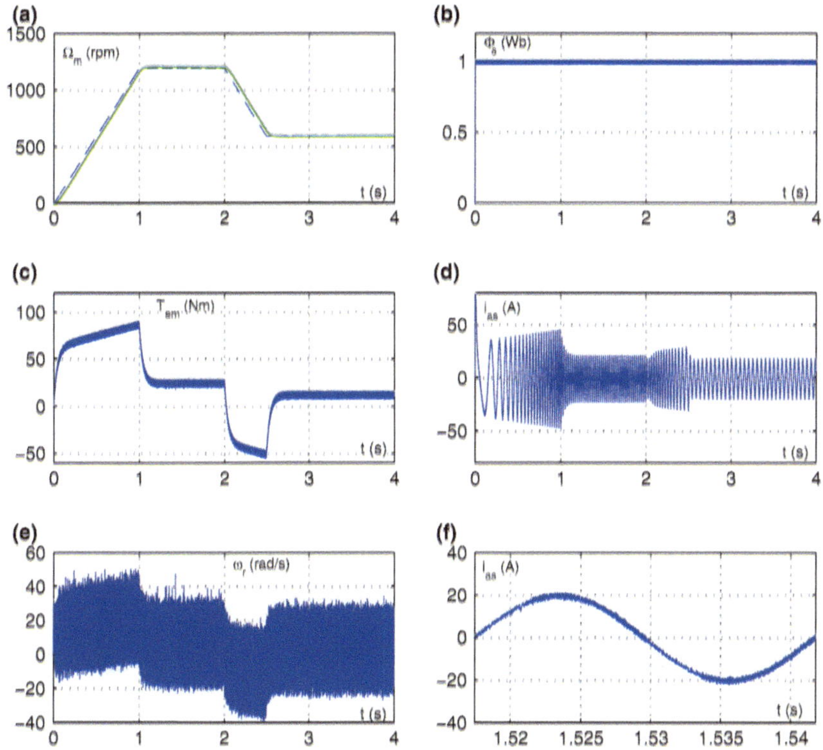

Fig. 14.5 Transient behavior of the IM under DTC-SVM using PI controllers with a load torque $T_L = K_1\Omega_m + K_2\Omega_m^2$, $K_1 = \frac{1}{2}K_{1n}$ and $K_2 = \frac{1}{2}K_{2n}$

This gives:

$$\text{Case 1 (Fig. 14.4)}: \; T_L = T_{L1} \; = T_n\left[\frac{\Omega_m}{\Omega_n}\right]$$

$$\text{Case 2 (Fig. 14.5)}: \; T_L = T_{L2} \; = \frac{1}{2}T_n\left[\frac{\Omega_m}{\Omega_n} + \left(\frac{\Omega_m}{\Omega_n}\right)^2\right] \; \leq T_{L1}$$

$$\text{Case 3 (Fig. 14.6)}: \; T_L = T_{L3} \; = \frac{1}{4}T_n\left[\frac{\Omega_m}{\Omega_n} + \left(\frac{\Omega_m}{\Omega_n}\right)^2\right] \; \leq T_{L2} \; \leq T_{L1}$$

It is to be noted that we have not considered a constant torque load, because PI controllers presents several difficulties and the system cannot follow its desired trajectory. Moreover, if the load is unknown, PI controllers give bad performances, and the system cannot reach its desired outputs (flux and speed).

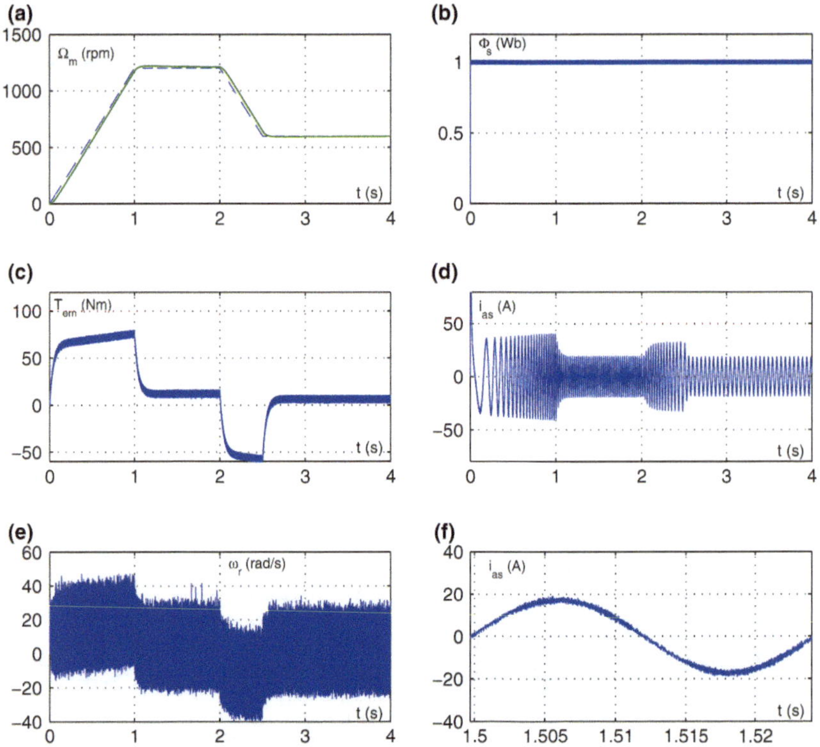

Fig. 14.6 Transient behavior of the IM under DTC-SVM using PI controllers with a load torque $T_L = K_1 \Omega_m + K_2 \Omega_m^2$, $K_1 = \frac{1}{4} K_{1n}$ and $K_2 = \frac{1}{4} K_{2n}$

14.5.1.2 Considering SM Controllers

Figures 14.7, 14.8, 14.9 and 14.10 present the evolution of the speed Ω_m, the flux $|\Phi_s|$, the electromagnetic torque T_{em}, the stator current i_{as}, one period of the stator current i_{as} for the steady state defined by the speed equal to 1200 rpm, and the control variable ω_r which represents the rotor pulsation. It is obvious that the flux $|\Phi_s|$ reaches its desired size rapidly, with very low ripples, and the speed follows its desired trajectory. Moreover, the electromagnetic torque has a good evolution with very low ripples, and the stator current presents sinusoidal variations for constant speeds. It is also obvious that torque and flux ripples are reduced compared to those given by PI controllers.

Figures 14.7, 14.8, 14.9 and 14.10 present the response of the system for different loads:

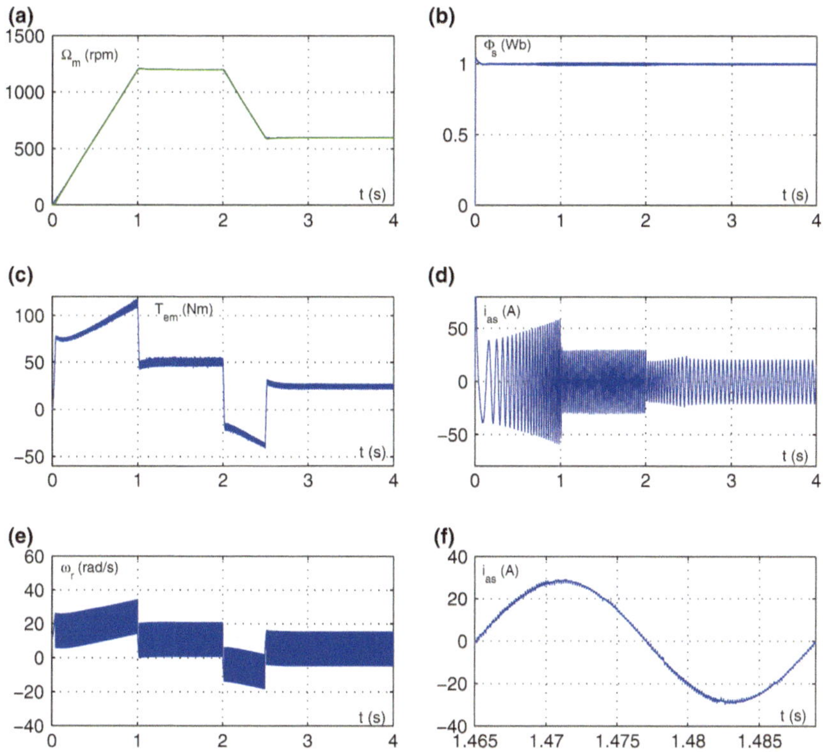

Fig. 14.7 Transient behavior of the IM under DTC-SVM using SM controllers with a linear load torque $T_L = K_1 \Omega_m$, $K_1 = K_{1n}$

$$\text{Case 1 (Fig. 14.7)}: \; [K_0,\; K_1,\; K_2] = [0,\; K_{1n},\; 0]$$

$$\text{Case 2 (Fig. 14.8)}: \; [K_0,\; K_1,\; K_2] = \frac{1}{2}[0,\; K_{1n},\; K_{2n}]$$

$$\text{Case 3 (Fig. 14.9)}: \; [K_0,\; K_1,\; K_2] = \frac{1}{3}[K_{0n},\; K_{1n},\; K_{2n}]$$

$$\text{Case 4 (Fig. 14.10)}: \; [K_0,\; K_1,\; K_2] = \frac{1}{6}[K_{0n},\; K_{1n},\; K_{2n}]$$

Cases 1, 2 and 3 give nominal load torques at the steady state around the nominal speed. However, it is well obvious that we have: $T_{L1} \leq T_{L2} \leq T_{L3} \leq T_{L4}$. This is why the electromagnetic torque and the stator current are larger for the first case, and they decrease for the second case, then for the third case, and then for the fourth case.

These figures show good several improvements compared to the case of PI controllers. The torque and the flux ripples have been reduced, and their evolutions become more smooth. Moreover, the control variable ω_r has a better shape and a smaller magnitude using SM controllers than those given by PI controllers.

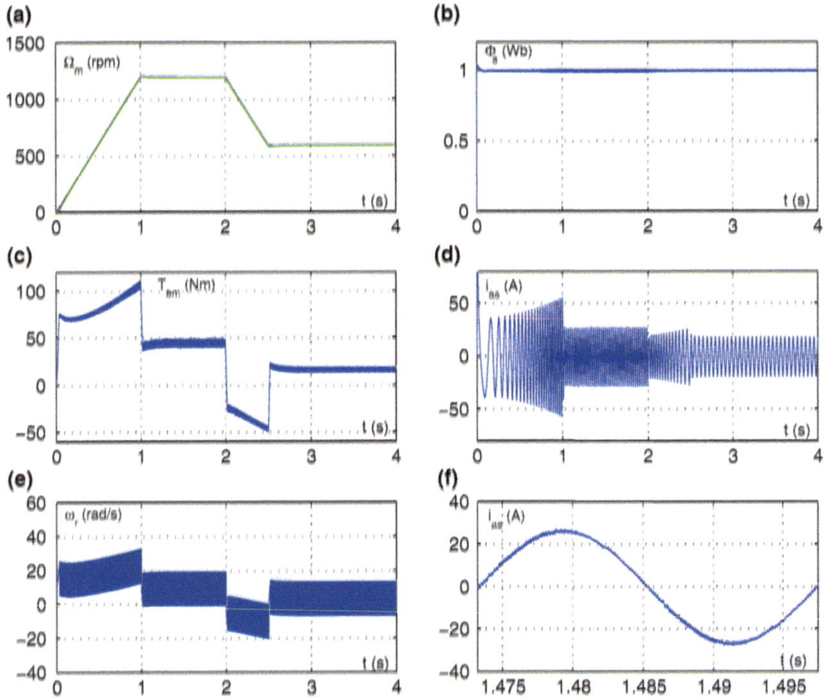

Fig. 14.8 Transient behavior of the IM under DTC-SVM using SM controllers with a load torque $T_L = K_1 \Omega_m + K_2 \Omega_m^2$, $K_1 = \frac{1}{2} K_{1n}$ and $K_2 = \frac{1}{2} K_{2n}$

14.5.2 Performance Criteria

Two comparison criteria have been selected to evaluate the effectiveness of the two DTC-SVM approaches under comparison.

First of all, consider the expression of the i_{as} current, around a steady state operating point, as follows:

$$i_{as}(t) = \Re e\left(\sum_{N=1}^{\infty} I_N \exp(jN\omega_s t)\right) \tag{14.41}$$

$|I_N|$ is the amplitude of the harmonic N, and $|I_1|$ is the amplitude of the fundamental.

The first criterion is the average total harmonic distortion (THD) of the stator current which is defined as follows:

$$\text{THD} = \frac{\sqrt{\sum_{N=2}^{\infty} |I_N|^2}}{|I_1|} \tag{14.42}$$

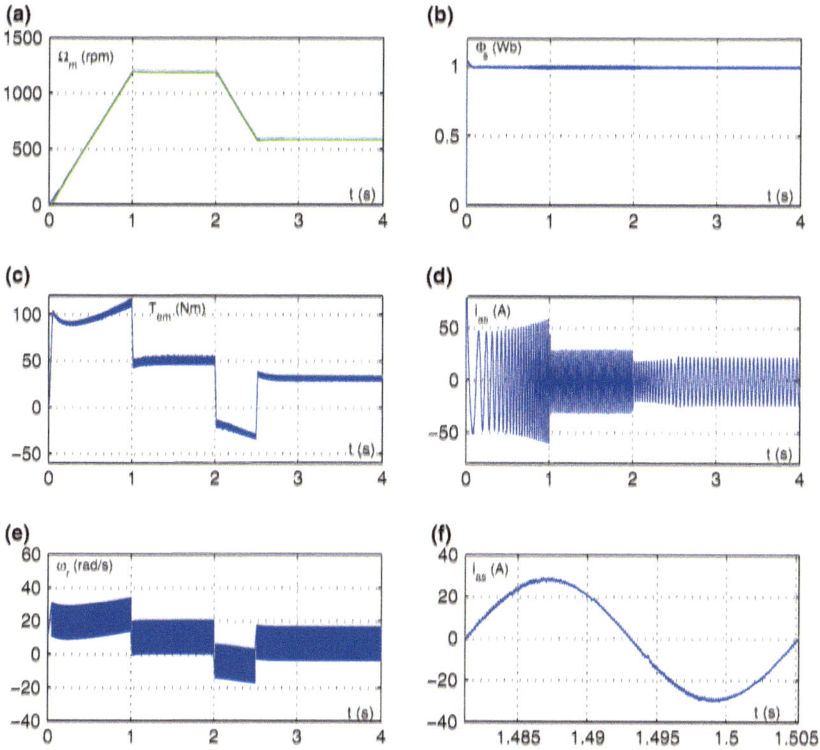

Fig. 14.9 Transient behavior of the IM under DTC-SVM using SM controllers with a load torque $T_L = K_0 + K_1\Omega_m + K_2\Omega_m^2$, $K_0 = \frac{1}{3}K_{0n}$, $K_1 = \frac{1}{3}K_{1n}$ and $K_2 = \frac{1}{3}K_{2n}$

The second comparison criterion translates the torque and the flux ripples around their steady state values. It can be expressed as the norm of the ratio of torque ripples by the torque mean, and the ratio of flux ripples by the flux mean, during one period:

$$\mathrm{T}_{RIP} = \left\| \frac{T_{em}(t) - T_{em,mean}}{T_{em,mean}} \right\| = \left\| \frac{T_{em}(t)}{T_{em,mean}} - 1 \right\| \tag{14.43}$$

$$\Phi_{RIP} = \left\| \frac{\Phi_s(t) - \Phi_{s,mean}}{\Phi_{s,mean}} \right\| = \left\| \frac{\Phi_s(t)}{\Phi_{s,mean}} - 1 \right\| \tag{14.44}$$

Two norms have been considered leading to two criteria describing the torque ripples and the flux ripples, based on the two well known norms (the norm 2 and the infinite norm):

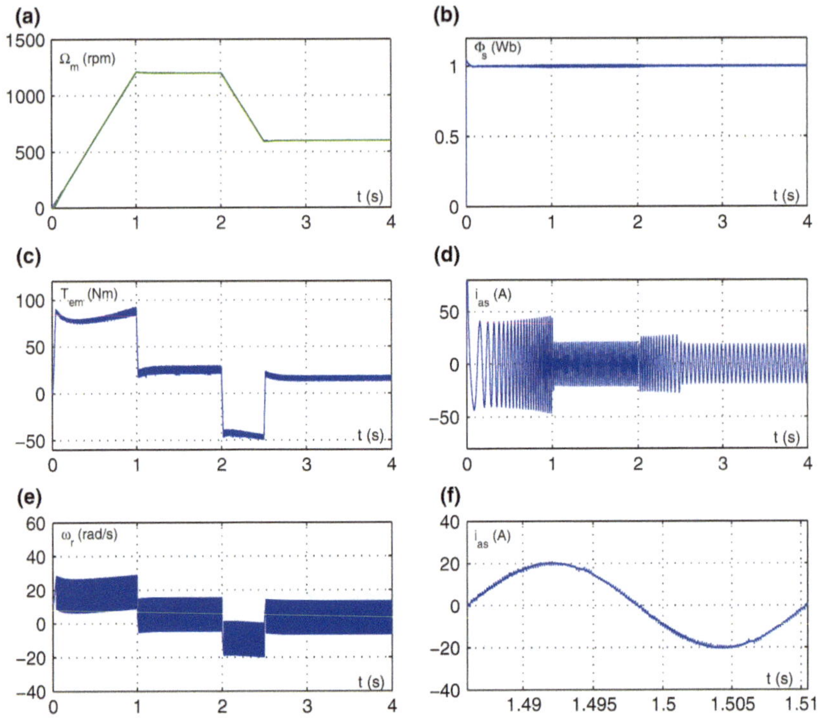

Fig. 14.10 Transient behavior of the IM under DTC-SVM using SM controllers with a load torque $T_L = K_0 + K_1 \Omega_m + K_2 \Omega_m^2$, $K_0 = \frac{1}{6} K_{0n}$, $K_1 = \frac{1}{6} K_{1n}$ and $K_2 = \frac{1}{6} K_{2n}$

$$\mathrm{T}_{RIP,2} = \sqrt{\frac{1}{T} \int_{t_0}^{t_0+T} \left[\frac{T_{em}(t) - T_{em,mean}}{T_{em,mean}} \right]^2 dt} \tag{14.45}$$

$$\mathrm{T}_{RIP,\infty} = \max_{t_0 \leq t < t_0+T} \left| \frac{T_{em}(t) - T_{em,mean}}{T_{em,mean}} \right| \tag{14.46}$$

$$\Phi_{RIP,2} = \sqrt{\frac{1}{T} \int_{t_0}^{t_0+T} \left[\frac{\Phi_s(t) - \Phi_{s,mean}}{\Phi_{s,mean}} \right]^2 dt} \tag{14.47}$$

$$\Phi_{RIP,\infty} = \max_{t_0 \leq t < t_0+T} \left| \frac{\Phi_s(t) - \Phi_{s,mean}}{\Phi_{s,mean}} \right| \tag{14.48}$$

Period T has been chosen equal to the stator period and time t_0 should be chosen in such away the system reaches its steady state for t larger than t_0.

We have chosen the first case PI1 and the first case SM1 detailed in the last paragraphs, for a linear nominal known load.

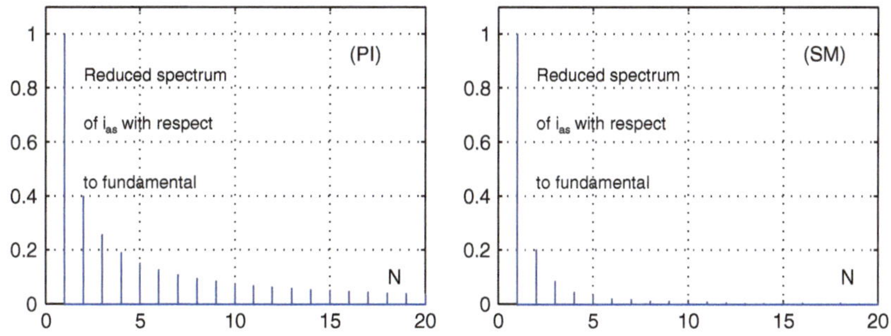

Fig. 14.11 Reduced spectrum of i_{as} with respect to fundamental, for DTC-SVM approach using PI controllers and for DTC-SVM approach using SM controllers

Table 14.2 Total harmonic distorsion of the stator current, flux ripples, and torque ripples

	Current	Flux ripples		Torque ripples	
	THD	$\Phi_{RIP,2}$	$\Phi_{RIP,\infty}$	$T_{RIP,2}$	$T_{RIP,\infty}$
PI1	0.6252	0.0149	0.0060	0.1799	0.0558
SM1	0.2316	0.0065	0.0015	0.1590	0.0401

Figure 14.11 presents the harmonics of the i_{as} current. It is obvious that sliding mode controllers yield less harmonics than PI controllers. This is also clear in the values of the total harmonic distorsion THD presented in Table 14.2. In this table, flux and torque ripples are computed. Table 14.2 shows that SM controllers give less ripples on the torque and the flux than those given by PI controllers.

14.5.3 Case of Ill-Known Loads Without Load Gain Estimators

To show that the used controllers are sensitive to load variations, we have considered that the load is the half of the nominal one. However, the controllers use parameters of the nominal load.

14.5.3.1 Considering PI Controllers

Figures 14.12 presents the evolution of the speed Ω_m, the flux $|\Phi_s|$, the electromagnetic torque T_{em}, the stator current i_{as}, one period of the stator current i_{as} for the steady state defined by the speed equal to 1200 rpm, and the control variable ω_r which represents the rotor pulsation.

The considered load is defined by:

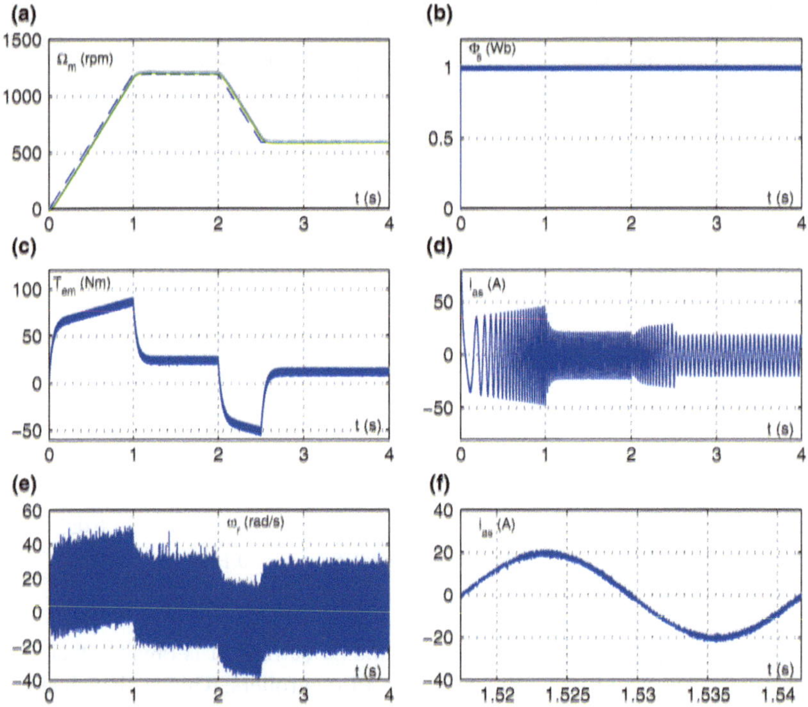

Fig. 14.12 Transient behavior of the IM under DTC-SVM using PI controllers with an ill-known load torque $T_L = K_1 \Omega_m$, $K_1 = \frac{1}{2} K_{1n}$

$$\text{Case 4 (Fig. 14.12)}: \; [K_0, \; K_1, \; K_2] = \frac{1}{2}[0, \; K_{1n}, \; 0]$$

However, the PI controller assumes that the load gain is K_{1n}.

It is obvious that the system performances have been affected by the fact that the controller ignores the exact load gains. A static error on the speed has appeared $\Delta\Omega_m \simeq 2$ rpm.

14.5.3.2 Considering SM Controllers

Figure 14.13 presents the evolution of the speed Ω_m, the flux $|\Phi_s|$, the electromagnetic torque T_{em}, the stator current i_{as}, one period of the stator current i_{as} for the steady state defined by the speed equal to 1200 rpm, and the control variable ω_r which represents the rotor pulsation.

The considered load is defined by:

$$\text{Case 5 (Fig. 14.13)}: \; [K_0, \; K_1, \; K_2] = \frac{1}{6}[K_{0n}, \; K_{1n}, \; K_{2n}]$$

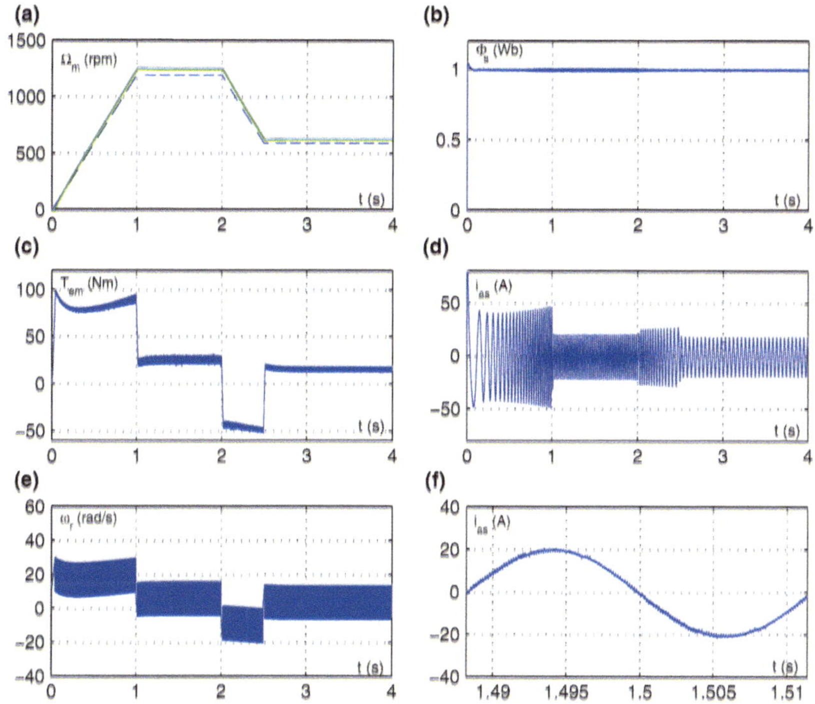

Fig. 14.13 Transient behavior of the IM under DTC-SVM using SM controllers with an ill-known load torque $T_L = K_0 + K_1\Omega_m + K_2\Omega_m^2$, $K_0 = \frac{1}{6}K_{0n}$, $K_1 = \frac{1}{6}K_{1n}$, $K_2 = \frac{1}{6}K_{2n}$

However, the SM controller assumes that the load gain is K_{1n}.

It is obvious that the system performances have been affected by the fact that the controller ignore the exact load gains. A static error on the speed has appeared $\Delta\Omega_m \simeq 6$ rpm.

14.5.4 *Case of Ill-Known Loads with Load Gain Estimators*

To check the robustness of the proposed adaptive approach, Figs. 14.14, 14.15 and 14.16 present the evolution of the speed Ω_m, the flux $|\Phi_s|$, the electromagnetic torque T_{em}, the stator current i_{as}, one period of the stator current i_{as} for the steady state defined by the speed equal to 1200 rpm, and the control variable ω_r which represents the rotor pulsation. These figures present the response of the system for three cases of unknown loads:

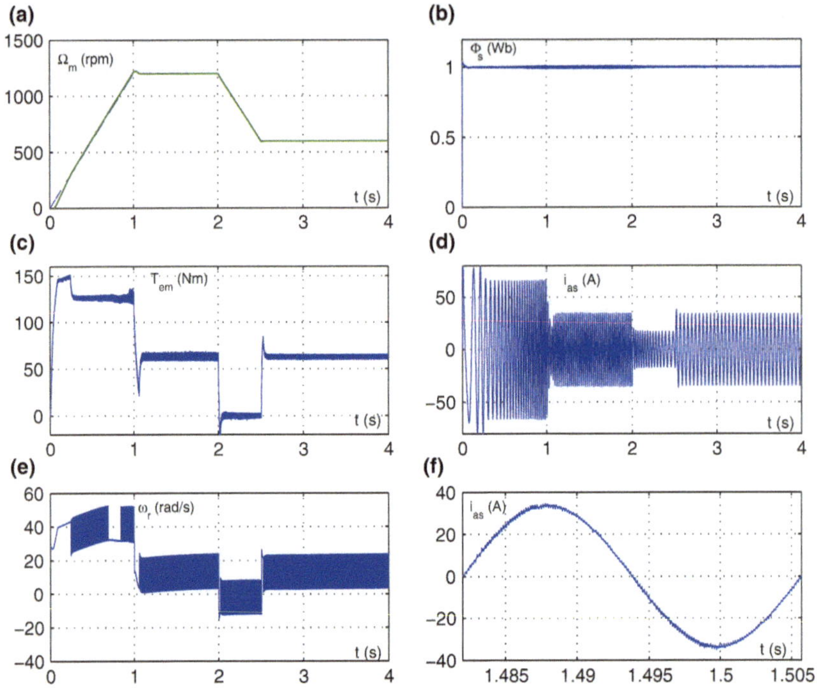

Fig. 14.14 Transient behavior of the IM under DTC-SVM using SM controllers with load gain estimators and for a constant torque load $T_L = K_0$, $K_0 = K_{0n}$

$$\text{Case 1 (Fig. 14.14)}: \ [K_0,\ K_1,\ K_2] = [K_{0n},\ 0,\ 0]$$
$$\text{Case 2 (Fig. 14.15)}: \ [K_0,\ K_1,\ K_2] = \frac{1}{3}[K_{0n},\ K_{1n},\ K_{2n}]$$
$$\text{Case 3 (Fig. 14.16)}: \ [K_0,\ K_1,\ K_2] = \frac{1}{6}[K_{0n},\ K_{1n},\ K_{2n}]$$

These figures show that the system performances are not affected by the fact that loads are unknown.

The evolution of the estimated load gains are presented in Figs. 14.17, 14.18 and 14.19. In these figures, estimated load gains converge to their actual values, and the electromagnetic torque reaches the demand on the load torque.

Despite the fact that the load is unknown, the adaptive approach, which computes the estimation of load gains, give very good performances. In fact, it is obvious that the flux $|\Phi_s|$ reaches its desired size rapidly, with very low ripples, and the speed follows its desired trajectory. Moreover, the electromagnetic torque has a good evolution with very low ripples, and the stator current presents sinusoidal variations for constant speeds.

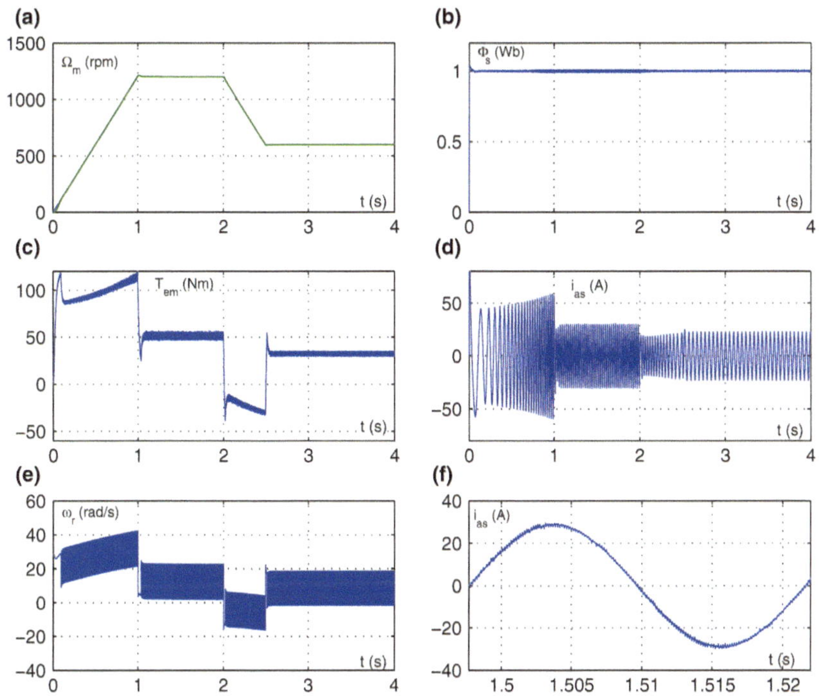

Fig. 14.15 Transient behavior of the IM under DTC-SVM using SM controllers with load gain estimators and for a load torque $T_L = K_0 + K_1\Omega_m + K_2\Omega_m^2$, $K_0 = \frac{1}{3}K_{0n}$, $K_1 = \frac{1}{3}K_{1n}$ and $K_2 = \frac{1}{3}K_{2n}$

14.6 Conclusion

A comparative study between two DTC-SVM approaches: (i) using PI controllers, and (ii) using sliding mode controllers, taking into account the effect of the load torque disturbances, has been developed in this paper. A discussion, using simulation results, with respect to load disturbances, considering constant, linear and quadratic load torques, has been presented. It has been found that the second approach with adaptive load gain, inserted in the control loop, offers best performances. Furthermore it exhibits a high capability to reject effects of the load disturbances. Moreover, it has been clearly shown the convergence of the adaptive load gains to their convenient values.

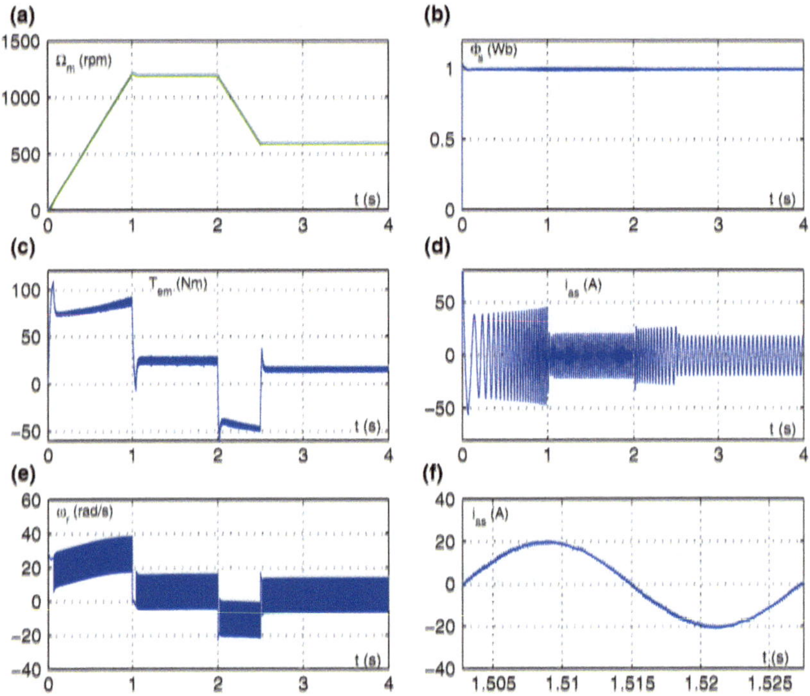

Fig. 14.16 Transient behavior of the IM under DTC-SVM using SM controllers with load gain estimators and for a load torque $T_L = K_0 + K_1 \Omega_m + K_2 \Omega_m^2$, $K_0 = \frac{1}{6} K_{0n}$, $K_1 = \frac{1}{6} K_{1n}$ and $K_2 = \frac{1}{6} K_{2n}$

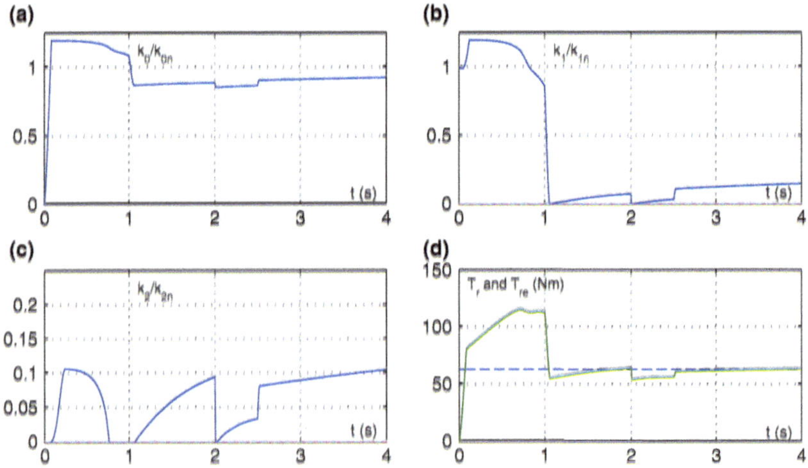

Fig. 14.17 The estimated gains evolution $\overline{K}_0$, $\overline{K}_1$ and $\overline{K}_2$, for a load torque $T_L = K_0 + K_1 \Omega_m + K_2 \Omega_m^2$, with: $K_0 = K_{0n}$, $K_1 = 0$ and $K_2 = 0$

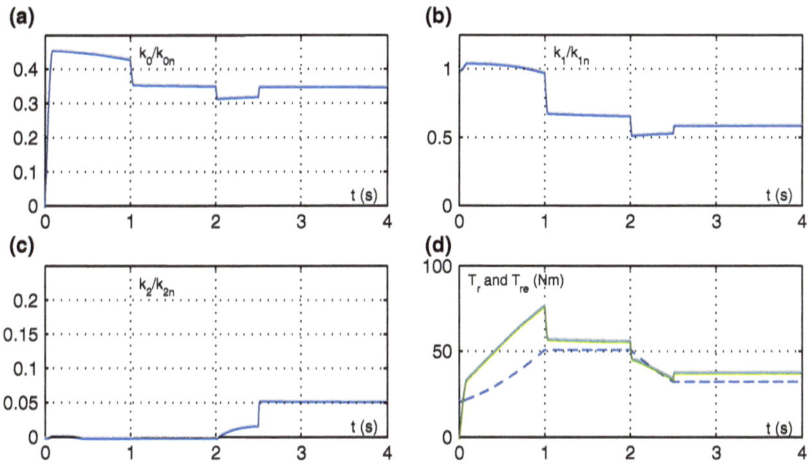

Fig. 14.18 The estimated gains evolution $\overline{K}_0$, $\overline{K}_1$ and $\overline{K}_2$, for a load torque $T_L = K_0 + K_1\Omega_m + K_2\Omega_m^2$, with: $K_0 = \frac{1}{3}K_{0n}$, $K_1 = \frac{1}{3}K_{1n}$ and $K_2 = \frac{1}{3}K_{2n}$

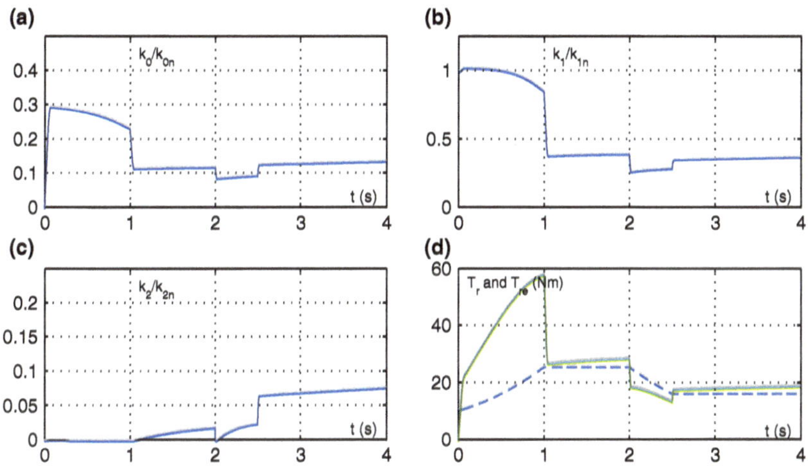

Fig. 14.19 The estimated gains evolution $\overline{K}_0$, $\overline{K}_1$ and $\overline{K}_2$, for a load torque $T_L = K_0 + K_1\Omega_m + K_2\Omega_m^2$, with: $K_0 = \frac{1}{6}K_{0n}$, $K_1 = \frac{1}{6}K_{1n}$ and $K_2 = \frac{1}{6}K_{2n}$

References

Ahammad, N., Khan, S. A., & Reddy, R. K. (2014). Novel DTC-SVM for an adjustable speed sensorless induction motor drive. *International Journal of Science Engineering and Advance Technology (IJSEAT).*, *2*(1), 31–36.

Allirani, S., & Jagannathan, V. (2014). Direct torque control technique in induction motor drives-a review. *Journal of Theoretical and Applied Information Technology*, *60*(3), 454–475.

Ben Salem, F., & Derbel, N. (2007). A sliding mode field oriented control of an induction machine operating with variable parameters. *Journal of Power and Energy Systems*, *27*(2), 205–212.

Ben Salem, F., & Derbel, N. (2014). Direct torque control of induction motors based on discrete space vector modulation using adaptive sliding mode control. *International Journal of Electric Power Components and Systems*, *42*(14), 1598–1610.

Ben Salem, F., & Derbel, N. (2016). Investigation of SM DTC-SVM performances of IM control considering load disturbances effects, in *IEEE 13th International Conference on Systems, Signals and Devices (SSD'16)*, Leipzig, Germany, March 21-24, 2016.

Ben Salem, F., & Masmoudi, A. (2005). On the reduction of the commutation frequency in dtc: A comparative study. *European Trans. on Electrical Power Engineering (ETEP)*, *15*(6), 571–584.

Ben Salem, F., & Masmoudi, A. (2007). A comparative analysis of the inverter switching frequency in Takahashi DTC strategy. *International Journal for Computation and Mathematics in Electrical and Electronic Engineering (COMPEL)*, *26*(1), 148–166.

Boucheta, A., Bousserhane, I. K., Hazzab, A., Sicard, P., & Fellah, M. K. (2012). Speed control of linear induction motor using sliding mode controller considering the end effects. *Journal of Electrical Engineering and Technology*, *7*(1), 34–45.

Bounadja, M., Belarbi, A., & Belmadani, B. (2009). A high performance space vector modulation-direct torque controlled induction machine drive based on stator flux orientation technique. *Advances in Electrical and Computer Engineering*, *9*(2), 28–33.

Carmelia, S., & Maurib, M. (2011). Direct torque control as variable structure control: Existence conditions verification and analysis. *Electric Power Systems Research*, *81*, 1188–1196.

Chaikhy, H., Khafallah, M., & Saad, A. (2011). Evaluation of two control strategies for induction machine. *International Journal of Computer Applications*, *35*(5), 571–584.

Chen, C. Y. (2009). Sliding mode controller design of induction motor based on space-vector pulsewidth modulation method. *International Journal of Innovative Computing, Information and Control*, *5*(10), 3603–3614.

Chlebis, P., Brandstetter, P., & Palacky, P. (2010). Direct torque control of induction motor with direct calculation of voltage vector. *Advances in Electrical and Computer Engineering Journal*, *4*, 17–22.

Habetler, T. G., Profumo, F., Pastorelli, M., & Tolbert, L. M. (1996). Direct torque control of induction machines using space vector modulation. *IEEE Transaction on Industry Applications*, *28*(5), 1045–1053.

Joseline Metilda, A., Arunadevi, R., Ramesh, N., & Sharmeela, C. (2011). Analysis of direct torque control using space vector modulation for three phase induction motor. *Recent Research in Science and Technology*, *3*(7), 37–40.

Mahmoudi, M. O., Madani, N., Ben Khouris, M. F., & Boudjemaa, F. (1999). Cascade sliding mode control of a field oriented induction machine drive. *The European Physical Journal, Applied Physics. EDP Sciences*, *7*(3), 225–277.

Rashag, H. F., Koh, S. P., Chong, K. H., Tiong, S. K., Tan, N. M. L., & Abdalla, A. N. (2013). High performance of space vector modulation direct torque control SVM-DTC based on amplitude voltage and stator flux angle. *Research Journal of Applied Sciences, Engineering and Technology*, *5*(15), 3934–3940.

Rashag, H. F., Tan, N. M. L., Koh, S. P., Abdalla, A. N., Chong, K. H., & Tiong, S. K. (2014). DTC-SVM based on PI torque and PI flux controllers to achieve high performance of induction motor. *Research Journal of Applied Sciences, Engineering and Technology*, *7*(4), 875–891.

Sadoui, R., & Meroufel, A. (2012). Performances comparative study of Field Oriented Control (FOC) and Direct Torque Control (DTC) of Dual Three Phase Induction Motor (DTPIM). *International Journal of circuits, Systems and Signal Processing*, *6*(2), 163–170.

Srirattanawichaikul, W., Kumsuwan, Y., & Premrudeepreechacharn, S. (2010). Reduction of torque ripple in direct torque control for induction motor drives using decoupled amplitude and angle of stator flux control. *ECTI Transaction on Electrical Engineering, Electronics, and Communications*, *8*(2), 187–196.

Takahashi, I., & Noguchi, T. (1986). A new quick-response and high-efficiency control strategy of an induction motor. *EEE Transaction Industry Application*, *16*, 820–827.

Utkin, V. (1993). Sliding mode control design principles and application to electric drives. *IEEE Transactions on Industry Applications.*, *40*(1), 23–36.

Veera, D., & Brahmananda Reddy, T. (2012). Implementation of discrete space vector modulation based direct torque control of induction motor for reduced ripple: a sliding mode control approach. *International Journal of Advanced Scientific and Technical Research*, *4*, 22–38.

Venkateswarlu, K., Sandeep, G., Srinivas, N., & Damodara Reddy, K. (2013). Speed sensorless sliding mode control of induction motor using simulink. *IOSR Journal of Electrical and Electronics Engineering (IOSR-JEEE)*, *6*(2), 50–56.

Yan, Z., Jin, C., & Utkin, V. I. (2000). Sensorless sliding-mode control of induction motors. *IEEE Transaction on Industrial Electronics*, *47*(6), 1286–1297.

Young, K. D., Utkin, V. I., & Ozguner, U. (1999). A control engineer's guide to sliding mode control. *IEEE Transaction on Control System Technology*, *7*(3), 328–342.

[illegible]

Chapter 15
An Enhanced High Order Sliding Mode based Method for Detecting Inter-Turn Short-Circuit Fault in Induction Machine with Decoupled Current Control

Amal Guezmil, Hanen Berriri, Anis Sakly and Mohamed Faouzi Mimouni

Abstract This chapter is devoted to the exploitation of high order sliding mode for fault detection of inter-turn short-circuit in induction machine. For this purpose, healthy and faulty induction machine models are firstly established for different operating conditions during open loop control. The faulty model show the impact of inter-turn short-circuit on induction machine variables. Secondly, a high order sliding mode is synthesized to design decoupled current controller for induction machine and a closed loop induction machine dynamics are analyzed. The fault detection algorithm is combined actual induction machine behavior given by faulty model including fault occurrence and the behavior given by high order sliding mode observer to provide sensitive residuals, which exploited such fault indicators. This theoretical research prove that high order sliding mode approach have good capabilities to ensure both decoupled current control and inter-turn short-circuit fault detection.

Keywords Induction machine · High order sliding mode · Decoupled current control · Inter-turn short-circuit · Fault detection

15.1 Introduction

Fault detection in electrical machines is a topic of increasing interest and importance in the field of control systems. Induction Machines (IM) are widely used in numerous application (Cao et al. 2012; Jurkovic et al. 2015; Sebastián and Peña-Alzola

---

A. Guezmil (✉) · H. Berriri · A. Sakly · M.F. Mimouni
Department of Electrical Engineering, National Engineering School of Monastir, University of Monastir, Monastir, Tunisia
e-mail: guezmil.amal@gmail.com

H. Berriri
e-mail: hanenberriri@yahoo.fr

A. Sakly
e-mail: sakly_anis@yahoo.fr

M.F. Mimouni
e-mail: mfaouzi.mimouni@enim.rnu.tn

N. Derbel et al. (eds.), *Applications of Sliding Mode Control*,
Studies in Systems, Decision and Control 79, DOI 10.1007/978-981-10-2374-3_15

2015; Sridharan and Krein 2014). However, owing to the electrical, mechanical, thermal, magnetic and environmental stresses, mechanical and electrical faults are unavoidable in IM (Siddique et al. 2005). Most recurrent faults are stator winding, rotor winding, bearings and other faults such as connection faults, shaft and coupling faults, etc (Kaikaa and Hadjami 2014). Several studies have shown that inter-turn short-circuit (ITSC) fault accounts for approximately 30–40 % of all possible faults kinds. ITSC fault is usually related to long-term thermal and subsequent deterioration of winding insulation (Nandi et al. 2005). Therefore, early detection of the ITSC fault is required to eliminate the undesirable IM behavior leading to unscheduled maintenance, process shutdown and huge unnecessary costs and delays.

Intensive researches have been carried out to solve the ITSC fault problem (Riera-Guasp et al. 2015). In Ahamed et al. (2014), Devi et al. (2014), Drif and Cardoso (2014), Liu et al. (2014), authors have been focused on the machine current signature analysis (MCSA) to solve the ITSC fault detection problem. This technique helps in the detection of such fault due to remarkable noise on the line current and transient operating conditions. Other works, such as Seshadrinath et al. (2014a) and Frosini et al. (2012), use vibration analysis and axial leakage flux-based techniques. However, these methods require the installation of additional sensors that are costly and sometimes inappropriate when the machine is operating in an adverse environment. Moreover some techniques are introduced using sequence component such as the zero sequence voltage or the negative sequence current component of the machine phase current (Gyftakis and Kappatou 2014). Another ITSC fault detection method using the air-gap torque, sensitive to asymmetric stator winding, is presented in Melero et al. (2003). Also, the increasing interest in artificial intelligence approaches (Filippetti et al. 2000) has led to the proliferation of new algorithms using fuzzy logic (Verma et al. 2014), genetic algorithms (Seshadrinath et al. 2014b) or artificial neural network (Bhavsar et al. 2014). A model-based approach remains widely used for fault detection area and especially for the ITSC fault detection (Ghazal and Poshtan 2011; Lu et al. 2011; Sellami et al. 2013; Toumi et al. 2012). Its main idea is to generate residual or set of residuals that should be sensitive to fault to be detected. These residuals result from a comparison between normal machine behavior and an abnormal behavior due to the fault occurrence.

In ITSC fault detection problem, the main goal is to develop an approach with a minimum knowledge about its parameters and constructional data, which are usually difficult to obtain when the machine is already installed and in operation. Following these goal, observer-based ITSC fault detection has received considerable interest in academic researches. In Lu et al. (2011), a Luenberger observer-based fault detection and diagnosis scheme is developed for double fed induction generator. This technique can not only provide a rapid detection when the ITSC fault occurs but also gives an accurate diagnosis of its position and level. Authors in Toumi et al. (2012) propose an adaptive observer to establish an online system able to detect the ITSC fault in IM. Based on faulty IM model, this observer can estimate the short-circuit turns fraction and other IM states. In Nohra (2013), μ analysis based observer followed by sliding mode estimator is developed to detect, isolate and estimate stator short-circuit fault.

Recently, there has been a special attention focus on sliding mode approach, where it is widely used for state estimation and fault detection in linear and non-linear systems due to the finite-time convergence, robustness with respect to uncertainties and the possibility of uncertainty estimation. In one of the first articles using sliding mode methods, Hermans and Zarrop (1996) are expended a sliding mode observer (SMO) for fault detection problem. Afterwards, several works have been treating the first order sliding mode (FOSM) observer (Edwards et al. 2000; Sellami et al. 2013). However, FOSM observer is limited by relative degree one requirement and the chattering phenomenon. The chattering is due to the inclusion of the sign function in the switching term. This is described as the appearance of oscillations of finite frequency and finite amplitude. These oscillations are caused by the high-frequency switching of the FOSM observer that can motivate no exhibited dynamics, like sensors and actuators dynamics, which are neglected in the system modeling. Generally, sensors and actuators dynamics are faster than the system one. The chattering phenomenon is also induced by uncertainties and perturbations that force system dynamics to leave the sliding mode. In order to reduce the chattering amplitude, a new generation of observers based on the high order sliding mode (HOSM) algorithms has been developed. In this chapter, HOSM observer is established, to estimate IM states which are not available and uncertain on one hand, and on other hand, to generate a set of residuals which close to zero when no fault is present and non-zero when fault occurs (Ashari et al. 2012).

A HOSM observer is established firstly for faulty IM in open loop control. Then, HOSM-based controllers have been investigated with decoupled control design (Kommuri et al. 2014, 2015). Indeed, HOSM current control is employed here to meet the following two purposes: (i) decoupled control of $(d-q)$ current for robust tracking and (ii) ITSC fault detection. The HOSM decoupled current control scheme is obtained using Proportional – Integral (PI) controller and the HOSM controller output. The nominal PI controller acts on the ideal system dynamics, devoid of coupling terms (Comanescu et al. 2008) and the HOSM controller generates an equivalent continuous approximation of the compensation voltages that remove the coupling terms in current dynamics deprived of the use of low-pass filtering (Comanescu 2009).

This chapter deals with ITSC fault detection scheme for IM using HOSM approach. Then, to show that impact of such fault on IM behavior, healthy and faulty IM models are established. Different fault scenarios are studied. Afterward, HOSM approach is used to avoid the estimation of three−phase stator current and rotor flux for ITSC fault detection and to eschew the decoupled control for robust tracking. A HOSM observer is developed firstly, to generate set of residuals that used for the fault detection technique. Secondly, the HOSM decoupled current controller is expanded to producc an equivalent continuous approximation of the compensation voltages that cancels out the coupling terms in the current dynamics. The present work investigated the effectiveness and feasibility of the use of HOSM approach for such fault detection.

15.2 Modeling of Induction Machine

15.2.1 Healthy IM Model

Under assumptions of linear magnetic circuits and balanced operating conditions, the three-phase voltage and current equations of the IM are expressed, in its natural reference frame $(a-b-c)$, as (15.1) and (15.2).

$$\begin{cases} [V^s_{abc}] = [R^s_{abc}][I^s_{abc}] + \frac{d}{dt}[\phi^s_{abc}] \\ [V^r_{abc}] = [R^r_{abc}][I^r_{abc}] + \frac{d}{dt}[\phi^r_{abc}] \end{cases} \tag{15.1}$$

Note that flux linkage may be expressed as,

$$\begin{cases} [\phi^s_{abc}] = [L^s_{abc}][I^s_{abc}] + [L^{sr}_{abc}][I^r_{abc}] \\ [\phi^r_{abc}] = [L^r_{abc}][I^r_{abc}] + [L^{sr}_{abc}]^\top [I^s_{abc}] \end{cases} \tag{15.2}$$

Using (15.2), $[\phi^s_{abc}]$ and $[\phi^r_{abc}]$ can be eliminated from (15.1), and the resulting equations can be written as:

$$\begin{cases} [V^s_{abc}] = [R^s_{abc}][I^s_{abc}] + [L^s_{abc}]\frac{d}{dt}[I^s_{abc}] + \frac{d}{dt}([L^{sr}_{abc}][I^r_{abc}]) \\ [V^r_{abc}] = [0] = [R^r_{abc}][I^r_{abc}] + [L^r_{abc}]\frac{d}{dt}[I^r_{abc}] + \frac{d}{dt}([L^{sr}_{abc}]^\top [I^s_{abc}]) \end{cases} \tag{15.3}$$

Accounting to (15.3), stator and rotor current dynamics can be expressed as:

$$\frac{d}{dt}\begin{bmatrix} [I^s_{abc}] \\ [I^r_{abc}] \end{bmatrix} = \begin{bmatrix} [L^s_{abc}] & [L^{sr}_{abc}] \\ [L^{sr}_{abc}]^T & [L^r_{abc}] \end{bmatrix}^{-1} \times \left(\begin{bmatrix} [V^s_{abc}] \\ [V^r_{abc}] \end{bmatrix} - \begin{bmatrix} [R^s_{abc}] & \omega\frac{d}{dt}[L^{sr}_{abc}] \\ \omega\frac{d}{dt}[L^{sr}_{abc}]^T & [R^r_{abc}] \end{bmatrix} \begin{bmatrix} [I^s_{abc}] \\ [I^r_{abc}] \end{bmatrix} \right) \tag{15.4}$$

where $[V^s_{abc}] = [V^s_a\ V^s_b\ V^s_c]^T$ and $[V^r_{abc}] = [0\ 0\ 0]^T$ indicate the stator and rotor voltage vectors respectively, $[I^s_{abc}] = [I^s_a\ I^s_b\ I^s_c]^T$ and $[I^r_{abc}] = [I^r_a\ I^r_b\ I^r_c]^T$ represent the stator and rotor current vectors respectively. ω is the rotor electrical angular velocity.

Resistance matrices $[R^s_{abc}]$ and $[R^r_{abc}]$ presented in (15.1) are:

$$[R^s_{abc}] = \begin{bmatrix} R_s & 0 & 0 \\ 0 & R_s & 0 \\ 0 & 0 & R_s \end{bmatrix}; [R^r_{abc}] = \begin{bmatrix} R_r & 0 & 0 \\ 0 & R_r & 0 \\ 0 & 0 & R_r \end{bmatrix}$$

where R_s and R_r are the stator and rotor resistances respectively.

The matrix comprising the stator and rotor self and mutual inductance $\left[L^{s}_{abc}\right]$ and $\left[L^{r}_{abc}\right]$ respectively, are given by (15.5):

$$\left[L^{s}_{abc}\right] = \begin{bmatrix} L_s + l_s & -\frac{1}{2}L_s & -\frac{1}{2}L_s \\ -\frac{1}{2}L_s & L_s + l_s & -\frac{1}{2}L_s \\ -\frac{1}{2}L_s & -\frac{1}{2}L_s & L_s + l_s \end{bmatrix}; \left[L^{r}_{abc}\right] = \begin{bmatrix} L_r + l_r & -\frac{1}{2}L_r & -\frac{1}{2}L_r \\ -\frac{1}{2}L_r & L_r + l_r & -\frac{1}{2}L_r \\ -\frac{1}{2}L_r & -\frac{1}{2}L_r & L_r + l_r \end{bmatrix} \quad (15.5)$$

where L_s and L_r are the stator and rotor magnetizing inductance of the IM respectively, l_s and l_r are the leakage inductance of the stator and rotor phase winding, respectively.

The matrix of stator and rotor mutual inductance referred to the stator side $\left[L^{sr}_{abc}\right]$ is given by (15.6) with M_{sr} is the mutual inductance.

$$\left[L^{sr}_{abc}\right] = \left[L^{rs}_{abc}\right] = M_{sr} \begin{bmatrix} \cos\theta & \cos(\theta + \frac{2\pi}{3}) & \cos(\theta + \frac{4\pi}{3}) \\ \cos(\theta + \frac{4\pi}{3}) & \cos\theta & \cos(\theta + \frac{2\pi}{3}) \\ \cos(\theta + \frac{2\pi}{3}) & \cos(\theta + \frac{4\pi}{3}) & \cos\theta \end{bmatrix} \quad (15.6)$$

The electromagnetic torque equation produced by an IM is given by:

$$C_{em} = \frac{1}{2}\left[I^{s}_{abc}\right]^{T}\left(\frac{d}{d\theta}\left[L^{sr}_{abc}\right]\right)\left[I^{s}_{abc}\right] \quad (15.7)$$

The rotor mechanical equation can be written as:

$$C_{em} = J\frac{d\omega}{dt} + f\omega + C_r \quad (15.8)$$

where, J is rotor inertia, f is the friction coefficient and C_r is the load torque.

15.2.2 Faulty IM Model

In this section, a three phase stator windings with a short circuit on phase a is considered. Figure 15.1 shows an ITSC occurring in phase a. A short circuit of turns introduces an extra winding in the machine winding structure. Therefore, the faulty IM can be modeled as a stator composed by four different winding as_1, as_2, bs and cs. Winding as_1 and as_2 represent the healthy and shorted turns winding respectively. An additional insulated winding is added and modeled by a resistance R_f. The current I_f represents the circulating current in the shorted turns. R_f value depends on the fault severity. A graver ITSC in the affected phase is got when the fault insulation resistance R_f decreases toward zero. It decreases from infinite toward zero in most materials. The number of short circuited turns is N_f out of the total number of turns N_s of the phase winding. The ratio of N_f to N_s (N_f $/N_s$) is indicated by the short circuit level k_{cc} and is remain between 0 and 1. This ratio present the fault severity (Lu et al. 2011).

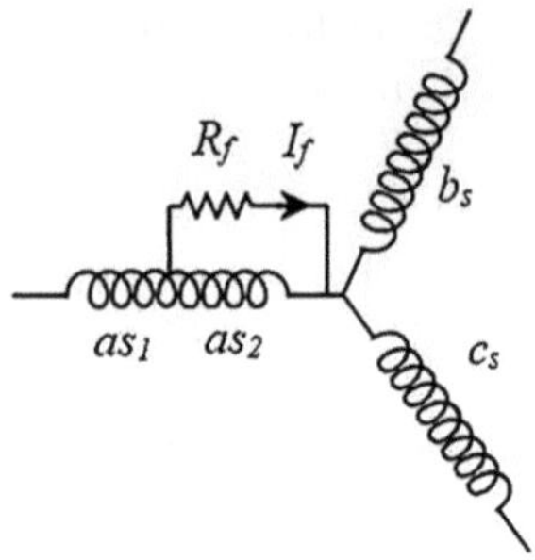

Fig. 15.1 Scheme of three phase stator winding with an ITSC fault in phase *a*

The IM voltage equations under ITSC fault conditions are written as:

$$\begin{cases} \left[V^s_{abcf}\right] = \left[R^s_{abcf}\right]\left[I^s_{abcf}\right] + \frac{d}{dt}\left[\phi^s_{abcf}\right] \\ \left[V^r_{abc}\right] = \left[R^r_{abc}\right]\left[I^r_{abc}\right] + \frac{d}{dt}\left[\phi^r_{abc}\right] \end{cases} \tag{15.9}$$

Knowing that:

$$\begin{cases} \left[\phi^s_{abcf}\right] = \left[L^s_{abcf}\right]\left[I^s_{abcf}\right] + \left[L^{sr}_{abcf}\right]\left[I^r_{abc}\right] \\ \left[\phi^r_{abc}\right] = \left[L^r_{abc}\right]\left[I^r_{abc}\right] + \left[L^{sr}_{abcf}\right]^\top\left[I^s_{abcf}\right] \end{cases} \tag{15.10}$$

Using (15.10), $\left[\phi^s_{abcf}\right]$ and $\left[\phi^r_{abc}\right]$ can be eliminated from (15.9) and the resulting equations can be written as:

$$\begin{cases} \left[V^s_{abcf}\right] = \left[R^s_{abcf}\right]\left[I^s_{abcf}\right] + \left[L^s_{abcf}\right]\frac{d}{dt}\left[I^s_{abcf}\right] + \frac{d}{dt}(\left[L^{sr}_{abcf}\right]\left[I^r_{abc}\right]) \\ \left[V^r_{abc}\right] = [0] = \left[R^r_{abc}\right]\left[I^r_{abc}\right] + \left[L^r_{abc}\right]\frac{d}{dt}\left[I^r_{abc}\right] + \frac{d}{dt}(\left[L^{sr}_{abcf}\right]^\top\left[I^s_{abcf}\right]) \end{cases} \tag{15.11}$$

Accounting to (15.10), stator and rotor current dynamics can be written as:

$$\frac{d}{dt}\begin{bmatrix} \left[I^s_{abcf}\right] \\ \left[I^r_{abc}\right] \end{bmatrix} = \begin{bmatrix} \left[L^s_{abcf}\right] & \left[L^{sr}_{abcf}\right] \\ \left[L^{sr}_{abcf}\right]^T & \left[L^r_{abc}\right] \end{bmatrix}^{-1} \times \left(\begin{bmatrix} \left[V^s_{abcf}\right] \\ \left[V^r_{abc}\right] \end{bmatrix} - \begin{bmatrix} \left[R^s_{abcf}\right] & \omega\frac{d}{dt}\left[L^{sr}_{abcf}\right] \\ \omega\frac{d}{dt}\left[L^{sr}_{abcf}\right]^T & \left[R^r_{abc}\right] \end{bmatrix} \begin{bmatrix} \left[I^s_{abcf}\right] \\ \left[I^r_{abc}\right] \end{bmatrix} \right) \tag{15.12}$$

with $\left[V^s_{abcf}\right] = \left[V^s_a \; V^s_b \; V^s_c \; 0\right]^T$ and $\left[I^s_{abcf}\right] = \left[I^s_a \; I^s_b \; I^s_c \; I_f\right]^T$ indicate the three−phase stator voltage and current vectors respectively, in presence of stator ITSC fault.

The resistance matrix $\left[R^s_{abcf}\right]$ of (15.9) is:

$$[R^s_{abcf}] = \begin{bmatrix} R_s & 0 & 0 & -k_{cc}R_s \\ 0 & R_s & 0 & 0 \\ 0 & 0 & R_s & 0 \\ k_{cc}R_s & 0 & 0 & -(k_{cc}R_s + R_f) \end{bmatrix}$$

All the stator–stator and stator–rotor mutual inductances will have an extra term due to this shorted loop. So far, the matrix comprising the stator self and mutual inductance's $[L^s_{abcf}]$ is given by (15.13):

$$[L^s_{abcf}] = \begin{bmatrix} L_s + l_s & -\frac{1}{2}L_s & -\frac{1}{2}L_s & -L_b \\ -\frac{1}{2}L_s & L_s + l_s & -\frac{1}{2}L_s & -\frac{1}{2}k_{cc}L_s \\ -\frac{1}{2}L_s & -\frac{1}{2}L_s & L_s + l_s & -\frac{1}{2}k_{cc}L_s \\ L_b & -\frac{1}{2}(1-k_{cc})L_s & -\frac{1}{2}(1-k_{cc})L_s & -(L_{sf} + l_{sf}) \end{bmatrix} \quad (15.13)$$

It is admitted that:

$$L_b = [L_{sf} + l_{sf} + L_s(1-k_{cc})k_{cc}];\ L_{sf} = k_{cc}^2 L_s;\ l_{sf} = k_{cc}^2 l_s$$

The matrix of stator and rotor mutual inductance's referred to the stator side $[L^{sr}_{abcf}]$ is given by (15.14):

$$[L^{sr}_{abcf}] = M_{sr}\begin{bmatrix} \cos\theta & \cos(\theta + \frac{2\pi}{3}) & \cos(\theta + \frac{4\pi}{3}) \\ \cos(\theta + \frac{4\pi}{3}) & \cos\theta & \cos(\theta + \frac{2\pi}{3}) \\ \cos(\theta + \frac{2\pi}{3}) & \cos(\theta + \frac{4\pi}{3}) & \cos\theta \\ -k_{cc}\cos\theta & -k_{cc}\cos(\theta + \frac{2\pi}{3}) & -k_{cc}\cos(\theta + \frac{4\pi}{3}) \end{bmatrix} \quad (15.14)$$

The electromagnetic torque equation produced by a faulty three phase IM can be written as:

$$C_{em} = \frac{1}{2}[I^s_{abcf}]^T \left(\frac{d}{d\theta}[L^{sr}_{abcf}]\right)[I^s_{abcf}] \quad (15.15)$$

15.2.3 *Impact of ITSC Fault*

Simulations are performed to evaluate an 1.5 kW IM in open loop. Its specifications and parameters are given in Table 15.1. For the simulation, the IM drive is operated in healthy and faulty mode.

In order to test the performance of the IM model in healthy mode, the short-circuit level K_{cc} is set to zero with an insulation resistance R_f fixed to 15 Ω. The load torque is chosen as constant, equal to 3 Nm, applied at $t = 1$ s. Then, an ITSC fault occurs the IM at $t = 1.5$ s ($K_{cc} = 0.20$ and $R_f = 0\,\Omega$). To test the fault severity, the short-circuit level K_{cc} increase from 0.20 to 0.30 with keeping $R_f = 0\,\Omega$ at $t = 2.5$ s.

Figures 15.2 and 15.3 show the three phase stator current (I^s_a, I^s_b, I^s_c). It is clear that the real stator current becomes unbalanced and the current in phase *a* is bigger than

Table 15.1 Induction machine parameters and its nominal values

Model parameter		
P_n	Output power	1.5 Kw
V_s	Stator voltage	220 V
N_p	Pole number	2
R_s	Stator resistance	5.272 Ω
R_r	Rotor resistance	4.282 Ω
L_s	Stator inductance	0.464 H
L_r	Rotor inductance	0.464 H
l_s	Stator leakage inductance	0.024 H
l_r	Rotor leakage inductance	0.024 H
M_{sr}	Mutual inductance	0.44 H
J	Inertia moment	0.048 kgm^2
f	Friction coefficient	0.003 Nms

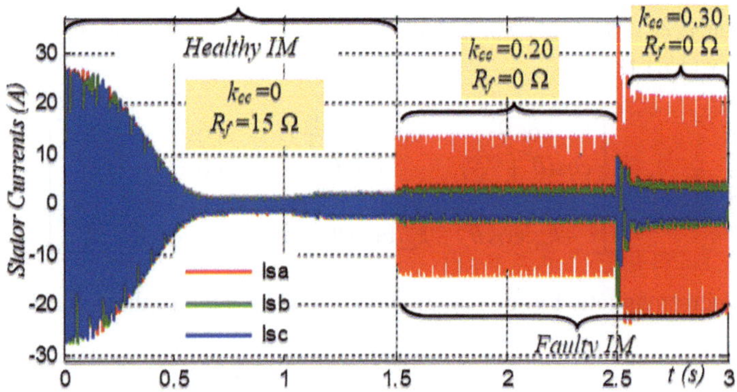

Fig. 15.2 Three phase stator currents (I_a^s, I_b^s, I_c^s) in healthy and faulty cases

the other two. The three-phase rotor current are presented in Fig. 15.4. As it shown, the three-phase rotor current $\left(I_a^r, I_b^r, I_c^r\right)$ increase when short circuit fault occurs. The short-circuit current I_f is shown in Fig. 15.5. This current present oscillations increase.

When the ITSC fault appears. Figures 15.6 and 15.7 show, respectively the rotor speed ω and electromagnetic and load torque (C_{em}, C_r) variations. The ITSC application causes a fluctuation in the rotor speed at high frequency.

It can be revealed that, with the increase of short-circuit level K_{cc}, the affected phase current I_a^s variation becomes more constables and the amplitude of I_a^s and I_f increase, as shown in Fig. 15.3.

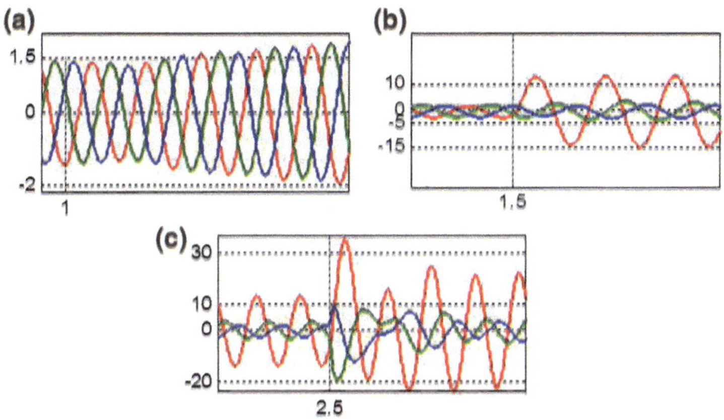

Fig. 15.3 Zoom of $(I_a^s,\ I_b^s,\ I_c^s)$, **a** $t = 1$ s, **b** $t = 1.5$ s and **c** $t = 2.5$ s

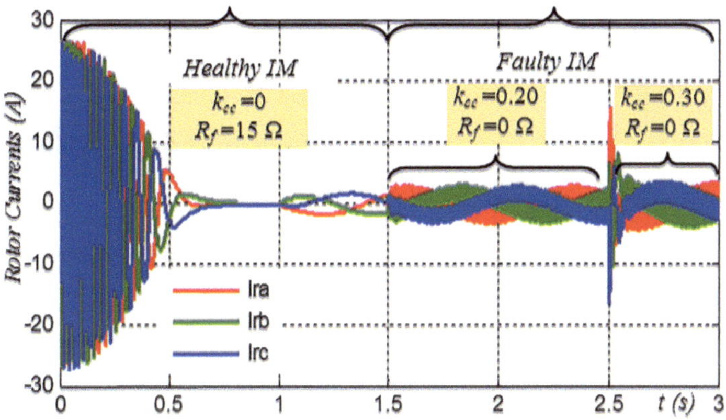

Fig. 15.4 Three phase rotor currents $(I_a^r,\ I_b^r,\ I_c^r)$ in healthy and faulty cases

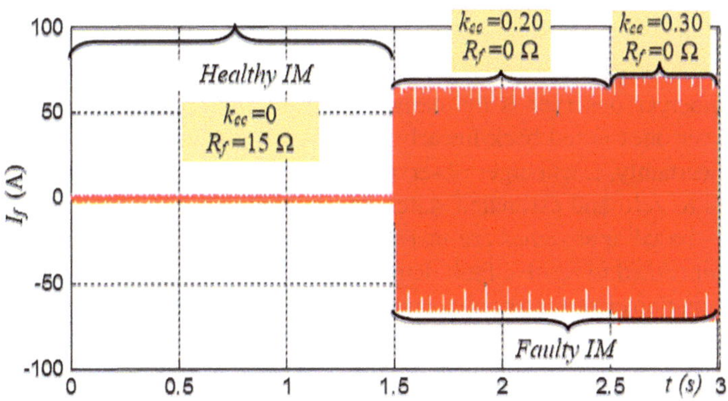

Fig. 15.5 Short-circuit current I_f in healthy and faulty cases

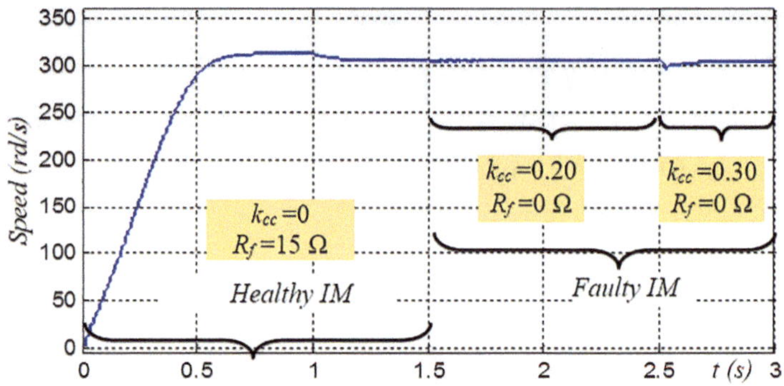

Fig. 15.6 Rotor speed ω in healthy and faulty cases

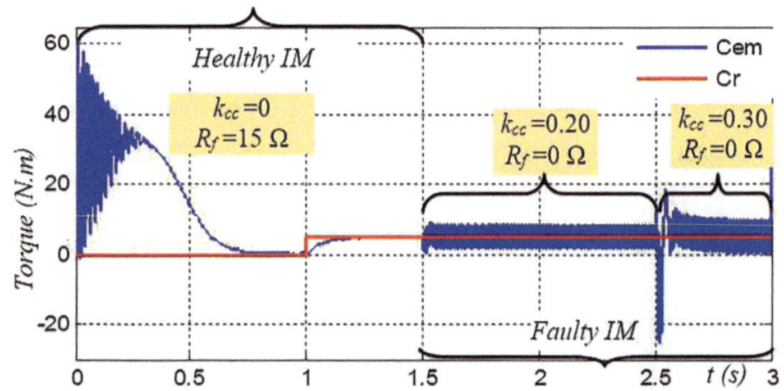

Fig. 15.7 Electromagnetic and load torque (C_{em}, C_r) in healthy and faulty cases

15.3 HOSM Observer

The purpose of this section is to design HOSM observer used for fault detection to generate residual signal on one hand and on the other hand, for control. Indeed, observer−based approach is a popular model based technique used in fault detection. Luenberger observer is the earliest. The difference between model outputs and observer ones is fed back linearly. However, in the presence of unknown signals or uncertainty, Luenberger observer is usually unable to force outputs estimation errors to zero and estimated states does not converge to system states. Other observers−based approaches are investigated (Lu et al. 2011; Toumi et al. 2012). SMO is concerned to resolve fault detection problem. This observer feeds back the output estimation error via a nonlinear switching term, provides an attractive solution to this issue. SMO is used in both linear and nonlinear systems with uncertainties. Provided a bound on the magnitude of the disturbances is known, the SMO can force the output estimation error to converge to zero in finite time, estimated states converge asymptotically to system states. Moreover, system disturbances can be reconstructed

(Shtessel et al. 2014b). SMO has also a unique property which starts from the fact that the introduction of a sliding motion forces the output of the observer to perfectly track the system measurements. In case of fault, the output of the SMO follows the system output. Indeed, the aim of SMO scheme is to brings states to a sliding surface and then they are bound to remain on sliding surface onwards (Aurora and Ferrara 2007). Thus, an important step in the design of the SMO is to introduce a proper sliding surface so that tracking errors and output deviations are reduced to a satisfactory level.

In the literature, SMO has been adopted by researchers (Alwi and Edwards 2014; Edwards et al. 2013; Mekki et al. 2015). Those papers are concerned on the use of sliding mode ideas for fault detection and reconstruction. Their main advantages in this domain are that they exhibit fundamental robustness against certain kinds of parameter variations and that they also enable faults and/or values of immeasurable system parameters to be reconstructed. Besides, the use of the FOSM observer introduces some drawbacks such as chattering effect, limited flexibility for the designer with a sliding function and constant gain as the error variable. Taking into account the chattering problem, HOSM observer is introduced by Levant (1993). This observer is the generalization of the FOSM observer and the observer input is performed such that it acts on higher derivatives of the sliding surface. It should be noted that different HOSM approaches have been presented in the literature such as the twisting (Guezmil et al. 2014), super-twisting (Zhao et al. 2014) and sub-optimal (Ferrara and Rubagotti 2009). Readers are referred for the extensive information about these methods to Shtessel et al. (2014a).

In the following section, authors delineate a HOSM observer, based on twisting algorithm, in order to observe the rotor flux modulus and reconstruct the three-phase stator current.

15.3.1 HOSM Observer Design

To simplify and clarify the development of HOSM observer, the healthy IM equations according to (15.4) are transformed to $(\alpha - \beta)$ reference frame as the following set of state variables equations:

$$\begin{cases} \frac{d}{dt}\left[I^s_{\alpha\beta}\right] = a_1\left[I_2\right]\left[I^s_{\alpha\beta}\right] + a_2\left[A\right]\left[\phi^r_{\alpha\beta}\right] + \left[C\right]\left[V^s_{\alpha\beta}\right] \\ \frac{d}{dt}\left[\phi^r_{\alpha\beta}\right] = b_1\left[I_2\right]\left[I^s_{\alpha\beta}\right] - \left[A\right]\left[I_2\right] \\ \frac{d}{dt}\omega = \frac{1}{J}(m\left[J_2\right]\left[\phi^r_{\alpha\beta}\right]\left[I^s_{\alpha\beta}\right]) - C_r - f\omega \end{cases} \tag{15.16}$$

with $\left[I^s_{\alpha\beta}\right] = \left[I^s_\alpha \;\; I^s_\beta\right]^T$, $\left[\phi^s_{\alpha\beta}\right] = \left[\phi^s_\alpha \;\; \phi^s_\beta\right]^T$ and $\left[V^s_{\alpha\beta}\right] = \left[V^s_\alpha \;\; V^s_\beta\right]^T$ are the two-phase stator current, the rotor flux and the two-phase stator voltage components, respectively, in the $(\alpha - \beta)$ reference frame. T_s and T_r are the stator and rotor time constant, respectively and N_p is the number of pole pairs.

Expressions of $[I_2]$; $[A]$; $[C]$; $[J_2]$ matrix, $a_1, a_2, \sigma, T_s, T_r, b_1, b_2, c$ and m are depend on IM parameters.

$$[I_2] = \begin{pmatrix} 1 & 0 \\ 0 & 1 \end{pmatrix};\ [A] = \begin{pmatrix} -b_2 & N_p\omega \\ -N_p\omega & -b_2 \end{pmatrix};\ [C] = \begin{pmatrix} c & 0 \\ 0 & c \end{pmatrix};\ [J_2] = \begin{pmatrix} 0 & -1 \\ 1 & 0 \end{pmatrix}$$

$$a_1 = \left(\frac{R_s}{\sigma L_s} + \frac{M_{sr}^2}{\sigma L_s L_r T_s}\right);\ a_2 = \frac{M_{sr}}{\sigma L_s L_r};\ \sigma = 1 - \frac{M_{sr}^2}{\sigma L_s L_r}$$

$$T_s = \frac{L_s}{R_s};\ T_r = \frac{L_r}{R_r};\ b_1 = \frac{M_{sr}}{T_r};\ b_2 = -\frac{1}{T_r};\ c = \frac{1}{\sigma L_s};\ m = \frac{N_p M_{sr}}{L_r}$$

From the IM healthy model (15.16), the HOSM observer, based on the method as proposed in Benderradji et al. (2012), will be introduced as follows:

$$\begin{cases} \frac{d}{dt}\left[\widehat{I^s_{\alpha\beta}}\right] = a_1\left[I_2\right]\left[\widehat{I^s_{\alpha\beta}}\right] + a_2\left[A\right]\left[\widehat{\phi^r_{\alpha\beta}}\right] + \left[C\right]\left[V^s_{\alpha\beta}\right] \\ \frac{d}{dt}\left[\widehat{\phi^r_{\alpha\beta}}\right] = b_1\left[I_2\right]\left[\widehat{I^s_{\alpha\beta}}\right] - \left[A\right]\left[I_2\right] + \left[\chi_{\alpha\beta}\right] \\ J\frac{d}{dt}\widehat{\omega} = \left(m\left[J_2\right]\left[\widehat{\phi^r_{\alpha\beta}}\right]\left[\widehat{I^s_{\alpha\beta}}\right]\right) - C_r - f\widehat{\omega} \end{cases} \quad (15.17)$$

with $\left[\widehat{I^s_{\alpha\beta}}\right] = \left[\widehat{I^s_\alpha}\ \widehat{I^s_\beta}\right]^T$ and $\left[\widehat{\phi^r_{\alpha\beta}}\right] = \left[\widehat{\phi^r_\alpha}\ \widehat{\phi^r_\beta}\right]^T$ are the estimated stator current and the rotor flux components, respectively. $\left[\chi_{\alpha\beta}\right] = \left[\chi_\alpha\ \chi_\beta\right]^T$ is the observer matrix gains to be designed.

Currents and flux estimation errors are defined as follow:

$$\begin{cases} \left[e_I\right] = \left[\widehat{I^s_{\alpha\beta}}\right] - \left[I^s_{\alpha\beta}\right] \\ \left[e_\phi\right] = \left[\widehat{\phi^r_{\alpha\beta}}\right] - \left[\phi^r_{\alpha\beta}\right] \end{cases} \quad (15.18)$$

The observation error dynamic is obtained from (15.16) and (15.17):

$$\begin{cases} \frac{d}{dt}\left[e_I\right] = a_1\left[A\right]\left[e_\phi\right] \\ \frac{d}{dt}\left[e_\phi\right] = b_1\left[I_2\right]\left[e_I\right] - \left[A\right]\left[e_\phi\right] + \left[\chi_{\alpha\beta}\right] \end{cases} \quad (15.19)$$

Let us select a sliding surface$[S]$ as follow:

$$[S] = \begin{bmatrix} S_1 \\ S_2 \end{bmatrix} = \frac{1}{a_2}\left[A\right]^{-1}\left[e_I\right] \quad (15.20)$$

The dynamic sliding surface is:

$$\frac{d}{dt}[S] = \frac{1}{a_2}[A]^{-1}\left(\frac{d}{dt}[e_I]\right) + \frac{1}{a_2}\left(\frac{d}{dt}[A]^{-1}\right)[e_I] \tag{15.21}$$

The derivative of ω is supposed constant compared to the derivative of $\left[I^s_{\alpha\beta}\right]$ and $\left[\phi^r_{\alpha\beta}\right]$. Consequently, $\left[A^{-1}\right]$ term is considered null and the dynamic of sliding surface become:

$$\begin{cases} \frac{d}{dt}[S] = \frac{1}{a_2}\left[A^{-1}\right]\left(\frac{d}{dt}[e_I]\right) = [e_\phi] \\ \frac{d^2}{dt^2}[S] = \frac{d}{dt}[e_\phi] = b_1[I_2][e_I] - [A][e_\phi] + [\chi_{\alpha\beta}] \end{cases} \tag{15.22}$$

The idea of using twisting approach to fault detection consists of using estimated variable to generate a residual.

$$[\chi_{\alpha\beta}] = \begin{cases} -\lambda_m \text{sign}[S] & \text{if } S\frac{d}{dt}(S) \leq 0 \\ -\lambda_M \text{sign}[S] & \text{if } S\frac{d}{dt}(S) > 0 \end{cases} \tag{15.23}$$

where λ_m and λ_M are appropriately designed positive constants that ensure the above condition defined in (15.24):

$$\begin{cases} -\lambda_m > |e_\phi| \\ -\lambda_M > \lambda_m + 2|e_\phi|_{max} \end{cases} \tag{15.24}$$

The system (15.23) evolves featuring a HOSM, after a finite time. Therefore,

$$\frac{d}{dt}(S) = \frac{d^2}{dt}(S) = 0 \tag{15.25}$$

15.3.2 HOSM Observer Simulation

In order to illustrate the dynamic behavior of the proposed observer, simulation tests were carried out in the IM used in Sect. 15.2.2. The ITSC fault, characterized by an insulation resistance R_f equal to 0 Ω and short-circuit level $K_{cc} = 0.20$, is applied at $t = 1.5$ s. The short-circuit level increase from 0.20 to 0.30 at $t = 2.5$ s. Assumed that the rotor speed ω, three-phase stator current $\left(I^s_a,\ I^s_b,\ I^s_c\right)$ and stator voltages $\left(V^s_a,\ V^s_b,\ V^s_c\right)$ are available.

The measured, estimated electromagnetic and load torque of the IM are depicted in Fig. 15.8. The rotor speed ω and his estimated one $\hat{\omega}$ are presented in Fig. 15.9.

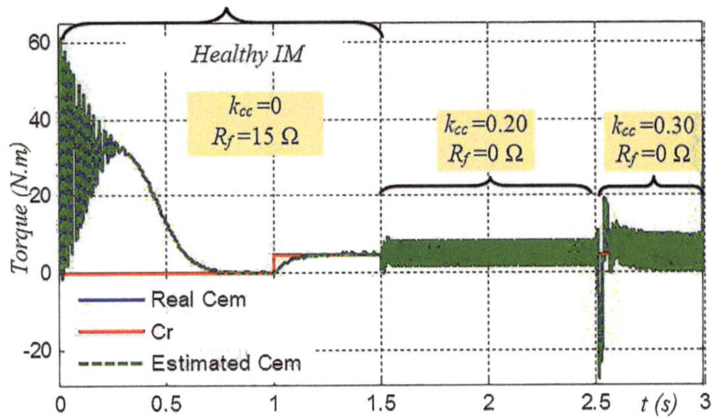

Fig. 15.8 Measured, estimated electromagnetic and load torque (C_{em}, $\hat{C}_{em}$ and C_r) in healthy and faulty cases, using HOSM observer

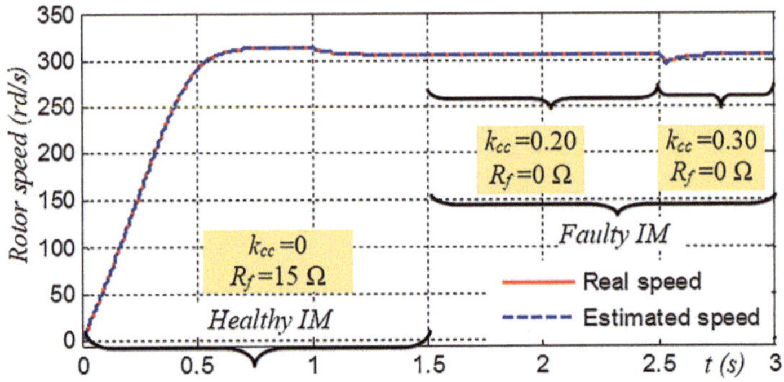

Fig. 15.9 Measured and estimated rotor speed (ω and $\hat{\omega}$) in healthy and faulty cases using HOSM observer

The observed rotor flux $\hat{\phi}_r$ is appeared in Fig. 15.10. Figure 15.11 shows the three-phase stator current$\left(I_a^s,\ I_b^s,\ I_c^s\right)$. After application of ITSC fault, the different IM variables present a small drop when increasing shorted turns level. From these simulation results, it can be seen that the proposed method can estimate the rotor flux quickly and actually.

15.4 Control Scheme and Decoupling

It is known that the techniques developed for IM fault detection in open loop drives cannot be used straight forward when the machine is included in control structure based on direct torque control and field oriented control (Ashari et al. 2012). Indeed, in

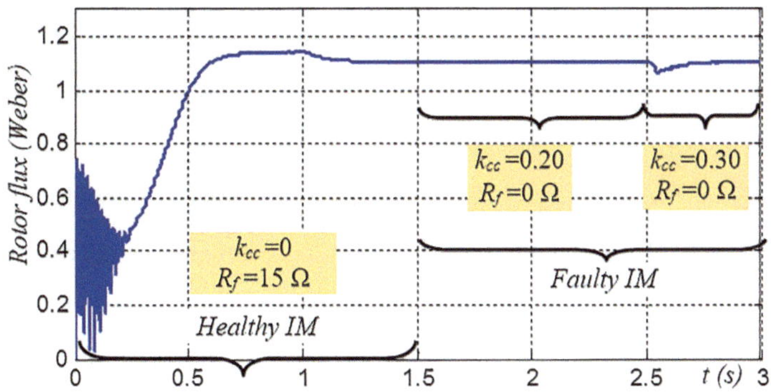

Fig. 15.10 Estimated rotor flux $\hat{\phi}_r$ in healthy and faulty cases, using HOSM observer

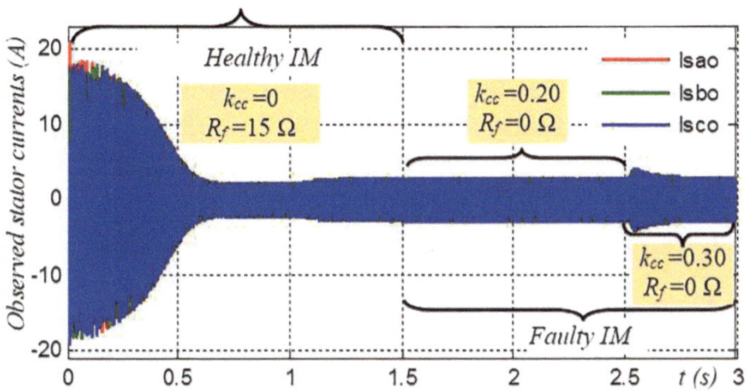

Fig. 15.11 Estimated stator currents ($\hat{I}_a^s$, $\hat{I}_b^s$, $\hat{I}_c^s$) in healthy and faulty cases, using HOSM observer

efficient operation, the first difficulty get up from, the coupling between the currents of d-axis and q-axis, respectively. This coupling effects result a change in one of the current and produces a transient disturbance in the other. As a result, the torque generated in IM is distorted. Another problem, related to synchronous current control, is the tuning of current loops. To overcome those problems, PI controllers are used as in Eker (2012).

15.4.1 Classical Decoupled Current Control for IM

The classical decoupled current control scheme is based on a conventional solution of synchronous frame current control using simple PI controllers.

By applying the Park transform to healthy IM model (15.4), current equations in synchronous reference frame $(d - q)$, are defined as:

$$\begin{cases} \dfrac{dI_d^s}{dt} = a_1 I_d^s - a_2 b_2 \phi_d^r + \omega N_p I_q^s + b_1 \dfrac{(I_q^s)^2}{\phi_d^r} + c V_d^s \\ \dfrac{dI_q^s}{dt} = a_1 I_q^s - a_2 N_p \omega \phi_d^r - N_p \omega I_d^s - b_1 \dfrac{I_d^s I_q^s}{\phi_d^r} + c V_q^s \end{cases} \tag{15.26}$$

where I_d^s, I_q^s, V_d^s and V_q^s are the stator current and voltage, respectively in the $(d-q)$ reference frame. ϕ_d^r is a magnitude of the rotor flux vector and ω is the rotor electrical speed. a_1, a_2, b_1 and b_2 are defined in Sect. 15.3.1.

The current dynamics ($\frac{dI_d^s}{dt}$ and $\frac{dI_q^s}{dt}$) are dependent on each other and also they have additional terms which are function of IM parameters, rotor flux magnitude and speed. To obtain a torque production, current controllers must provide both regulation and decoupling.

Figure 15.12 shows the classical decoupled current control scheme using PI controllers. The current controllers are purposed according to machine dynamics and must satisfy the control requirements for a certain type of reference inputs. PI current controllers regulate currents with zero steady-state error and have proportional gains K_d, K_q and integral times T_d, T_q. Compensation voltages are looking to cancel the additional machine dynamics and unknown terms. Compensation terms are dependent on the machine speed and are given by (15.27):

$$\begin{cases} V_{dcomp}^s = -\dfrac{1}{c}\left(N_p \omega I_q^s + b_1 \dfrac{(I_q^s)^2}{\phi_d^r}\right) \\ V_{qcomp}^s = -\dfrac{1}{c}\left(a_2 N_p \omega \phi_d^r + N_p \omega I_d^s + b_1 \dfrac{I_d^s I_q^s}{\phi_d^r}\right) \end{cases} \tag{15.27}$$

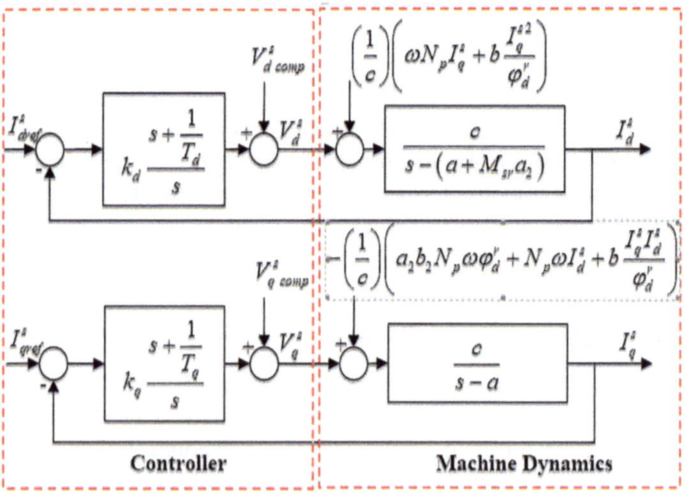

Fig. 15.12 Scheme based on PI controllers for IM decoupled current control

The objective of this section is to design a robust current controllers, which can decouple dependent currents (I_d^s and I_q^s). For instance, decoupled control of current I_d^s is achieved by inserting (15.27) in (15.26), this control cancels the unknown disturbances that depends on I_q^s. If the cancellation takes place, the overall dynamics of I_d^s is simplified and it is easy to design the corresponding PI controller gains.

In order to implement the current control scheme, the following conditions must be validate.

(1) Reference currents I_{dref}^s and I_{qref}^s, real currents I_d^s and I_q^s must be fed in PI controllers to generate V_d^s and V_q^s. These four currents are available. Reference currents are generated by structures of speed and flux controllers. I_d^s and I_q^s are measured.
(2) Terms V_{dcomp}^s and V_{qcomp}^s must be computed (and added to I_d^s and I_q^s). This computation requires the knowledge of four things: machine parameters, stator currents, the IM speed ω, and the flux magnitude ϕ_d^r.

Generally, the scheme in Fig. 15.12 can be relatively be implement. In this case, variables ω and ϕ_d^r are not available. Consequently, the control algorithm would need a flux estimator for the sole purpose of being able to compute the voltages V_{dcomp}^s and V_{qcomp}^s. Therefore it is difficult to compute the compensation voltages (15.27) in control schemes. Because of this disadvantage, some implementations use only PI controllers (the compensation voltages are omitted). These implementations cannot regulate the current if the reference input is a step function. Since the machine accelerates or decelerates, the speed is a ramp function and a PI alone cannot reject the extra terms in Fig. 15.12 and is unable to regulate. The steady-state errors of currents can be reduced by increasing proportional gains of the PI. However, the steady state current errors can be reduced by increasing the proportional gains of the PI controllers (Comanescu 2010). Also, the $(d - q)$ axis currents will be coupled. As a result, changing the reference current in one axis produces a transient disturbance in the other axis. This results in a transient distortion in the torque.

15.4.2 HOSM Decoupled Current Control Design

The selection of the PI current controllers gain, presented in the previous sections, is not easy and is usually subject to continuous adjustments. Equivalent control approach is used in solution based on the high order IM model. Stability of the closed-loop system is proved using direct Lyapunov approach. Let us note that the unknown disturbance terms of model dynamics in (15.26) act like extra dynamics. The idea is to use sliding mode controller on both the d-axis and q-axis to cancel the extra dynamics of the IM. In the following, a HOSM controller will be designed to cancel the disturbance dynamics.

The current dynamics in (15.26) can be expressed as:

$$\begin{cases} \frac{d}{dt} I_d^s = a_1 I_d^s + a_2 b_2 \phi_d^r + \Gamma_d\left(\omega, I_q^s, \phi_d^r\right) + c V_d^s \\ \frac{d}{dt} I_q^s = a_1 I_q^s + \Gamma_q\left(\omega, I_d^s, I_q^s, \phi_d^r\right) + c V_q^s \end{cases} \tag{15.28}$$

where $\Gamma_d\left(\omega, I_q^s, \phi_d^r\right)$ and $\Gamma_q\left(\omega, I_d^s, I_q^s, \phi_d^r\right)$ are the cross coupling and unknown terms for the d-axis and q-axis, respectively.

$$\begin{cases} \Gamma_d\left(\omega, I_q^s, \phi_d^r\right) = \omega N_p I_q^s + b_1 \frac{(I_q^s)^2}{\phi_d^r} \\ \Gamma_q\left(\omega, I_d^s, I_q^s, \phi_d^r\right) = -a_2 N_p \omega \phi_d^r - N_p \omega I_d^s - b_1 \frac{I_d^s I_q^s}{\phi_d^r} \end{cases} \tag{15.29}$$

The disturbance terms consist of some known terms and some unknown terms which result in coupling the current dynamics. The objective is to design a controller that performs decoupling action by canceling the coupling and unknown terms. The control input provided to the IM comprises of nominal control and disturbance rejection control.

The nominal machine dynamics which is devoid of the disturbances can be expressed as:

$$\begin{cases} \frac{d}{dt} I_d^s = a_1 I_d^s + c V_{d0}^s \\ \frac{d}{dt} I_q^s = a_1 I_q^s + c V_{q0}^s \end{cases} \tag{15.30}$$

The nominal control inputs (V_{d0}^s; V_{q0}^s) are generated by the use of PI controllers as shown in Fig. 15.13.

The nominal control inputs (V_{d0}^s; V_{q0}^s) are generated by the use of PI controllers as shown in Fig. 15.13. The output of the nominal machine dynamics are the fictitious currents (I_{dsim}^s, I_{qsim}^s), called model currents. Note that measured currents (I_d^s, I_q^s) are not used in the generation of nominal control inputs (V_{d0}^s; V_{q0}^s). This currents are only used by the proposed HOSM controllers to provide the tracking of the reference currents. However, it is atypical control scheme because the error term (($I_{ref} - I_{real}$)) does not appear directly in Fig. 15.13 like in a classical decoupled control scheme shown in Fig. 15.12. To achieve the rejection of the unknown terms $\Gamma_d\left(\omega, I_q^s, \phi_d^r\right)$ and $\Gamma_q\left(\omega, I_d^s, I_q^s, \phi_d^r\right)$, HOSM controllers are employed to control inputs ((ψ_d^s, ψ_d^s)).

For the design of HOSM controllers, the following sliding surfaces are selected:

$$\begin{cases} S_d(t) = I_d^s - I_{dref}^s \\ S_q(t) = I_q^s - I_{qref}^s \end{cases} \tag{15.31}$$

where I_d^s and I_q^s are the actual currents, I_{dref}^s and I_{qref}^s are the reference currents.

Applying the same design principles as for variable structure control, the controller trajectories are constrained to evolve after a finite time on the sliding manifold by the use of a discontinuous output injection signal. Hence, the sliding motion provides

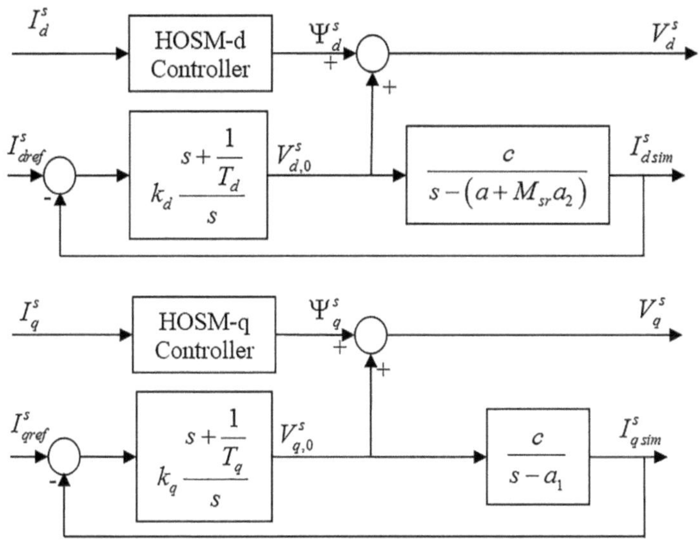

Fig. 15.13 Proposed HOSM decoupled current control scheme

a cancellation of extra terms of the system states. Dynamics of sliding surfaces, employing the modified HOSM algorithm (Moreno et al. 2008), can be obtained for the d-axis and q-axis, respectively, as:

$$\begin{cases} \dot{S}_d = a_1 I_d^s - a_2 b_2 \phi_d^r + \Gamma_d\left(\omega, I_q^s, \phi_d^r\right) + cV_d^s(t) - \frac{d}{dt}\left(I_{dref}^s\right) \\ \dot{S}_d = a_1 I_d^s - a_1 I_{dref}^s - \frac{d}{dt}\left(I_{dref}^s\right) - a_2 b_2 \phi_d^r + \Gamma_d\left(\omega, I_q^s, \phi_d^r\right) + cV_d^s(t) \\ V_d^s(t) = -K_1 \Phi_1(S_d) - K_2 \int \Phi_2(S_d)\,dt - K_3 S_d \\ \Phi_1(S_d) = |S_d|^{\frac{1}{2}} \operatorname{sign}(S_d) \\ \Phi_2(S_d) = \operatorname{sign}(S_d) \end{cases} \tag{15.32}$$

$$\begin{cases} \dot{S}_q = a_1 I_q^s + \Gamma_q\left(\omega, I_d^s, I_q^s, \phi_d^r\right) + cV_q^s(t) - \frac{d}{dt}\left(I_{qref}^s\right) \\ \dot{S}_q^s = a_1 I_q^s + a_1 I_{qref}^s - \frac{d}{dt}\left(I_{qref}^s\right) + \Gamma_q\left(\omega, I_d^s, I_q^s, \phi_d^r\right) + cV_q^s(t) \\ V_q^s(t) = -K_1 \Phi_1\left(S_q\right) - K_2 \int \Phi_2\left(S_q\right) dt - K_3 S_q \\ \Phi_1\left(S_q\right) = |S_q|^{\frac{1}{2}} \operatorname{sign}\left(S_q\right) \\ \Phi_2\left(S_q\right) = \operatorname{sign}\left(S_q\right) \end{cases} \tag{15.33}$$

and define terms of $\dot{S}_d$ and $\dot{S}_q$ as:

$$\begin{cases} \Omega_d = -a_1 I_{dref}^s - \frac{d}{dt}\left(I_{dref}^s\right) - a_2 b_2 \phi_d^r + \Gamma_d\left(\omega, I_q^s, \phi_d^r\right) \\ \Omega_q = a_1 I_{qref}^s - \frac{d}{dt}\left(I_{qref}^s\right) + \Gamma_q\left(\omega, I_d^s, I_q^s, \phi_d^r\right) \end{cases} \tag{15.34}$$

where $V_d^s(t)$ and $V_q^s(t)$ are the sliding mode terms, K_1, K_2 and K_3 are properly designed positive constants.

In the sliding mode, $S_d = \dot{S}_d = 0$ and $S_q = \dot{S}_q = 0$. The reduced order dynamics of the system (15.28) can be obtained as

$$\begin{cases} 0 = -a_2 b_2 \phi_d^r + \Omega_d - c\left(K_2 \int \Phi_2(S_d)\, dt\right) \\ 0 = \Omega_q - c\left(K_2 \int \Phi_2(S_q)\, dt\right) \end{cases} \tag{15.35}$$

The estimated robust terms correspond to the continuous compensation voltages (V_{dcomp}^s and V_{qcomp}^s) and are given by

$$\begin{cases} \psi_d = \frac{1}{c}\left(a_1 I_{dref}^s - \frac{d}{dt} I_{dref}^s + a_2 b_2 \phi_d^r\right) - K_2\left(\int \Phi_2(S_d)\, dt\right) \\ \quad = -\frac{1}{c}\left(N_p \omega I_q^s + b_1 \frac{(I_q^s)^2}{\phi_d^r}\right) \\ \psi_q = \frac{1}{c}\left(a_1 I_{qref}^s + \frac{d}{dt} I_{qref}^s\right) - K_2\left(\int \Phi_2(S_q)\, dt\right) \\ \quad = \frac{1}{c}\left(a_2 N_p \omega \phi_d^r + N_p \omega I_d^s + b_1 \frac{I_d^s I_q^s}{\phi_d^r}\right) \end{cases} \tag{15.36}$$

As a summarize, the control voltages (V_d^s, V_q^s) in (15.28) are generated by summing two components as

$$\begin{cases} V_d^s = V_{d0}^s + \psi_d \\ V_q^s = V_{q0}^s + \psi_q \end{cases} \tag{15.37}$$

The terms (V_{d0}^s, V_{q0}^s) are the nominal inputs while $(d - q)$ are the continuous estimations of the unknown compensation voltages.

15.4.2.1 Convergence in the Sliding Mode

The HOSM controller employed in this paper provides finite-time convergence and chattering free estimation of compensation voltages. For illustration, the d-axis controller and prove the finite time convergence to the selected sliding surface S_d is considered. The sliding surface dynamics in (15.31) can be written as:

$$\dot{S} = a_1 I_d^s + a_2 b_2 \phi_d^r + \Omega_d + c V_d^s \tag{15.38}$$

Here $a_2 b_2 \phi_d^r$ is a bounded quantity which comprises of known system parameters and the flux. The boundlessness of this term can be established as $a_2 b_2 \phi_d^r \leq p \phi_d^r$.

The terms of in (15.32) and (15.33) are functions of the physical quantities ω, I_q^s, ϕ_d^r and other known system parameters. Hence, the boundlessness of this function can also be established at least locally as:

$$\left| \Omega_d + a_2 b_2 \frac{d}{dt} \phi_d^r \right| \leq \rho_d \tag{15.39}$$

For some positive constant ρ_d. The above condition (15.41) is not restrictive since ω, I_q^s and ϕ_d^r are continuous on a compact set.

Theorem 1 *With the condition (15.41) satisfied, the controller (15.32) will ensure that the sliding surfaces* (S_d, S_q) *converges to zero in finite time.*

Proof For simplicity of exposition, we only prove the convergence of sliding surface. With the robust terms of the controller defined as in and the perturbations bounded as in (15.41), the convergence of the sliding surface dynamics can be proved similar to Moreno et al. (2008) by consideration of the following Lyapunov function:

$$V(S_d) = \xi^{\top} Q \xi \tag{15.40}$$

where $\xi = \left[\Phi_1 (S_d)\ S_d\ \int \Phi_2 (S_d)\, dt\right]^{\top}$ and $Q = Q^{\top} > 0$ is a positive definite matrix defined as

$$Q = \frac{1}{2}\begin{bmatrix} (4K_2 + K_1^2) & K_1K_3 & -K_1 \\ K_1K_3 & (1 + K_3^2) & -K_3 \\ -K_1 & -K_3 & 2 \end{bmatrix} \tag{15.41}$$

The considered Lyapunov function satisfies

$$\lambda_{min} ||\xi||_2^2 \leq V (S_d) \leq \lambda_{max} ||\xi||_2^2 \tag{15.42}$$

where λ_{max} represents the maximum eigen value and λ_{min} represents the minimum singular value. The time derivative along the trajectories of the system can be obtained as,

$$\dot{V} (S_d) = 2\xi^T Q \dot{\xi} \tag{15.43}$$

On further simplification, $\dot{V} (S_d)$ can be now written as:

$$\dot{V} (S_d) = -\frac{1}{|S_d|}\xi^T Q_1 \xi - \xi^T Q_2 \xi \tag{15.44}$$

Where

$$Q_1 = \begin{bmatrix} q_{11} & 0 & -\frac{K_1^2}{2} \\ 0 & q_{12} & -\frac{3K_1K_2}{2} \\ -\frac{K_1^2}{2} & -\frac{3K_1K_3}{2} & \frac{K_1}{2} \end{bmatrix}; Q_2 = \begin{bmatrix} q_{21} & 0 & -\frac{K_1\gamma}{4} \\ 0 & q_{22} & q_{23} \\ -\frac{K_1\gamma}{4} & q_{23} & K_3 \end{bmatrix}$$

with

$$\begin{cases} q_{11} = (2K_2 + K_1^2)\frac{K_1}{2} - K_3\rho_d \\ q_{12} = (2 + 5K_3^2)\frac{K_1}{2} - \frac{3K_1K_3\gamma}{2} \\ q_{21} = (K_2 + 2K_1^2) K_3 - K_3\rho_d - \gamma\left(\frac{K_1^2}{2} + 2K_2\right) \\ q_{22} = K_3 (K_3^2 + 1) - \gamma (K_3 + 2) \\ q_{23} = K_3^2 - \frac{\gamma K_3}{2} \end{cases}$$

Similar to the arguments in Moreno et al. (2008), it can be shown that if the gains K_1, K_2 and K_3 satisfy the following inequalities.

$$\begin{cases} K_1 > \left(\dfrac{2K_2\gamma + K_3\rho_d - K_2K_3}{2K_3 - \frac{\gamma}{2}}\right)^{0.5} \\ K_2 > \dfrac{2\rho_d - K_1^2}{2} \\ K_3 > \dfrac{\gamma\left(\frac{K_1^2}{2}\right) + 2K_2}{K_2 + 2K_1^2 - \rho_d} \end{cases} \tag{15.45}$$

Then, it can be shown from Eq. (15.46) that

$$\dot{V}(S_d) = -\frac{1}{|S_d|^{0.5}}\lambda_{min}Q_1||\xi||^2 - \lambda_{min}Q_2||\xi||^2 \tag{15.46}$$

$$|S_d|^{0.5} \leq ||\xi|| \leq \frac{V^{0.5}(S_d)}{\lambda_{min}^{0.5}Q}$$

It can be thus show that

$$\dot{V}(S_d) \leq \beta_1 V^{0.5}(S_d) - \beta_2 V(S_d) \tag{15.47}$$

where

$$\beta_1 = \frac{\lambda_{min}^{0.5}(Q)\,\lambda_{min}(Q_1)}{\lambda_{max}(Q)};\ \beta_2 = \frac{\lambda_{min}(Q_2)}{\lambda_{max}(Q)}$$

with the proper selection of the gains, K_i, $i = 1; 2; 3$; such that (15.47) is satisfied ensures that $\dot{V}(S_d)$ is negative definite. Thus the sliding surface is reached in finite time and maintained thereafter. Similarly, the convergence analysis for the q-axis current can be established.

As a conclusion, the HOSM current controllers scheme is developed in order to decouple the dynamics of the $(d - q)$ currents. It consists on a PI and an HOSM controller on each axis. The nominal PI control acts on the ideal plant dynamics, devoid of coupling terms. The HOSM current control is an attractive alternative to classic decoupled current control because it does not use the speed estimator. The HOSM play the same role as the compensation voltages of the classic method.

15.4.2.2 Closed Loop HOSM Decouple Current Control Simulations and Analysis

Simulations are performed with same parameters of IM as Sect. 15.2. The HOSM decoupled current control considers that the speed ω and the rotor flux ϕ_d^r are accessible for feedback and IM parameters are known exactly.

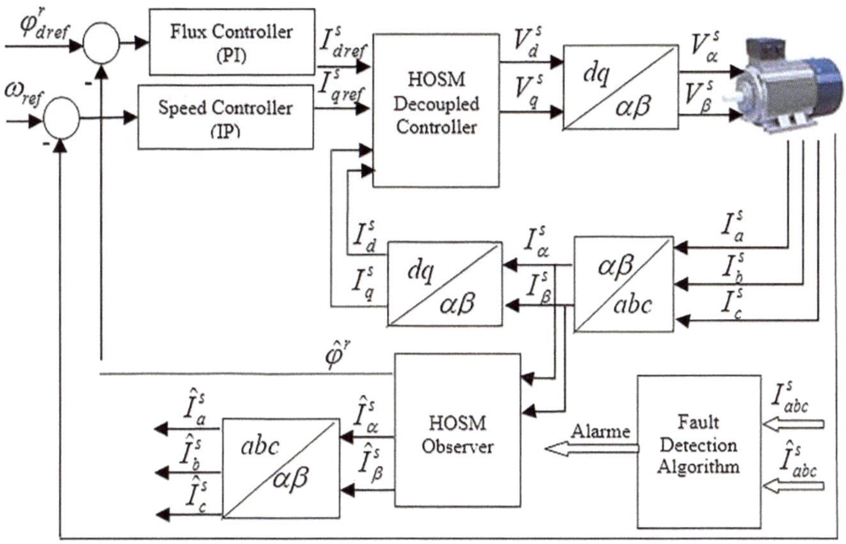

Fig. 15.14 Proposed HOSM decoupled current control scheme

Table 15.2 HOSM decoupled current control and observer parameters

Control parameters			Observer parameters	
K_1	K_2	K_3	λ_m	λ_M
2	200	50	0.1	3

Besides, the rotor flux is inaccessible for measurement. To avoid this problem, the HOSM observer established in Sect. 15.3 is taken on. The developed control law supposes that only three-phase stator current and voltage are available. All the observer initial conditions are set to zero.

Figure 15.14 illustrates the block diagram of HOSM decoupled current control scheme using HOSM observer. The rotor speed ω and flux ϕ_d^r are compared with its reference ($\omega_{ref} = 314$ rd/s and $\phi_{dref}^r = 0.9$ Wb) and adjusted by the IP and PI controller, respectively. The output of this regulators is considered to be the reference quadratic and direct currents, respectively. Two phase I_d^s and I_q^s currents are measured. These measurements feed the Clark transformation module. The outputs of this projection are designated I_α^s and I_β^s. These current components are the inputs of the park transformation that gives the current in the $d - q$ rotating reference frame. The I_d^s and I_q^s components are compared to the references I_{qref}^s and direct I_{dref}^s, respectively. The outputs of the current regulators are I_{dref}^s and I_{qref}^s. This outputs are applied to the inverse park transformation to obtain and, which are the components of the stator vector voltage in the $(\alpha - \beta)$ stationary orthogonal reference frame. These are the inputs of IM. For the implementation of the HOSM decoupled currents controller, the sliding mode gains are selected as in Table 15.2.

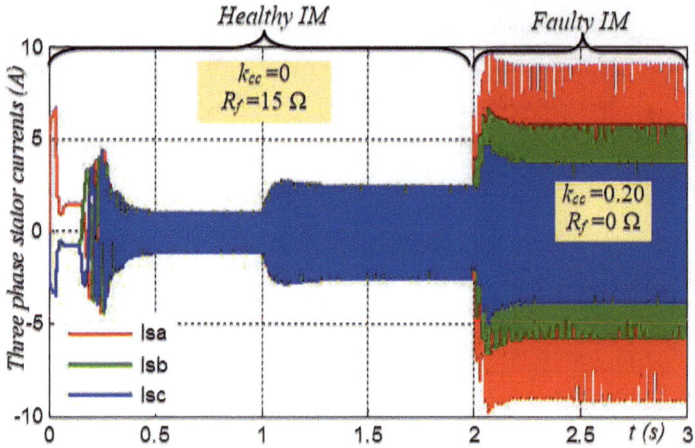

Fig. 15.15 Stator currents (I_a^s, I_b^s, I_c^s) evaluation of HOSM decoupled controller with ITSC fault

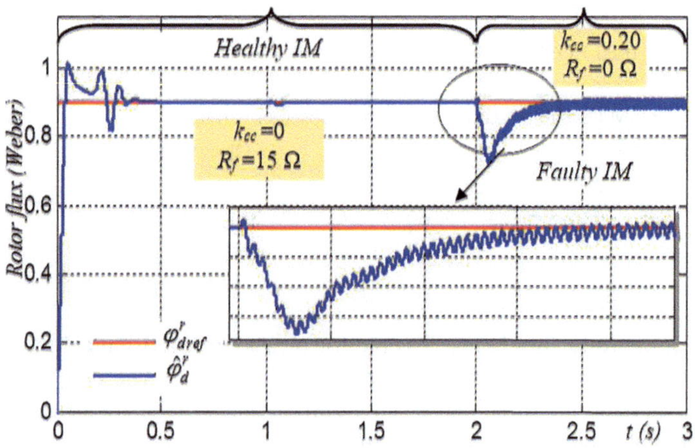

Fig. 15.16 Rotor flux ϕ_d^r evaluation of HOSM decoupled controller with ITSC fault

Figures 15.15, 15.16, 15.17 and 15.18 show responses of HOSM decoupled current control for IM with an ITSC fault in phase a. From these numerical simulations, it can noticed that the proposed controller (nominal control) force the speed and d-axis components of the estimated rotor flux trajectory converge to her desired references with good dynamics and present robustness compared to the load torque disturbance (Figs. 15.16 and 15.17) in healthy and faulty conditions. It is also seen that electromagnetic torque, presented in Fig. 15.18, tracks the load torque. Figure 15.15 shows a balanced three phase stator current with relative phase difference of $\frac{2\pi}{3}$ rad. Therefore, HOSM decoupled current performances with control scheme based on HOSM observer works very well with the present flux estimation technique.

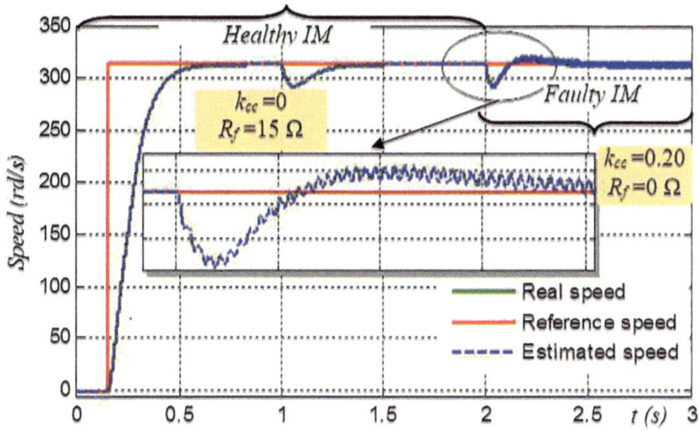

Fig. 15.17 Rotor speed ω evaluation of HOSM decoupled controller with ITSC fault

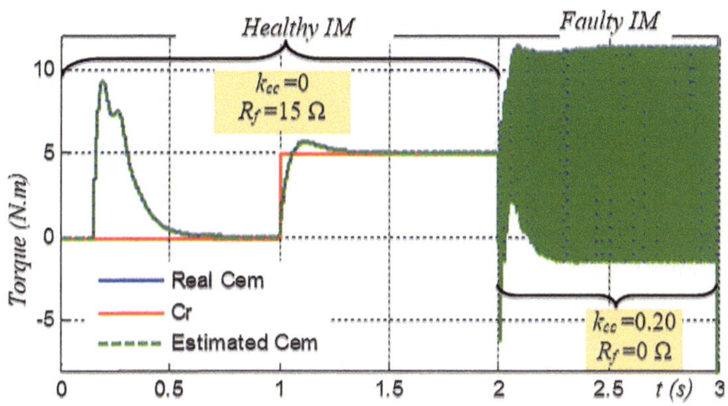

Fig. 15.18 Electromagnetic and load torque (C_{em}, C_r) evaluation of HOSM decoupled controller with ITSC fault

To simulate the sudden occurrence of stator ITSC fault, the insulation resistance R_f become equal to zero and the short circuit level K_{cc} increase from zero to 0.20 at $t = 1.5$ s. When fault occurs the IM, it can be early observed that the three-phase stator current become unbalanced and the current is bigger than the other two. The speed shows up oscillations around of his reference value (Fig. 15.18). The rotor flux deviates from his reference resulting in a coupling between the torque and flux and after few time it tracks its reference with an oscillatory behavior (Fig. 15.17). The electromagnetic torque present a crucial fluctuation which caused by the asymmetry of the stator currents.

15.5 Fault Detection and Residual Generation

In this section, a fault detection based residual generation is presented. This scheme is established using a mathematical model of IM under considered faulty conditions (15.14). Its implementation requires the measurement of three-phase stator current, rotor speed, voltages and the knowledge of nominal IM parameters. The fundamental purpose of a fault detection scheme is to generate an alarm when a fault occurs and to pin-point the source.

Most existing fault detection schemes in the literature are concerned with the design of so-called residuals. These residual signals are used as alarms to indicate the occurrence of a fault, and if properly designed, give information from which the source of the fault may be identified. In analytic redundancy approaches, the residuals are the difference between measurement outputs and reconstructed ones, as shown in Fig. 15.19. Residuals are wanted to be identified such that they are close to zero when the IM behavior in healthy mode, and promptly become nonzero when the associated ITSC fault occurs. However, by appropriate HOSM observer, developed in Sect. 15.3, stator currents can be reconstructed with an estimation error exponentially decaying to zero.

In this context, to obtain residuals, the HOSM observer outputs are compared with those from the healthy and faulty models developed in Sect. 15.2. The out-puts of these models are considered as measures from the real machine. Residual is chosen such an absolute value of the difference between measured variable and his estimate one by HOSMO. The generated residuals in the case of fault absence and parametric variations are established in the aim to define threshold.

Four residuals can be deduced agree with three phase stator current and speed:

$$F_i\left(I_a^s\right) = |I_a^s - \hat{I}_a^s| \tag{15.48}$$

$$F_i\left(I_b^s\right) = |I_b^s - \hat{I}_b^s| \tag{15.49}$$

$$F_i\left(I_c^s\right) = |I_c^s - \hat{I}_c^s| \tag{15.50}$$

The simplest fault detection strategy could be sought as follows (Pilloni et al. 2013).

$$\begin{cases} if\ F_i \leq \epsilon & \text{then IM is healthy} \\ if\ F_i > \epsilon & \text{then IM is faulty} \end{cases} \tag{15.51}$$

A good choice of threshold is used to detect ITSC fault while avoiding false alarms. In fact, only the variables variations due to the ITSC fault are considered and not those due to parametric change.

To illustrate the previously presented results, even in the presence of ITSC fault and step wise variations of the load torque, simulation analysis of the responses of HOSM decoupled current control shown in Sect. 15.3 were carried out. Figures 15.20, 15.21 and 15.22, given by (15.48), (15.49) and (15.50), in the case of ITSC fault in

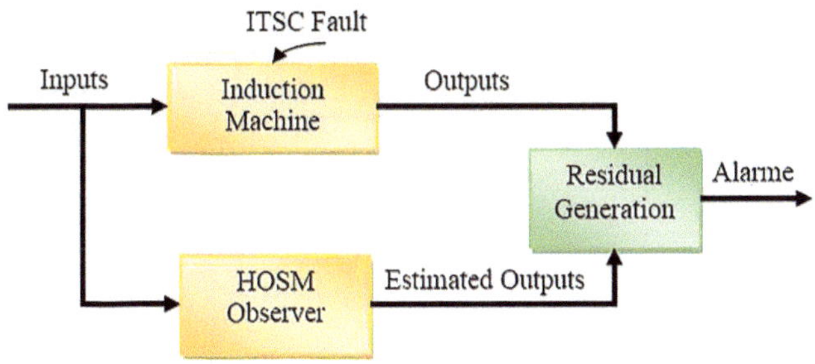

Fig. 15.19 A general model-based fault detection scheme

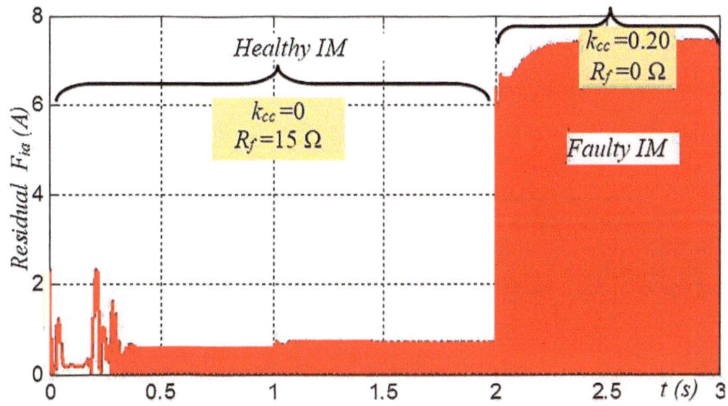

Fig. 15.20 Evaluation of $F_i(I_a^s)$ of faulty IM

current phase a at $t = 1.5$ s. The fault is characterized by fraction of shorted turns of 0.20 of the phase and insulation resistance $R_f = 0$. Any adaptive thresholding technique can be utilized for unsupervised online fault detection and identification. When residuals exceed thresholds, the overtaking is caught by fault indicators that indicate fault presence. The highest value of the stator currents recorded by fault indicator I_a^s, I_b^s and I_c^s corresponds to the stator phase affected by the ITSC. The analyze of residuals shows that all residuals are affected and exceed their maximum value. However, only the first residual (15.48) according to affected current phase a increases more than others residuals. That allows the isolation of affected phase.

As a conclusion, the effectiveness of this proposed observer based fault detection scheme is validated through simulation tests performed by different operation conditions. As it is shown, the fault can be rapidly detected by observing the modulus of three phase stator currents residuals.

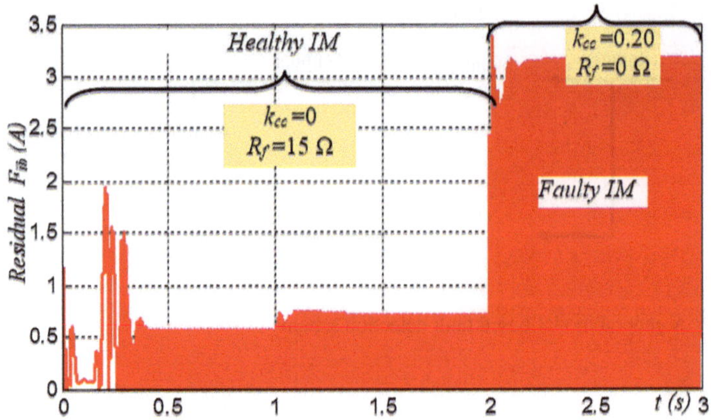

Fig. 15.21 Evaluation of $F_i(I_b^s)$ of faulty IM

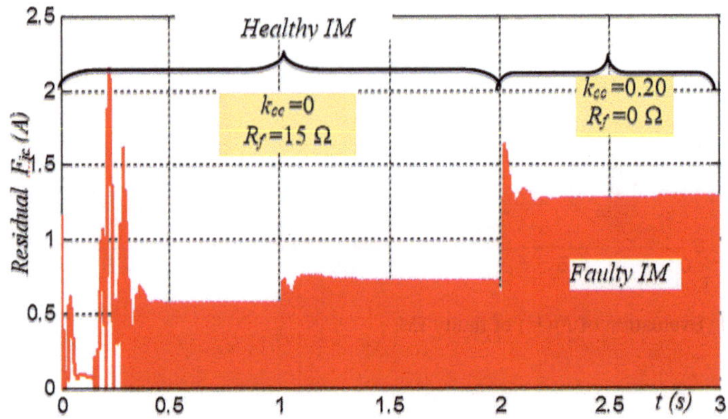

Fig. 15.22 Evaluation of $F_i(I_c^s)$ of faulty IM

15.6 Conclusion

In this chapter, HOSM approach is exploited for ITSC fault detection in IM. To show that impact of such fault on IM behavior, healthy and faulty IM models are established. Different fault scenarios and operation conditions are studied. An observer based method in close loop decoupled current control is presented. The decoupled current control method based on HOSM is designed to ensure the robust tracking performance, to reduce the chattering phenomenon and provide an accurate ITSC fault detection. This fault detection method is based on HOSM observer used to rotor flux estimation and residual generation. Simulation results are presented to highlight the performance and the validity of the developed scheme. Analysis of residuals allows the identification of affected phase and show the sensitivity to ITSC fault occurrence.

References

Ahamed, S. K., Sarkar, A., Mitra, M., & Sengupta, S. (2014). Novel approach for detection of inter-turn short circuit of induction motor's stator winding through envelope analysis. In *International Conference on Electrical and Computer Engineering (ICECE)* (pp. 457–460). IEEE.

Alwi, H., & Edwards, C. (2014). Robust fault reconstruction for linear parameter varying systems using sliding mode observers. *International Journal of Robust and Nonlinear Control*, *24*(14), 1947–1968.

Ashari, A. E. E., Nikoukhah, R., & Campbell, S. L. (2012). Active robust fault detection in closed-loop systems: Quadratic optimization approach. *IEEE Transactions on Automatic Control*, *57*(10), 2532–2544.

Aurora, C., & Ferrara, A. (2007). A sliding mode observer for sensorless induction motor speed regulation. *International Journal of Systems Science*, *38*(11), 913–929.

Benderradji, H., Benamor, A., Chrifi-Alaoui, L., Bussy, P., & Makouf, A. (2012). Second order sliding mode induction motor control with a new Lyapunov approach. In *9th International Multi-Conference on Systems, Signals and Devices (SSD)* (pp. 1–6). IEEE.

Bhavsar, R. C., Patel, R., Bhalja, B., et al. (2014). Condition monitoring of induction motor using artificial neural network. In *Annual International Conference on Emerging Research Areas: Magnetics, Machines and Drives (AICERA/iCMMD)* (pp. 1–6). IEEE.

Cao, W., Mecrow, B. C., Atkinson, G. J., Bennett, J. W., & Atkinson, D. J. (2012). Overview of electric motor technologies used for more electric aircraft (MEA). *IEEE Transactions on Industrial Electronics*, *59*(9), 3523–3531.

Comanescu, M. (2009). An induction-motor speed estimator based on integral sliding-mode current control. *IEEE Transactions on Industrial Electronics*, *56*(9), 3414–3423.

Comanescu, M. (2010). Cascaded EMF and speed sliding mode observer for the nonsalient PMSM. In *IECON 2010-36th Annual Conference on IEEE Industrial Electronics Society* (pp. 792–797). IEEE.

Comanescu, M., Xu, L., & Batzel, T. D. (2008). Decoupled current control of sensorless induction-motor drives by integral sliding mode. *IEEE Transactions on Industrial Electronics*, *55*(11), 3836–3845.

Devi, N. R., Sarma, D., & Rao, P. (2014). Detection and identification of stator inter-turn faults in three-phase induction motor in presence of supply unbalance condition. In *IEEE International Conference on Power Electronics, Drives and Energy Systems (PEDES)* (pp. 1–4). IEEE.

Drif, M., & Cardoso, A. J. M. (2014). Stator fault diagnostics in squirrel cage three-phase induction motor drives using the instantaneous active and reactive power signature analyses. *IEEE Transactions on Industrial Informatics*, *10*(2), 1348–1360.

Edwards, C., Spurgeon, S. K., & Patton, R. J. (2000). Sliding mode observers for fault detection and isolation. *Automatica*, *36*(4), 541–553.

Edwards, C., Alwi, H., & Menon, P. P. (2013). Applications of sliding observers for FDI in aerospace systems. *Advances in Sliding Mode Control* (pp. 341–360). New York: Springer.

Eker, İ. (2012). Second-order sliding mode control with PI sliding surface and experimental application to an electromechanical plant. *Arabian Journal for Science and Engineering*, *37*(7), 1969–1986.

Ferrara, A., & Rubagotti, M. (2009). A sub-optimal second order sliding mode controller for systems with saturating actuators. *IEEE Transactions on Automatic Control*, *54*(5), 1082–1087.

Filippetti, F., Franceschini, G., Tassoni, C., & Vas, P. (2000). Recent developments of induction motor drives fault diagnosis using AI techniques. *IEEE Transactions on Industrial Electronics*, *47*(5), 994–1004.

Frosini, L., Borin, A., Girometta, L., & Venchi, G. (2012). A novel approach to detect short circuits in low voltage induction motor by stray flux measurement. In *XXth International Conference on Electrical Machines (ICEM)* (pp. 1538–1544). IEEE.

Ghazal, M., & Poshtan, J. (2011). Robust stator winding fault detection in induction motors. In *2nd Power Electronics, Drive Systems and Technologies Conference (PEDSTC)* (pp. 163–168). IEEE.

Guezmil, A., Berriri, H., Sakly, A., & Mimouni, M. (2014). Backstepping control for induction machine with high order sliding mode observer and unknown inputs observer: A comparative study. In *International Conference on Electrical Sciences and Technologies in Maghreb (CISTEM)* (pp. 1–8). IEEE.

Gyftakis, K. N., & Kappatou, J. C. (2014). The zero-sequence current as a generalized diagnostic mean in δ-connected three-phase induction motors. *IEEE Transactions on Energy Conversion, 29*(1), 138–148.

Hermans, F. J., & Zarrop, M. (1996). Sliding mode observers for robust sensor monitoring. In *Proceedings of the 13th IFAC World Congress* (pp. 211–216).

Jurkovic, S., Rahman, K. M., Morgante, J. C., & Savagian, P. J. (2015). Induction machine design and analysis for general motors e-assist electrification technology. *IEEE Transactions on Industry Applications, 51*(1), 631–639.

Kaikaa, M. Y., & Hadjami, M. (2014). Effects of the simultaneous presence of static eccentricity and broken rotor bars on the stator current of induction machine. *IEEE Transactions on Industrial Electronics, 61*(5), 2452–2463.

Kommuri, S. K., Rath, J. J., Veluvolu, K. C., & Defoort, M. (2014). An induction motor sensor fault detection and isolation based on higher order sliding mode decoupled current controller. In *European Control Conference (ECC)* (pp. 2945–2950). IEEE.

Kommuri, S. K., Rath, J. J., Veluvolu, K. C., Defoort, M., & Soh, Y. C. (2015). Decoupled current control and sensor fault detection with second-order sliding mode for induction motor. *IET Control Theory and Applications, 9*(4), 608–617.

Levant, A. (1993). Sliding order and sliding accuracy in sliding mode control. *International Journal of Control, 58*(6), 1247–1263.

Liu, H., Huang, J., Cheng, M., Hou, Z., & Zhao, L. (2014). Stator inter-turn fault detection for the converter-fed induction motor based on the adjacent-current phase-shift. In *17th International Conference on Electrical Machines and Systems (ICEMS)* (pp. 981–987). IEEE.

Lu, Q., Breikin, T., & Wang, H. (2011). Modelling and fault diagnosis of stator inter-turn short circuit in doubly fed induction generators. In *Proceedings of the IFAC. Milan, Italy.*

Mekki, H., Benzineb, O., Boukhetala, D., Tadjine, M., & Benbouzid, M. (2015). Sliding mode based fault detection, reconstruction and fault tolerant control scheme for motor systems. *ISA Transactions.*

Melero, M., Cabanas, M., Rojas, C., Orcajo, G., Cano, J., & Solares, J. (2003). Study of an induction motor working under stator winding inter-turn short circuit condition. In *4th IEEE International Symposium on Diagnostics for Electric Machines, Power Electronics and Drives, 2003. SDEMPED* (pp. 52–57). IEEE.

Moreno, J., Osorio, M., et al. (2008). A Lyapunov approach to second-order sliding mode controllers and observers. In *47th IEEE Conference on Decision and Control, 2008. CDC* (pp. 2856–2861). IEEE.

Nandi, S., Toliyat, H., Li, X., et al. (2005). Condition monitoring and fault diagnosis of electrical motors—A review. *IEEE Transactions on Energy Conversion, 20*(4), 719–729.

Nohra, C. (2013). Online stator and rotor fault diagnosis in induction machines by H_∞ observer and sliding mode estimator. In *25th Chinese Control and Decision Conference (CCDC)* (pp. 3269–3274). IEEE.

Pilloni, A., Pisano, A., Riera-Guasp, M., Puche-Panadero, R., & Pineda-Sanchez, M. (2013). Fault detection in induction motors. *AC electric motors control: Advanced design techniques and applications* (pp. 275–309).

Riera-Guasp, M., Antonino-Daviu, J., Capolino, G.-A., et al. (2015). Advances in electrical machine, power electronic, and drive condition monitoring and fault detection: State of the art. *IEEE Transactions on Industrial Electronics, 62*(3), 1746–1759.

Sebastián, R., & Peña-Alzola, R. (2015). Control and simulation of a flywheel energy storage for a wind diesel power system. *International Journal of Electrical Power and Energy Systems*, *64*, 1049–1056.

Sellami, T., Berriri, H., Jelassi, S., & Mimouni, M. F. (2013). Sliding mode observer-based fault-detection of inter-turn short-circuit in induction motor. In *14th International Conference on Sciences and Techniques of Automatic Control and Computer Engineering (STA)* (pp. 524–529). IEEE.

Seshadrinath, J., Singh, B., & Panigrahi, B. (2014a). Vibration analysis based interturn fault diagnosis in induction machines. *IEEE Transactions on Industrial Informatics*, *10*(1), 340–350.

Seshadrinath, J., Singh, B., & Panigrahi, B. K. (2014b). Incipient interturn fault diagnosis in induction machines using an analytic wavelet-based optimized Bayesian inference. *IEEE Transactions on Neural Networks and Learning Systems*, *25*(5), 990–1001.

Shtessel, Y., Edwards, C., Fridman, L., & Levant, A. (2014a). Second-order sliding mode controllers and differentiators. *Sliding mode control and observation* (pp. 143–182). New York: Springer.

Shtessel, Y., Edwards, C., Fridman, L., & Levant, A. (2014b). *Sliding mode control and observation*. New York: Springer.

Siddique, A., Yadava, G., & Singh, B. (2005). A review of stator fault monitoring techniques of induction motors. *IEEE Transactions on Energy Conversion*, *20*(1), 106–114.

Sridharan, S., & Krein, P. T. (2014). Induction motor drive design for traction application based on drive-cycle energy minimization. In *Twenty-Ninth Annual IEEE Applied Power Electronics Conference and Exposition (APEC)* (pp. 1517–1521). IEEE.

Toumi, D., Boucherit, M., & Tadjine, M. (2012). Observer-based fault diagnosis and field oriented fault tolerant control of induction motor with stator inter-turn fault. *Archives of Electrical Engineering*, *61*(2), 165–188.

Verma, A., Sarangi, S., & Kolekar, M. (2014). Stator winding fault prediction of induction motors using multiscale entropy and grey fuzzy optimization methods. *Computers and Electrical Engineering*, *40*(7), 2246–2258.

Zhao, L., Huang, J., Liu, H., Li, B., & Kong, W. (2014). Second-order sliding-mode observer with online parameter identification for sensorless induction motor drives. *IEEE Transactions on Industrial Electronics*, *61*(10), 5280–5289.

Chapter 16
Sliding Mode Control Applied to Electrohydraulic System

Emna Kolsi Gdoura and Moez Feki

Abstract In this chapter a control law is designed to accurately control the rod position of hydraulic servo system. In fact, due to its having a nonlinear model, the hydraulic servo system is not accurately stabilized by a proportional controller and suffers from wind up phenomenon when applying the PI controller. To overcome the problems encountered by the action of these linear controllers, we propose two ways. The first, consists in adding a simple anti-windup algorithm to the PI controller while the second consists in designing a sliding mode controller with an integral and realizable reference compensation. Both methods are applied to a symmetrical and non symmetrical pistons and have showed an accurate position control in addition to having a short settling time. Finally, to circumvent the problem of unmeasurable pressures, we proposed a sliding mode observer to estimate them from the position of the piston. The efficiency of the proposed scheme is illustrated using numerical simulations.

Keywords Hydraulic servo system · PI control · Anti-windup · Sliding mode control · Actuator saturation · Realizable reference

16.1 Introduction

Electrohydraulic systems generate hydraulic power in response to the electrical signal inputs. Their scope of application extends to the field of precision control of systems such as robotics, heavy manipulators, etc. Therefore, they have been widely used in industrial applications. However, this type of systems has a nonlinear behavior due to the pressure-flow characteristics inside the control valves and the friction that acts on the actuator and moving parts of the valves which they make the design of

E. Kolsi Gdoura (✉) · M. Feki
Control and Energy Management Laboratory, National Engineering School of Sfax, University of Sfax, Sfax, Tunisia
e-mail: kolsiemna@gmail.com

M. Feki
e-mail: moez.feki@enig.rnu.tn

N. Derbel et al. (eds.), *Applications of Sliding Mode Control*,
Studies in Systems, Decision and Control 79, DOI 10.1007/978-981-10-2374-3_16

a controller with state feedback a hard task. In a wide range of applications where electrohydraulic systems are involved, the output force is required to follow a given specific reference or the piston is required to achieve a given specific position. This requires the design of a state feedback control law.

The control of electrohydraulic systems was treated by several researchers and it reveals to be a hard task. To solve this problem, different methods of control have been used such as PID controllers (Alleyne et al. 1998) and the input/output linearizing controller (Feki 2001). But these methods do not take into account the uncertainties and lead to controllers that sacrifice the performance and the robustness in favor of simplicity. Therefore, several types of variable structure controls were applied to the electrohydraulic systems.

Recently, nonlinear control methods and intelligent methods have been used to control hydraulic systems taking into account the presence of unmodeled dynamics, parameter uncertainties and external disturbances. Adaptive controllers have been proposed in (Ahn and Dinh 2009; Knohl and Unbehauen 2000; Yao et al. 2000) where the uncertainty parameters are linear. To handle the case where the unknown parameters are nonlinear as the volume of the cylinder, a controller using the adaptive control theory and the backstepping technique has been developed (Guan and Pan 2008; Sirouspour and Salcudean 2000). In order to improve the control quality of the loading system while eliminating or reducing the disturbance, a prediction model "grey" associated with a fuzzy PID controller is proposed to achieve a predetermined force (Truong and Ahn 2009). PID controllers have also been improved using fuzzy methods (Mihajlov et al. 2002), genetic algorithm (Aly 2011) and zero-pole placement method (Jian-jun et al. 2012) to obtain accurate position control.

The variable structure control (VSC) is another controller that has been widely used to deal with systems with uncertain parameters (Kolsi-Gdoura et al. 2013; Khebbache and Tadjine 2013; Rossomando 2014). In (Hwang and Lan 1994), a time varying switching gain, a second order relation between the sliding surface and uncertainties, and a boundary layer of the sliding surface, is used to deal with the position control. In (Chen et al. 2005), the position control was processed using (VSC) with varying boundary layers to improve tracking performance by reducing the width of the boundedness and reducing the chattering effect by increasing the width the boundedness.

In this chapter, we will design a simple controller that can reach the reference position in the presence of uncertain parameters, disturbance and saturation of the actuator. To achieve this aim, we first present the effects of proportional control (P) and a proportional integral control (PI). Subsequently, we present a sliding mode controller enhanced by an integral surface and a realizable reference to reduce the reaching time to the sliding surface and to obtain a shorter settling time. The different types of control will be applied on two different mathematical models of the electrohydraulic system the first concerns a symmetrical piston and the second model concerns a non-symmetrical piston. Numerical simulation results are presented to illustrate the effectiveness of the proposed control methods.

16.2 Model of the Electrohydraulic System

Figure 16.1 presents a hydraulic servo system.

If we consider $S_1 \neq S_2$, Then we obtain the dynamical system represented by the differential equations (16.1) (Feki 2001).

$$\dot{P}_1 = \frac{B}{V_0 + S_1 y}\Big(Q_1 - S_1 v\Big), \tag{16.1}$$

$$\dot{P}_2 = \frac{B}{V_0 - S_2 y}\Big(Q_2 + S_2 v\Big), \tag{16.2}$$

$$\dot{v} = \frac{1}{m + m_0}(S_1 P_1 - S_2 P_2 - bv - k_l y), \tag{16.3}$$

$$\dot{y} = v, \tag{16.4}$$

The first two equations of the adopted model is obtained using the Bernoulli law of the fluid flow (Merritt 1967) and the last two equations are obtained using Newton's second law of motion. where the flow rates Q_1 and Q_2 are:

$$Q_1 = \begin{cases} ku\sqrt{P_s - P_1} + \frac{\alpha}{1+\gamma u}(P_s + P_r - 2P_1) & \text{if } u \geq 0 \\ ku\sqrt{P_1 - P_r} + \frac{\alpha}{1-\gamma u}(P_s + P_r - 2P_1) & \text{if } u < 0 \end{cases} \tag{16.5}$$

$$Q_2 = \begin{cases} -ku\sqrt{P_2 - P_r} + \frac{\alpha}{1+\gamma u}(P_s + P_r - 2P_1) & \text{if } u \geq 0 \\ -ku\sqrt{P_s - P_2} + \frac{\alpha}{1-\gamma u}(P_s + P_r - 2P_1), & \text{if } u < 0 \end{cases} \tag{16.6}$$

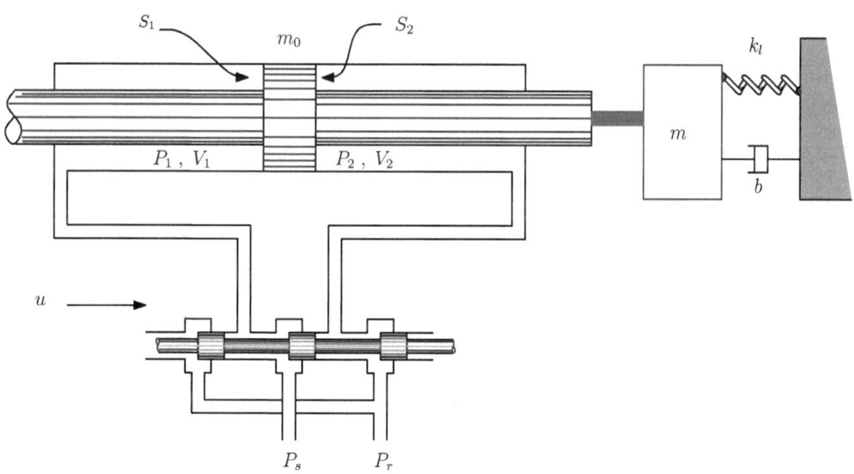

Fig. 16.1 Hydraulic servo system

P_1 and P_2 present respectively the pressure inside the first and the second chamber of the cylinder, v and y present respectively the velocity and the position of the rod. m is the mass of the load and m_0 is the mass of the rod, V_1 and V_2 present respectively the volume of the first and the second chambers of the cylinder, S_1 and S_2 present respectively the section surface of the first and the second sides of the piston, P_r denotes the return pressure (atmospheric pressure) and P_s denotes the supply pressure(pressure of the pump). b is the friction coefficient and k_l is the spring stiffness constant. B denotes the effective bulk modulus of the fluid and u represents the control signal. The value of the various parameters of system are given in Appendix A.

If we consider a symmetrical piston i.e. as $S_1 = S_2$ (noted S), then we can define the variable $\Delta P = P_1 - P_2$ so $\Delta\dot{P} = \dot{P}_1 - \dot{P}_2$. After simplification we obtain the model represented by the system of equations (16.7)

$$\Delta\dot{P} = \frac{4B}{V_t}\Big(ku\sqrt{P_d - \text{sign}(u)\Delta P} - \frac{\alpha}{1+\gamma|u|}\Delta P - Sv\Big), \tag{16.7}$$

$$\dot{v} = \frac{1}{m+m_0}(S\Delta P - bv - k_l y), \tag{16.8}$$

$$\dot{y} = v, \tag{16.9}$$

where $P_d = P_s - P_r$. From this model, we can deduce that the system is highly non-linear with respect to the state vector and also not affine with respect to the control signal u. Thus, the design of a control law is not a simple task.

16.3 Position Control of a Symmetric System

In this section, we will adopt the model of the electrohydraulic system expressed by the Eq. (16.10):

$$\dot{x}_1 = \frac{4B}{V_t}\Big(ku\sqrt{P_d - \text{sign}(u)x_1} - \frac{\alpha}{1+\gamma|u|}x_1 - Sx_2\Big), \tag{16.10}$$

$$\dot{x}_2 = \frac{1}{m+m_0}(Sx_1 - bx_2 - k_l x_3), \tag{16.11}$$

$$\dot{x}_3 = x_2 + d, \tag{16.12}$$

where x_1 is the pressure difference $x_1 = P_1 - P_2$. We suppose that:

$$|P_1 - P_2| = |x_1| < P_d = P_s - P_r.$$

x_2 and x_3 represent respectively the velocity and the displacement of the piston.

We consider that the system is under the effect of several mismatched perturbations. Especially, the spring constant is used with an uncertainty of up to 20 % of its nominal value and the piston velocity undergoes a constant perturbation d. Finally, due to practical limitation, the input signal $u(t)$ which is simply the current injected to the actuator is restricted to a maximum value $|u(t)| \leq u_{\max}$.

It is always interesting to start by checking the effect of proportional and PI controllers. Indeed, a command that does not show significant advantages compared to PI will always be excluded under application. In the following, the aim of the control is to move the piston from its central position considered as the origin to a reference position noted x_{ref}.

16.3.1 Proportional Control

The control signal of a proportional controller is given by:

$$u_p = k_0(x_{3ref} - x_3) \tag{16.13}$$

where x_{3ref} is the reference position to be attained and $k_0 = 0.05$. The application of this control law leads to the following result:

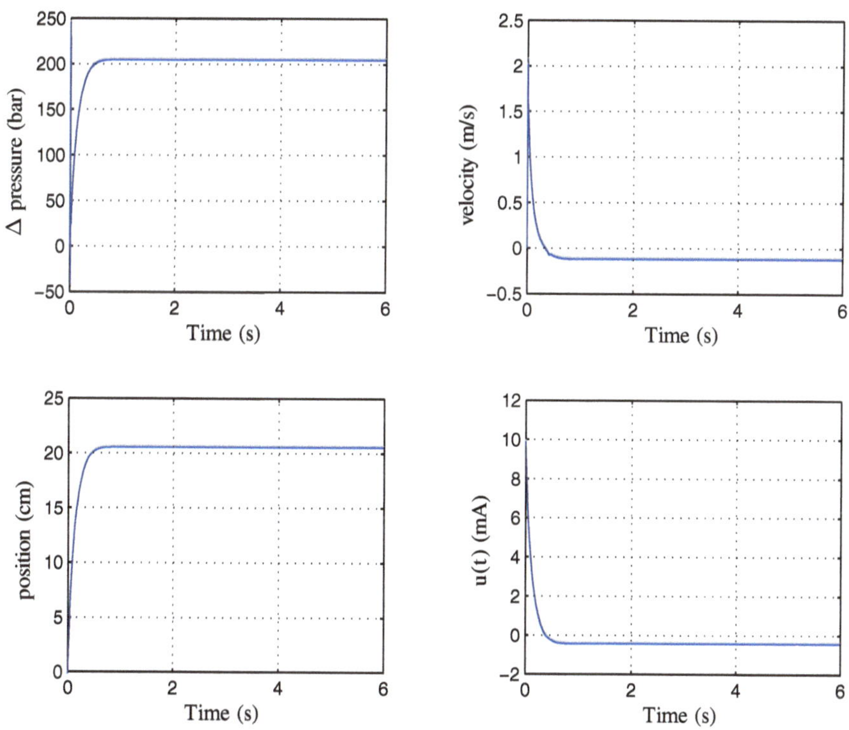

Fig. 16.2 Position control of the electrohydraulic system under the proportional controller: SSE = 3.55 %, $T_{s5\,\%} = 0.37$ s

From Fig. 16.2, we notice that the proportional controller yields to steady state error of 3.55 % and a 5 % settling time of $T_{s5\,\%} = 0.37$ s.

16.3.2 Proportional Integral Control

The proportional integral control signal is expressed as follows

$$u_{pi} = k_0(x_{3ref} - x_3) + k_i \int_0^t (x_{3ref} - x_3)dt \tag{16.14}$$

By applying this control law to our model with $k_0 = 0.05$ and $k_i = 0.05$, we obtain the results shown in Fig. 16.3.

From these figures, it can be deduced that the proportional integral controller eliminates the steady state error but the settling time becomes too long due to wind-up phenomenon, in fact the settling time becomes 2.31 s.

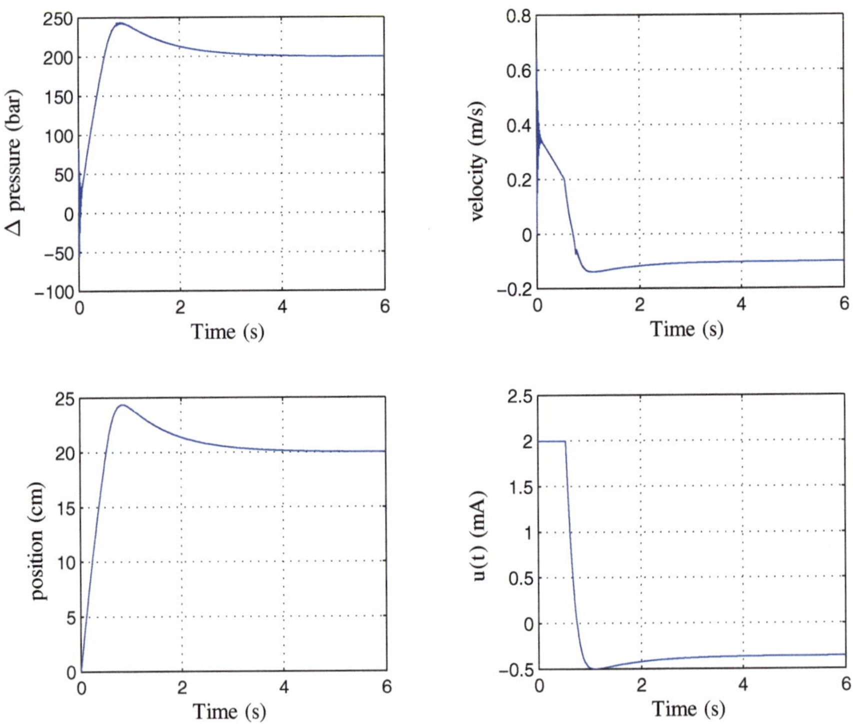

Fig. 16.3 Position control of the electrohydraulic system under the PI controller: SSE = 0, $T_{s5\,\%} = 2.31$ s

To improve these results and achieve good performances, we will apply in the first part the anti wind-up approach to this model and in the second part we will deal with the sliding mode method.

16.3.3 Anti Wind-Up Approach

From the previous section, we remark that when applying the proportional integral controller to our system the output takes a long time to eliminate the steady state error, this is due the wind-up phenomenon of the integrator that maintains the integration of tracking error especially if the input is saturated. So, to overcome this problem we think about the use of an anti wind-up approach (Fig. 16.4).

In this section we will use an anti wind-up with back calculation:

- The anti-windup scheme has no effect when the actuator is not saturating ($e_t(t) = 0$)
- The time constant T_t determines how quickly the integrator of the PI controller is reset
- If the actual output $u(t)$ of the actuator in not measurable, we can use a mathematical model of the actuator. For instance $e_t(t) = v(t) - \text{sat}(v(t))$.

The application of this type of anti wind-up leads to the results in Fig. 16.5.

We can remark from Fig. 16.5 that the anti wind-up reduces the effect of the integral and so we obtain a reduced settling time compared to PI controller without anti wind-up.

Now, in the following section we will use another type of controller which is the sliding mode controller that can help to obtain a good performance of the electrohydraulic system.

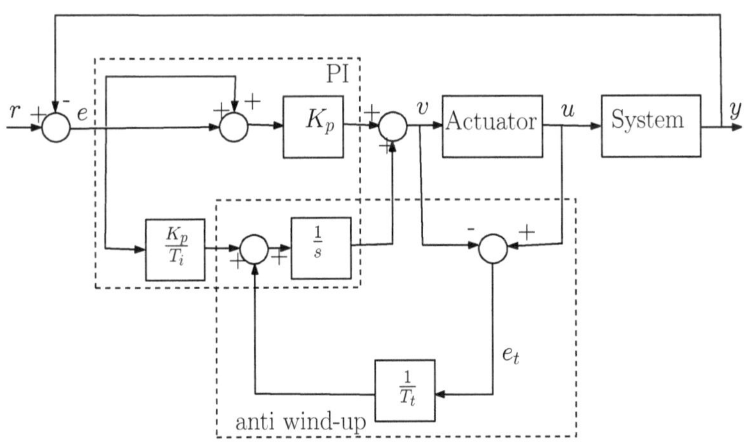

Fig. 16.4 Anti wind-up using back calculation

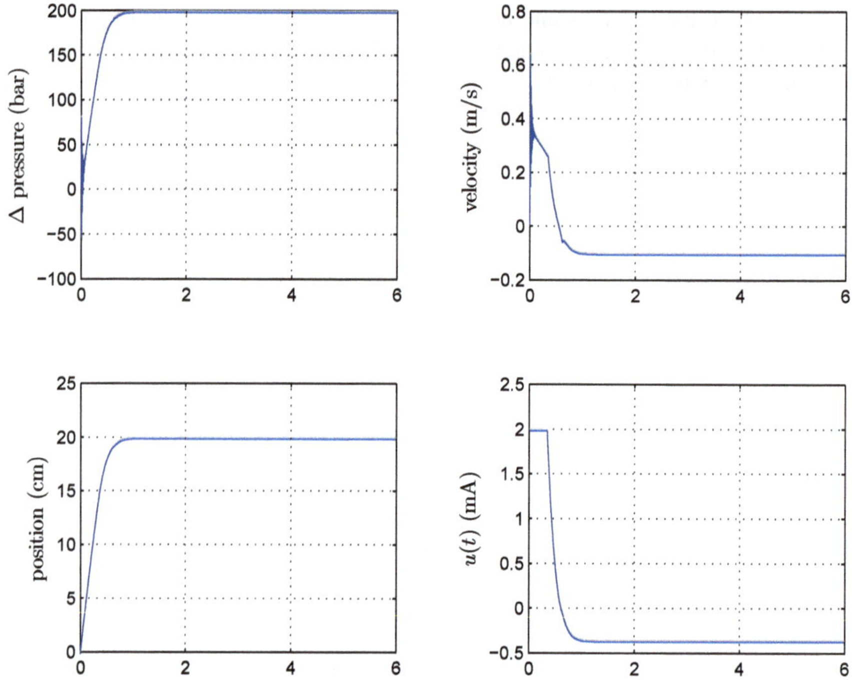

Fig. 16.5 Position control of the hydraulic servo system under the PI controller with anti wind-up: SSE = 0, $T_{s5\,\%} = 0.59\,\text{s}$

16.3.4 Sliding Mode Control

In this section, we suggest a sliding mode controller by conjunction of an integral controller that helps to attenuate the constant mismatched perturbation effect. We propose the expression of a simple bang-bang control law given by:

$$u = -u_{\max}\text{sign}(\sigma(x)) \tag{16.15}$$

where $\sigma(x) = 0$ is the sliding surface. By conjunction of the integral state equation defined as follows:

$$\dot{x_4} = x_{3ref} - x_3 \tag{16.16}$$

we obtain a four dimensional system. So, the sliding surface is chosen as:

$$\sigma(x) = C_4x_4 + C_3(x_{3ref} - x_3) + C_2x_2 + x_1 \tag{16.17}$$

where the constants C_4, C_3 and C_2 are calculated such that the behavior becomes asymptotically stable when the system is confined to the sliding surface $\sigma(x)$.

To restrict the behavior of the system to $\sigma(x)$, we should satisfy the attractivity condition of $\sigma(x)\dot{\sigma}(x) < 0$.

Using (16.15) and (16.17), the attractivity condition becomes:

$$\sigma(x)\dot{\sigma}(x) = \begin{cases} \sigma(x)\Big(C_4(x_{3ref} - x_3) - C_3 x_2 + \frac{C_2}{m+m_0}(Sx_1 - bx_2 - k_l x_3)\Big) \\ +\frac{4B}{V_t}\Big(-ku_{\max}|\sigma(x)|\sqrt{P_d - x_1} - \frac{\alpha}{1+\gamma u_{\max}} x_1\sigma(x) - Sx_2\sigma(x)\Big) \\ \qquad \text{if } \sigma(x) < 0 \\ \sigma(x)\Big(C_4(x_{3ref} - x_3) - C_3 x_2 + \frac{C_2}{m+m_0}(Sx_1 - bx_2 - k_l x_3)\Big) \\ +\frac{4B}{V_t}\Big(-ku_{\max}|\sigma(x)|\sqrt{P_d + x_1} - \frac{\alpha}{1+\gamma u_{\max}} x_1\sigma(x) - Sx_2\sigma(x)\Big) \\ \qquad \text{if } \sigma(x) > 0 \end{cases}$$

Thus, to satisfy the attractivity condition, we should choose C_4, C_3 and C_2 such that:

- if $\sigma(x) < 0$:

$$\left|\frac{\Big(C_4(x_{3ref} - x_3) - C_3 x_2 + C_2\Big(\frac{Sx_1 - bx_2 - k_l x_3}{m+m_0}\Big)\Big) - \frac{4B}{V_t}\Big(\alpha x_1 - Sx_2\Big)}{\frac{4Bk}{V_t}\sqrt{P_d - x_1}}\right| \le u_{\max}$$

- if $\sigma(x) > 0$:

$$\left|\frac{\Big(C_4(x_{3ref} - x_3) - C_3 x_2 + C_2\Big(\frac{Sx_1 - bx_2 - k_l x_3}{m+m_0}\Big)\Big) - \frac{4B}{V_t}\Big(\alpha x_1 - Sx_2\Big)}{\frac{4Bk}{V_t}\sqrt{P_d + x_1}}\right| \le u_{\max}$$

Furthermore, in sliding mode the system is confined to the sliding surface, so we have $\sigma(x) = 0$. Therefore, in closed loop the system can be considered as three dimensional linear system (the nonlinearity exists only in the dynamics of x_1) with the characteristic equation:

$$s^3 + \frac{SC_2 + b}{m + m_0}s^2 - \frac{SC_3 - k_l}{m + m_0}s - \frac{SC_4}{m + m_0} = 0 \qquad (16.18)$$

Eventually, using the method of pole placement and imposing a stable multiple pole at $s = -\lambda$, we can determine the control parameters:

$$C_2 = \frac{3\lambda(m + m_0) - b}{S} \qquad (16.19)$$

$$C_3 = \frac{k_l - 3\lambda^2(m + m_0)}{S} \qquad (16.20)$$

$$C_4 = -\frac{\lambda^3(m + m_0)}{S} \qquad (16.21)$$

Now, by analyzing the behavior of the sliding mode controller, we can note that if the system is initiated at the origin of the state space then $\sigma(0) = C_3 x_{3ref}$. So, if we chooseλ such that $C_3 > 0$ then the control signal u will be negative for all $t \in [0, t_1]$ where t_1 is the necessary time for the system to attain the sliding surface for the first time. During this interval, the actuator moves in the opposite direction of the positive reference. If the sliding condition is satisfied at $t = t_1$ then the system will slide to the reference point. Otherwise, the system will pass through the sliding surface until the sliding condition is satisfied before it slides to the reference point.

If we choose λ such that $\lambda \geq \lambda_0 = \sqrt{\frac{k_l}{3(m+m_0)}}$ then the control signal u will be positive until the system attains the sliding surface. During this interval, the actuator may exceed the reference value which lead to an overshoot, before moving to the reference position while sliding on the surface $\sigma(x)$.

In Fig. 16.6, we show the system behavior with the actuator which moves in the opposite direction to reach $x_3 = -15$ cm before reaching the reference position $x_3 = 20$ cm with a settling time equal to $T_{s5\,\%} = 2.87$ s. Figure 16.7 shows the system behavior with the fastest settling time $T_{s5\,\%} = 0.88$ s obtained by placing the poles at $\lambda = 14.5$. The system has an undershoot of 30 % but no overshoot. By the choice

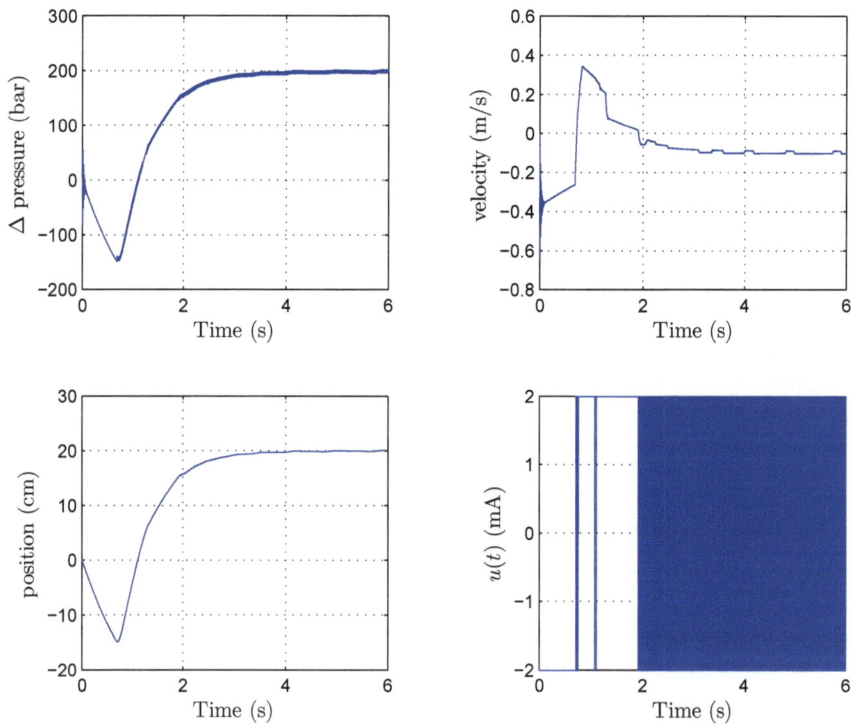

Fig. 16.6 Position control of the electrohydraulic system under the sliding mode controller: $\lambda = 10$, $T_{s5\,\%} = 2.87$ s

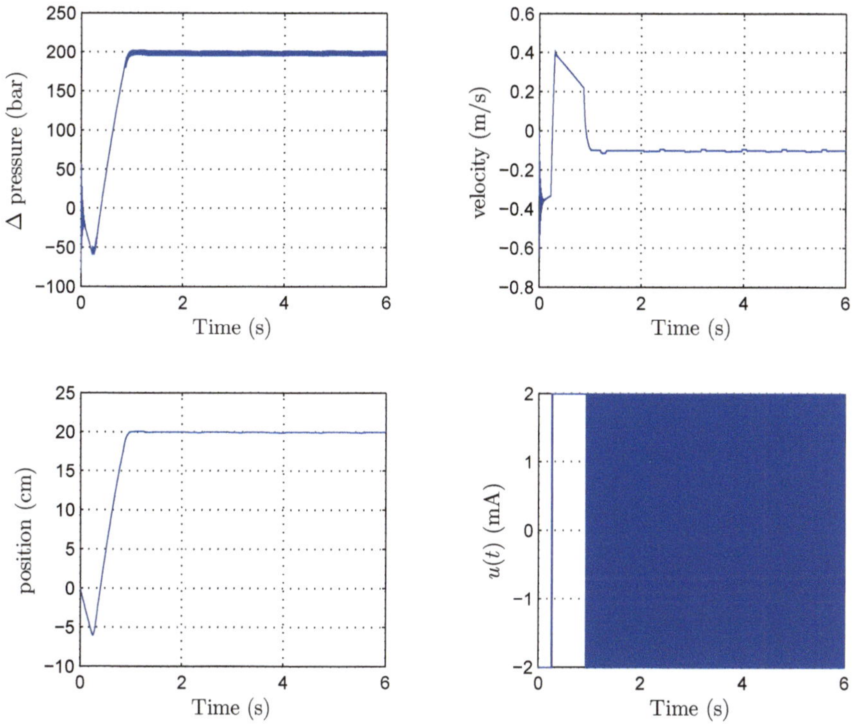

Fig. 16.7 Position control of the electrohydraulic system under the sliding mode controller: $\lambda = 14.5$

of $\lambda = \lambda_0$, the system behavior has a settling time equal to $T_{s5\,\%} = 1.05\,\text{s}$ and an overshoot of 35.75 % as shown in Fig. 16.8.

By analyzing the system behavior under the action of the sliding mode controller, we deduce that the reaching phase is the defective part in the behavior of the system, therefore, it should be improved. To tackle this problem, we propose a modification of the controller in terms of the reference noted as the realizable reference method (Hanus 1987).

16.3.5 Sliding Mode Control with "Realizable Reference"

We consider the sliding surface Eq. (16.17), the method is to determine the realizable reference at each instance, that the surface equation is always satisfied. Next, we modify the expression of the sliding surface such that the system is behaving on the surface.

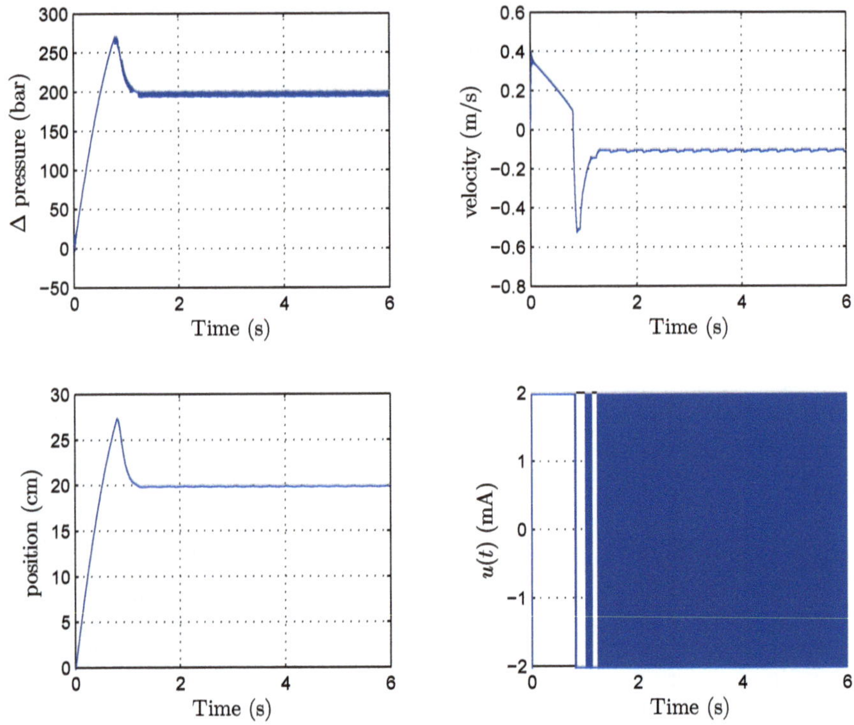

Fig. 16.8 Position control of the electrohydraulic system under the sliding mode controller: $\lambda = \lambda_0$

Plainly, with the realizable reference r_{ref} we have $\sigma(x) = 0$, then substituting this reference in (16.17) we obtain:

$$0 = C_4x_4 + C_3(r_{ref} - x_3) + C_2x_2 + x_1 \tag{16.22}$$

Subtracting (16.22) from (16.17), it follows that:

$$r_{ref} = x_{3ref} - \frac{1}{C_3}\sigma(x) \tag{16.23}$$

Next, substituting the reference signal by the realizable reference in the dynamics of the integral state we obtain:

$$\dot{x_4} = x_{3ref} - \frac{1}{C_3}\sigma(x) - x_3 \tag{16.24}$$

Finally, we get:

$$\dot{\sigma}(x) = \begin{cases} -\frac{C_4}{C_3}\sigma(x) + C_4(x_{3ref} - x_3) - C_3 x_2 + \frac{C_2}{m+m_0}(Sx_1 - bx_2 - k_l x_3) \\ -\frac{4B}{V_t} k u_{\max} \text{sign}(\sigma(x))\sqrt{P_d - x_1} - \frac{4B}{V_t}\left(\frac{\alpha}{1+\gamma u_{\max}} x_1 + Sx_2\right) \\ \hfill \text{if } \sigma(x) < 0 \\ -\frac{C_4}{C_3}\sigma(x) + C_4(x_{3ref} - x_3) - C_3 x_2 + \frac{C_2}{m+m_0}(Sx_1 - bx_2 - k_l x_3) \\ -\frac{4B}{V_t} k u_{\max} \text{sign}(\sigma(x))\sqrt{P_d + x_1} - \frac{4B}{V_t}\left(\frac{\alpha}{1+\gamma u_{\max}} x_1 + Sx_2\right) \\ \hfill \text{if } \sigma(x) > 0 \end{cases}$$

wherefrom one can deduce that if $\frac{C_4}{C_3} > 0$ then the dynamics of $\sigma(x)$ are stable and makes the sliding surface more attractive. Knowing that $C_4 < 0$ so we should choose λ such that the constant C_3 is also negative, equivalently, we should choose $\lambda > \lambda_0$.

The behavior of system under the action of a sliding mode controller with the realizable reference is delineated in Fig. 16.9. The settling time was significantly

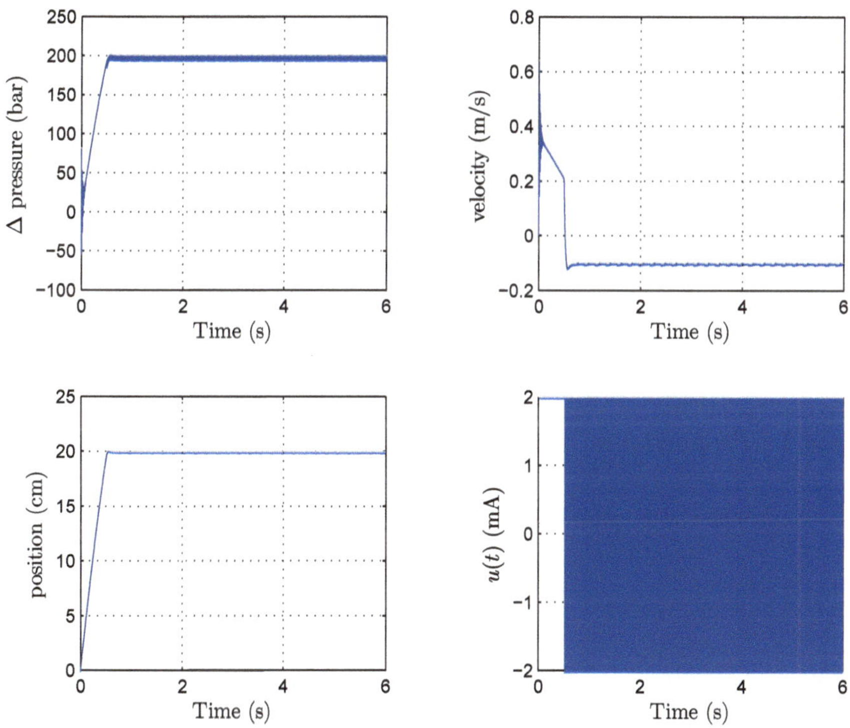

Fig. 16.9 Position control of the electrohydraulic system under the sliding mode controller with realizable reference: $\lambda = 50$

reduced to $T_{s5\,\%} = 0.48$ s while we get a zero steady state error and no overshoot. Clearly, this controller outperforms all the previous designed controllers.

16.4 Position Control of a Non-symmetric System

In this section, we will adopt the model of the electrohydraulic system defined by (16.1).

We will repeat the same steps detailed in the previous sections to a model of a non-symmetric system. In a first step, we will check the effect of the proportional and the proportional integral controller then we will apply the sliding mode control with conjunction of the "realizable reference" method.

16.4.1 Proportional Control

The control signal of the proportional controller is given by:

$$u_p = k_0(y_{ref} - y) \tag{16.25}$$

where y_{ref} is the reference position to be attained and $k_0 = 0.5$. In addition, the actuator constraint is described by the saturation function:

$$u_{sat} = \frac{1}{2}(|u + u_{\max}| - |u - u_{\max}|) \tag{16.26}$$

Applying this control law to our model we obtain the result in Fig. 16.10

From this simulation, we notice that the proportional controller yields to steady state error (SSE) of 9.1 % and a 5 % settling time of $T_{s5\,\%} = 0.6$ s.

16.4.2 Proportional Integral Control

The control law of the proportional integral control is given by (16.27)

$$u_{pi} = k_0(y_{ref} - y) + k_i \int_0^t (y_{ref} - y)dt \tag{16.27}$$

with $k_0 = 0.5$ and $k_i = 0.5$. The application of this control law to the electrohydraulic system leads to the result shown in Fig. 16.11.

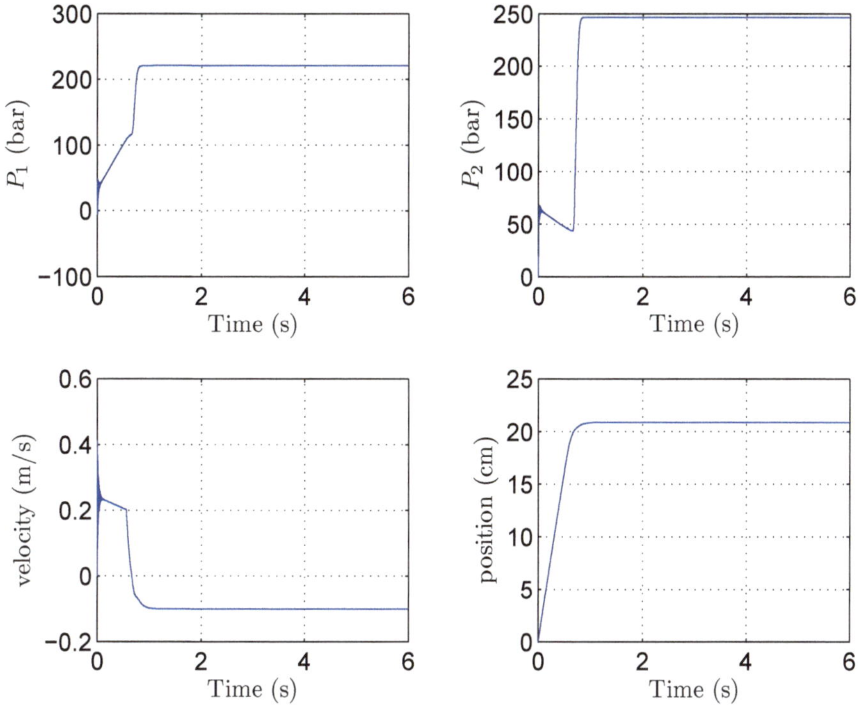

Fig. 16.10 Position control of the hydraulic servo system under the proportional controller: SSE = 9.1 %, $T_{s5\,\%} = 0.6\,\mathrm{s}$

From the simulation the PI controller eliminates the SSE, but the settling time becomes too long due to wind-up phenomenon, in fact the settling time becomes 2.64 s. To improve these results and achieve good performance, we will apply in the first part the anti wind-up approach to this model and in the second part we will deal with the sliding mode approach.

16.4.3 Anti Wind-Up Approach

In this section we will use the same type of anti wind-up applied to the symmetric system. So, applying the back calculation anti wind-up to the non-symmetric system leads to the results delineated in Fig. 16.12.

From Fig. 16.12 we can note that the effect of the integral is reduced and the settling time is reduced to $T_{s5\,\%} = 0.75\,\mathrm{s}$.

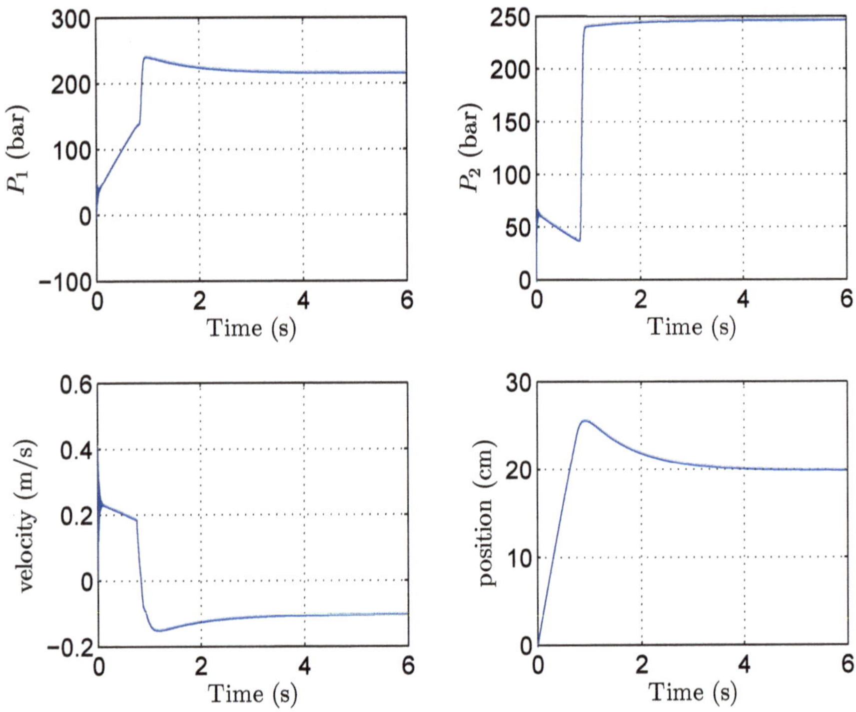

Fig. 16.11 Position control of the hydraulic servo system under the PI controller: SSE = 0, $T_{s5\,\%} = 2.64$ s

16.4.4 *Sliding Mode Control*

To improve the performance of the electrohydraulic system to obtain accurate and fast position control, we suggest in this section to design a sliding mode controller with an integral surface which helps to attenuate the effect of the constant disturbances having the following expression:

$$u = -u_{\max}\text{sign}(\sigma) \tag{16.28}$$

σ is the sliding surface defined by

$$\sigma = C_4\tilde{y}_{int} + C_3\tilde{y} + C_2\tilde{v} + C_1\tilde{P}_2 + \tilde{P}_1 \tag{16.29}$$

with

$$\tilde{P}_1 = P_1 - P_{1e},$$
$$\tilde{P}_2 = P_2 - P_{2e},$$

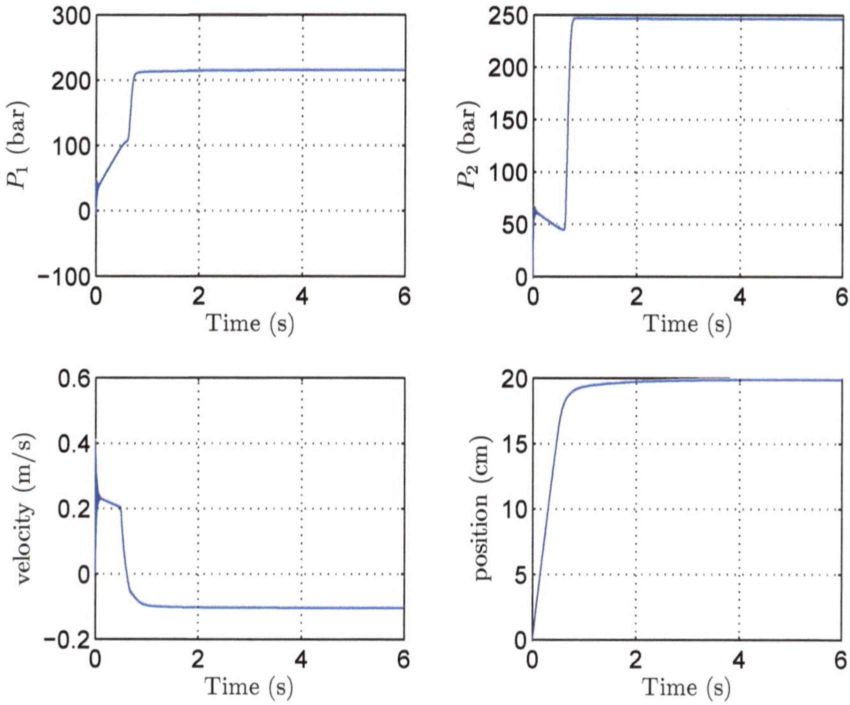

Fig. 16.12 Position control of the hydraulic servo system under the PI controller with anti wind-up: SSE = 0, $T_{s5\%} = 0.75\,\mathrm{s}$

$$\tilde{v} = v - v_e,$$
$$\tilde{y} = y - y_{ref}$$
$$\tilde{y}_{int} = \int \tilde{y}dt,$$

where $(P_{1e}, P_{2e}, v_e, y_{ref})$ is the unperturbed system equilibrium point when the reference position y_{ref} is reached, and C_4, C_3, C_2 and C_1 are the constants which must be chosen to ensure an asymptotic stability of the system when it is behaving in sliding mode, that is when $\sigma = 0$.

To limit the behavior of the system to the sliding surface we must satisfy the attractivity condition $\sigma\dot{\sigma} < 0$. Using (16.28) and (16.29), the attractivity condition becomes:

$$\sigma\dot{\sigma} = \begin{cases} \sigma\Big(C_4 y + C_3 v + \frac{C_2}{m+m_0}\Big(S_1 P_1 - S_2 P - bv - k_l y\Big)\Big) \\ +\frac{C_1 B}{V_0 - S_2 y}\Big(-ku_{\max}|\sigma|\sqrt{P_2 - P_r} + \frac{\alpha}{1+\gamma u_{\max}}(P_s + P_r - 2P_2)\sigma + S_2 v\sigma\Big) \\ +\frac{B}{V_0 + S_1 y}\Big(ku_{\max}|\sigma|\sqrt{P_s - P_1} + \frac{\alpha}{1+\gamma u_{\max}}(P_s + P_r - 2P_1)\sigma - S_1 v\sigma\Big) \\ \qquad\qquad\qquad\qquad \text{if } \sigma < 0 \\ \sigma\Big(C_4 y + C_3 v + \frac{C_2}{m+m_0}\Big(S_1 P_1 - S_2 P_2 - bv - k_l y\Big)\Big) \\ +\frac{C_1 B}{V_0 - S_2 y}\Big(-ku_{\max}|\sigma|\sqrt{P_s - P_2} + \frac{\alpha}{1+\gamma u_{\max}}(P_s + P_r - 2P_2)\sigma + S_2 v\sigma\Big) \\ +\frac{B}{V_0 + S_1 y}\Big(ku_{\max}|\sigma|\sqrt{P_1 - P_r} + \frac{\alpha}{1+\gamma u_{\max}}(P_s + P_r - 2P_1)\sigma - S_1 v\sigma\Big) \\ \qquad\qquad\qquad\qquad \text{if } \sigma > 0 \end{cases}$$

Thus, to satisfy the attractivity condition, we should choose C_4, C_3, C_2 and C_1 such that:

- if $\sigma < 0$:

$$\Big|C_4 y + C_3(\tilde{v} + v_e) + \frac{C_2}{m + m_0}(S_1(\tilde{P}_1 + P_{1e})$$
$$-S_2(\tilde{P}_2 + P_{2e}) - b(\tilde{v} + v_e) - k_l(\tilde{y} + y_{ref}))\Big|$$
$$\times \frac{1}{\frac{C_1 Bk}{V_0 - S_2(\tilde{y}+y_{ref})}\sqrt{(\tilde{P}_2 + P_{2e}) - P_r} - \frac{Bk}{V_0 + S_1(\tilde{y}+y_{ref})}\sqrt{P_s - (\tilde{P}_1 + P_{1e})}}$$
$$+ \frac{\Big|\frac{C_1 B}{V_0 - S_2(\tilde{y}+y_{ref})}\Big(\alpha(P_s + P_r - 2(\tilde{P}_2 + P_{2e})) + S_2(\tilde{v} + v_e)\Big)\Big|}{\frac{C_1 Bk}{V_0 - S_2(\tilde{y}+y_{ref})}\sqrt{(\tilde{P}_2 + P_{2e}) - P_r} - \frac{Bk}{V_0 + S_1(\tilde{y}+y_{ref})}\sqrt{P_s - (\tilde{P}_1 + P_{1e})}}$$
$$+ \frac{\Big|\frac{B}{V_0 + S_1(\tilde{y}+y_{ref})}\Big(\alpha(P_s + P_r - 2(\tilde{P}_2 + P_{2e})) - S_1(\tilde{v} + v_e)\Big)\Big|}{\frac{C_1 Bk}{V_0 - S_2(\tilde{y}+y_{ref})}\sqrt{(\tilde{P}_2 + P_{2e}) - P_r} - \frac{Bk}{V_0 + S_1(\tilde{y}+y_{ref})}\sqrt{P_s - (\tilde{P}_1 + P_{1e})}} \le u_{\max}$$

- if $\sigma > 0$:

$$\Big|C_4 y + C_3(\tilde{v} + v_e) + \frac{C_2}{m + m_0}(S_1(\tilde{P}_1 + P_{1e})$$
$$-S_2(\tilde{P}_2 + P_{2e}) - b(\tilde{v} + v_e) - k_l(\tilde{y} + y_{ref}))\Big|$$
$$\frac{1}{\frac{C_1 Bk}{V_0 - S_2(\tilde{y}+y_{ref})}\sqrt{P_s - (\tilde{P}_2 + P_{2e})} - \frac{Bk}{V_0 + S_1(\tilde{y}+y_{ref})}\sqrt{(\tilde{P}_1 + P_{1e}) - P_r}}$$

$$+\frac{\left|\frac{C_1 B}{V_0-S_2(\tilde{y}+y_{ref})}\Big(\alpha(P_s+P_r-2(\tilde{P}_2+P_{2e}))+S_2(\tilde{v}+v_e)\Big)\right|}{\frac{C_1 Bk}{V_0-S_2(\tilde{y}+y_{ref})}\sqrt{(\tilde{P}_2+P_{2e})-P_r}-\frac{Bk}{V_0+S_1(\tilde{y}+y_{ref})}\sqrt{P_s-(\tilde{P}_1+P_{1e})}}$$
$$+\frac{\left|\frac{B}{V_0+S_1(\tilde{y}+y_{ref})}\Big(\alpha(P_s+P_r-2(\tilde{P}_2+P_{2e}))-S_1(\tilde{v}+v_e)\Big)\right|}{\frac{C_1 Bk}{V_0-S_2(\tilde{y}+y_{ref})}\sqrt{(\tilde{P}_2+P_{2e})-P_r}-\frac{Bk}{V_0+S_1(\tilde{y}+y_{ref})}\sqrt{P_s-(\tilde{P}_1+P_{1e})}}\le u_{\max}$$

Clearly, there are different choices of C_4, C_3, C_2 and C_1 such that the condition of the attainability is satisfied. However, there is an additional and necessary condition that should also be satisfied which is the asymptotic stability of the system in sliding mode; that is when the system is confined to $\sigma = 0$. In this case, we have:

$$\tilde{P}_1 = -C_1\tilde{P}_2 - C_2\tilde{v} - C_3\tilde{y} - C_4\tilde{y}_{int} \tag{16.30}$$

thus the system dynamics can be expressed as follows:

$$\dot{\tilde{y}}_{int} = \tilde{y}, \tag{16.31}$$
$$\dot{\tilde{y}} = \tilde{v}, \tag{16.32}$$
$$\dot{\tilde{v}} = \frac{1}{m+m_0}\Big(S_1(-C_1\tilde{P}_2 - C_2\tilde{v} - C_3\tilde{y} - C_4\tilde{y}_{int}) \tag{16.33}$$
$$-S_2\tilde{P}_2 - b(\tilde{v}+v_e) - k_l(\tilde{y}+y_{ref})\Big), \tag{16.34}$$
$$\dot{\tilde{P}}_2 = \frac{B}{V_0 - S_2(\tilde{y}+y_{ref})}\Big(Q_2 + S_2(\tilde{v}+v_e)\Big), \tag{16.35}$$

Since at the equilibrium we have $S_1P_{1e} - S_2P_{2e} - bv_e - k_l y_{ref} = 0$ and we may choose

$$C_1 = -\frac{S_2}{S_1} \tag{16.36}$$

then we obtain

$$\dot{\tilde{y}}_{int} = \tilde{y}, \tag{16.37}$$
$$\dot{\tilde{y}} = \tilde{v}, \tag{16.38}$$
$$\dot{\tilde{v}} = \frac{1}{m+m_0}\Big(-S_1C_4\tilde{y}_{int} - (S_1C_3+k_l)\tilde{y} - (S_1C_2+b)\tilde{v}\Big), \tag{16.39}$$
$$\dot{\tilde{P}}_2 = \frac{B}{V_0 - S_2(\tilde{y}+y_{ref})}\Big(Q_2 + S_2(\tilde{v}+v_e)\Big), \tag{16.40}$$

Clearly, after application of a control signal $u(x_1, x_2)$, we obtain an autonomous system in the triangular form. Indeed (16.37) is represented as:

$$\dot{x}_1 = f_1(x_1), \tag{16.41}$$

$$\dot{x}_2 = f_2(x_1 + x_{1e}, x_2 + x_{2e}), \tag{16.42}$$

with $x_1 = (\tilde{y}_{int}, \tilde{y}, \tilde{v})$ and $x_2 = \tilde{P}_2$. Using proposition 1, we can prove the stability of system (16.37) which is actually limited to behave on the sliding surface.

Proposition 1 *Consider the dynamic system defined by* (16.41). *Assume that* $x_1 = 0$ *is an exponentially stable equilibrium for (16.41-a) and* $\dot{x}_2 = f_2(x_{1e}, x_2 + x_{2e})$ *is an exponentially bounded system. Moreover, assume that* $f_2(x_1 + x_{1e}, x_2 + x_{2e})$ *is Lipschitz w.r.t.* x_1 *and* x_2 *with constants* γ_1 *and* γ_2 *respectively. Then if* $\|x_{1e}\| \leq \gamma_3$, *then* $\lim_{t\to\infty} x_1(t) = 0$ *and* $\|x_2\| < \infty$.

Proof By assumption of exponential stability of (16.41a), we obtain: $\lim_{t\to\infty} x_1(t) = 0$, if $u(x_1, x_2)$ is bounded. So, we should show that $\|x_2\| < \infty$.

The exponential boundedness of $\dot{x}_2 = f_2(x_{1e}, x_2 + x_{2e})$ implies that there is $V(x_2)$ such that the following hold outside a ball of radius R for some positive constants $a_1,\ a_2,\ a_3$ and a_4 (Sastry and Isidori 1989).

$$a_1\|x_2\|^2 \leq V(x_2) \leq a_2\|x_2\|^2, \tag{16.43}$$

$$\frac{dV}{dx_2} f_2(x_{1e}, x_2 + x_{2e}) \leq -a_3\|x_2\|^2, \tag{16.44}$$

$$\left\|\frac{dV}{dx_2}\right\| < a_4\|x_2\|. \tag{16.45}$$

Combining (16.42b) and (16.44b) leads to

$$\begin{aligned}
\dot{V} &= \frac{dV}{dx_2} f_2(x_1 + x_{1e}, x_2 + x_{2e}) \\
&\leq -a_3\|x_2\|^2 + \frac{dV}{dx_2}.\big(f_2(x_1 + x_{1e}, x_2 + x_{2e}) - f_2(x_{1e}, x_2 + x_{2e})\big), \\
&\leq -a_3\|x_2\|^2 + a_4\|x_2\|.\gamma_1\gamma_3,
\end{aligned}$$

so

$$\dot{V} \leq 0 \quad \text{for} \quad \|x_2\| \geq \frac{a_4\gamma_1\gamma_3}{a_3}. \tag{16.46}$$

Using (16.43) and (16.46), it follows that any trajectory of $x_2(t)$ that starts at a finite value $x_2(0)$ will finish by entering in a ball of radius $\bar{R} = \max\{R, \frac{a_4\gamma_1\gamma_3}{a_3}\}$ so $\|x_2\| < \infty$ and $\lim_{t\to\infty} x_1(t) = 0$ is achieved. □

Now, to apply the above result to the controlled servo system (16.37) and to prove that is stable and should attain its reference value, we can easily deduce that the first three equations form a linear subsystem with a characteristic equation given by:

$$s^3 + \frac{S_1C_2 + b}{m + m_0}s^2 + \frac{S_1C_3 + k_I}{m + m_0}s + \frac{S_1C_4}{m + m_0} = 0 \tag{16.47}$$

Using the method of the pole placement and imposing a stable multiple pole at $s = -\lambda\ (\lambda > 0)$, we can determine the control parameters:

$$C_2 = \frac{3\lambda(m + m_0) - b}{S_1} \tag{16.48}$$

$$C_3 = \frac{3\lambda^2(m + m_0) - k_l}{S_1} \tag{16.49}$$

$$C_4 = -\frac{\lambda^3(m + m_0)}{S_1} \tag{16.50}$$

The fourth equation forms the second subsystem, we can deduce that this subsystem is not globally Lipschitz since it contains the square root term. However, since the system variables are not expected to behave in the vicinity of their physical limits, then a local Lipschitz condition is satisfied. Now, it remains to prove that the second subsystem expressed by (16.51) is exponentially bounded.

$$\dot{\tilde{P}}_2 = \begin{cases} \frac{B}{V_0 - S_2 y_{ref}}\left[\frac{\alpha}{1+\gamma u_{\max}}(P_s + P_r - 2\tilde{P}_2) + S_2 v_e - k u_{\max}\sqrt{\tilde{P}_2 + P_{2e} - P_r}\right] \\ \qquad \text{if } u \geq 0 \\ \frac{B}{V_0 - S_2 y_{ref}}\left[\frac{\alpha}{1+\gamma u_{\max}}(P_s + P_r - 2\tilde{P}_2) + S_2 v_e + k u_{\max}\sqrt{P_s - \tilde{P}_2 - P_{2e}}\right] \\ \qquad \text{if } u < 0 \end{cases} \tag{16.51}$$

From the equation bellow we obtain:

$$P_{2e} = \begin{cases} P_r & \text{if } u \geq 0 \\ P_s & \text{if } u < 0 \end{cases}$$

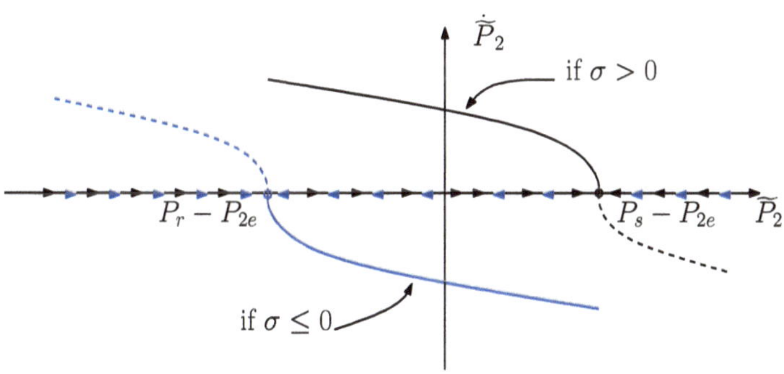

Fig. 16.13 Graphical analysis to prove the exponential boundedness of $\tilde{P}_2$

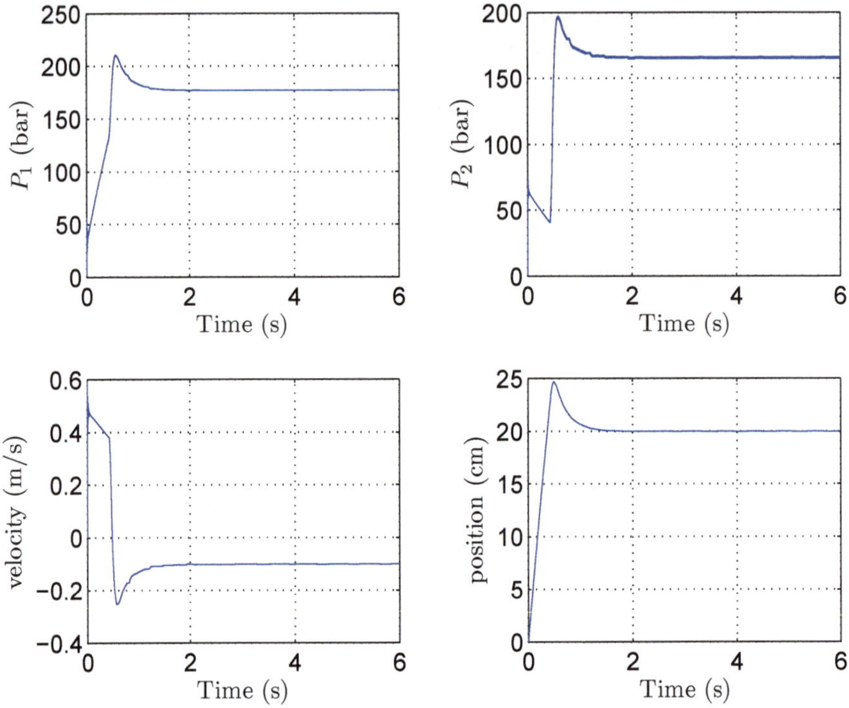

Fig. 16.14 Position control of the hydraulic servo system under the integral surface sliding mode controller: $\lambda = 15$

Clearly, the subsystem (16.51) is a system that toggles among two systems and each has its own equilibrium point. To prove the exponential boundedness, a graphical analysis is used as shown on Fig. 16.13.

In blue color, we present $\dot{\widetilde{P}}_2$ versus $\widetilde{P}_2$ when $\sigma \leq 0$ where the solid part represent the effective part of the vector field and the dashed part is not effective since the pressure in the system cannot drop beyond the return pressure. The arrows on the $\widetilde{P}_2$ axis point to the left if $\dot{\widetilde{P}}_2 < 0$ that is the pressure is decreasing and they point to the right if $\dot{\widetilde{P}}_2 > 0$ that is the pressure is increasing. Using similar argument for the case of $\sigma > 0$ shown in black curve and arrows, we may easily deduce that when the pressure is initiated in the interval $P_2(0) \in (P_r, P_s)$ then $P_2(t)$ remains in that interval for all subsequent time which proves the boundedness of the second subsystem.

Finally using Proposition 1, we can deduce that the reference position is reached by the servo system under the action of a sliding control signal with an integral surface. Figure 16.14 represents the behavior of the controlled system states in presence of the constant disturbance $d = 0.1$ and the uncertainty in the spring constant of the order of 20 % and with closed loop poles placed at $\lambda = 15$. It is noted that the reference position has been reached with a settling time $T_{s5\,\%} = 0.91$ s and an overshoot of about 25 %. The rod velocity has been stabilized at $v = -0.1$ to compensate the

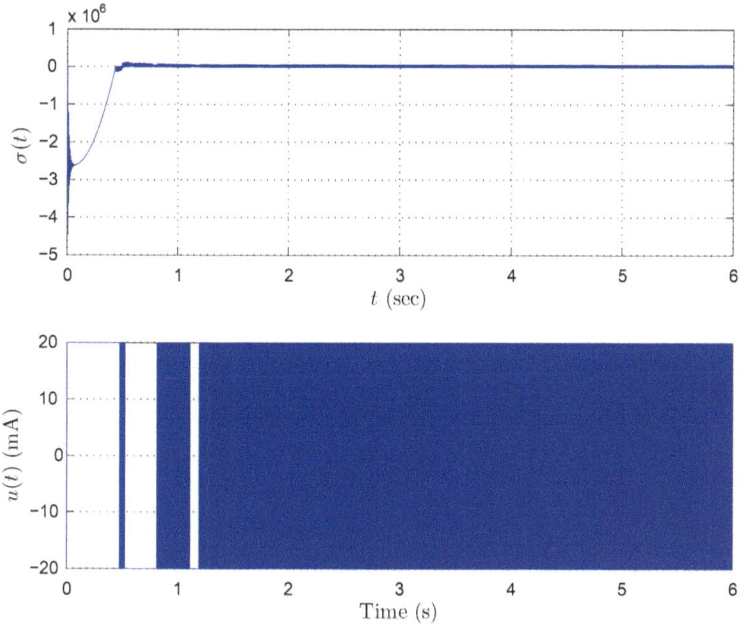

Fig. 16.15 Sliding surface $\sigma(t)$ and the control signal $u(t)$ used to assess the hydraulic servo system

disturbance d and therefore the physical velocity of the system is zero that is the rod is at rest. We can also note that the pressures in the chambers of the piston evolved within the interval (P_r, P_s). Finally, the control signal used to evaluate the hydraulic servo system is shown in Fig. 16.15.

16.4.5 Sliding Mode Control with Realizable Reference

To improve the system behavior such as eliminating the overshoot of the system and reducing the settling time, we can think of decreasing the time needed for the system to reach the sliding surface.

To do this, we consider the realizable reference that can be reached by the control at each instant, and use this information so that the system behaves as if it is in sliding mode at each instant of the transient time.

The realizable reference is noted by r_{ref}. With the realizable reference, the sliding mode is supposed to be achieved so $\sigma = 0$, then substituting in (16.29) we obtain:

$$0 = C_4 \tilde{y}_{int} + C_3(y - r_{ref}) + C_2 \tilde{v} + C_1 \tilde{P}_2 + \tilde{P}_1 \tag{16.52}$$

Subtracting (16.52) from (16.29), we obtain:

$$r_{ref} = y_{ref} + \frac{1}{C_3}\sigma \tag{16.53}$$

Next, we change the reference by the realizable reference in the dynamics of the integral state to obtain:

$$\dot{y}_{int} = y - y_{ref} - \frac{1}{C_3}\sigma = \tilde{y} - \frac{1}{C_3}\sigma \tag{16.54}$$

Finally, we obtain:

$$\dot{\sigma} = \begin{cases} -\frac{C_4}{C_3}\sigma + C_4(y - y_{ref}) + C_3 v + \frac{C_2}{m+m_0}\Big(S_1P_1 - S_2P - bv - k_l y\Big) \\ +\frac{C_1 B}{V_0 - S_2 y}\Big(- k u_{\max}\mathrm{sign}(\sigma)\sqrt{P_2 - P_r} + \frac{\alpha}{1+\gamma u_{\max}}(P_s + P_r - 2P_2) + S_2 v\Big) \\ +\frac{B}{V_0 + S_1 y}\Big(k u_{\max}\mathrm{sign}(\sigma)\sqrt{P_s - P_1} + \frac{\alpha}{1+\gamma u_{\max}}(P_s + P_r - 2P_1) - S_1 v\Big) \\ \qquad\qquad\qquad\qquad\qquad\qquad \text{if } \sigma < 0 \\ -\frac{C_4}{C_3}\sigma + C_4(y - y_{ref}) + C_3 v + \frac{C_2}{m+m_0}\Big(S_1P_1 - S_2P_2 - bv - k_l y\Big) \\ +\frac{C_1 B}{V_0 - S_2 y}\Big(- k u_{\max}\mathrm{sign}(\sigma)\sqrt{P_s - P_2} + \frac{\alpha}{1+\gamma u_{\max}}(P_s + P_r - 2P_2) + S_2 v\Big) \\ +\frac{B}{V_0 + S_1 y}\Big(k u_{\max}\mathrm{sign}(\sigma)\sqrt{P_1 - P_r} + \frac{\alpha}{1+\gamma u_{\max}}(P_s + P_r - 2P_1) - S_1 v\Big) \\ \qquad\qquad\qquad\qquad\qquad\qquad \text{if } \sigma > 0 \end{cases}$$

We can deduce that $\dot{\sigma}$ is expressed now as a function of σ in addition to the expression obtained earlier. So to satisfy the attractivity of the sliding surface σ we should satisfy the following relation:

$$\frac{C_4}{C_3} > 0 \tag{16.55}$$

Knowing that $C_4 > 0$, so we can choose λ such that C_3 is also positive, that is:

$$\lambda > \sqrt{\frac{k_l}{3(m + m_0)}} \tag{16.56}$$

Figure 16.16 delineates the behavior of the system states when it is controlled by the integral sliding controller using the integral state and being modified using the realizable reference as indicated in (16.54). To ensure the stability condition (16.56), the closed loop poles are placed at $\lambda = 90$. We can deduce that the reference position has been attained with a settling time $T_{s5\,\%} = 0.35$ s and an overshoot of about 1.5 %. The velocity of the rod has been again stabilized at $v = -0.1$ to compensate the disturbance d. We can also note that the pressures in the chambers of the piston evolved in the interval (P_r, P_s). Finally, the control signal used to evaluate the

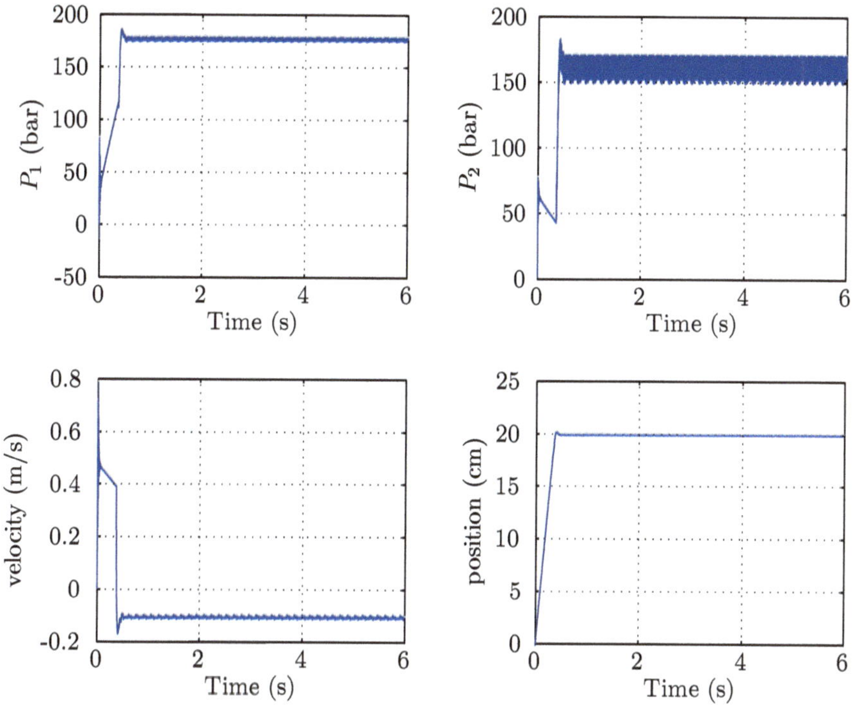

Fig. 16.16 Position control of the hydraulic servo system under the integral surface sliding mode controller: $\lambda = 90$

electrohydraulic system is shown in Fig. 16.17. Clearly, this controller exceeds all controllers previously designed in terms of performance.

We should mention here that we did not address the chattering phenomenon because our aim was to construct a simple controller expression. Nevertheless, we can reduce the chattering phenomenon by substituting the sign function by a smooth saturation function $\tanh(A\sigma)$ where A is a gain. We can note that when A is very large then $\tanh(A\sigma) \simeq \text{sign}(\sigma)$. Figure 16.18 represents the behavior of the system states when the controller uses a smooth function. We can note that the chattering phenomenon is decreasing especially in the pressure and velocity behaviors. Figure 16.19 represents the behavior of the controller and the sliding surface when the controller uses a smooth function.

16.4.6 Sliding Mode Observer Design

From the previous sections, we can notice that the controller conceived uses implicitly all four state variables through the sliding surface definition. Nevertheless, the

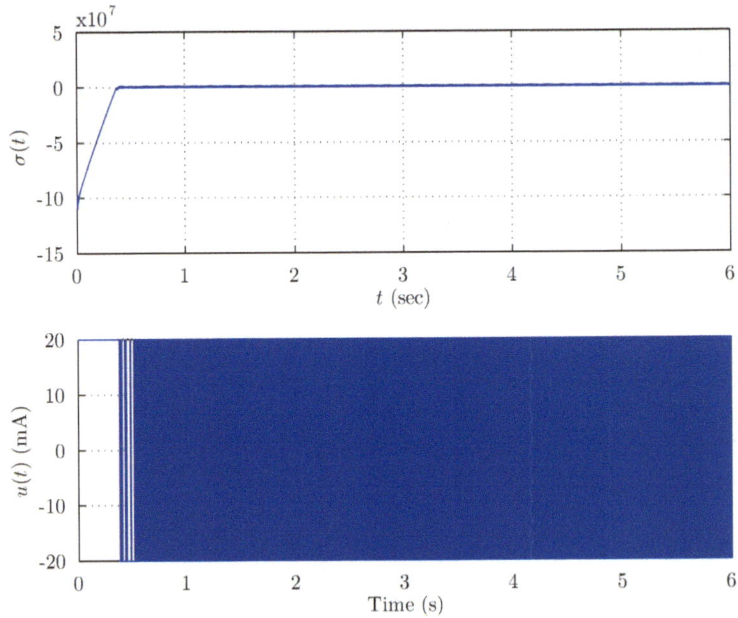

Fig. 16.17 Sliding surface $\sigma(t)$ and the control signal $u(t)$ used to assess the hydraulic servo system

measure of the pressures P_1 and P_2 is a costly task and needs high technology to eliminate the additional leakage. To overcome this problem, we propose in this section to design a sliding mode observer that can estimate the required states and used them to construct the sliding surface.

Before starting the observer design, it is interesting to note that when the rod position y is the measured output, then we can easily verify that the hydraulic servo system described by its model (16.1) is not fully observable. Indeed, the pressures P_1 and P_2 are not observable, nevertheless, the difference $E = S_1P_1 - S_2P_2$ is itself observable if it is considered as a single state. Referring to the choice of $C_1 = -\frac{S_2}{S_1}$ given in (16.36), and to the expression of the sliding surface (16.29), then it is deduced that we only need to estimate the expression $E = S_1P_1 - S_2P_2$ to be able to construct the sliding surface. Therefore, we choose to use a sliding mode observer due to its robustness and the possibility of estimating the term without knowing the full model. We here consider the reduced order model of the hydraulic servo system:

$$\dot{E} = f(P_1, P_2, v, y) \tag{16.57}$$

$$\dot{v} = \frac{1}{m + m_0}(E - bv - (k_l + \delta k_l)y) \tag{16.58}$$

$$\dot{y} = v + d \tag{16.59}$$

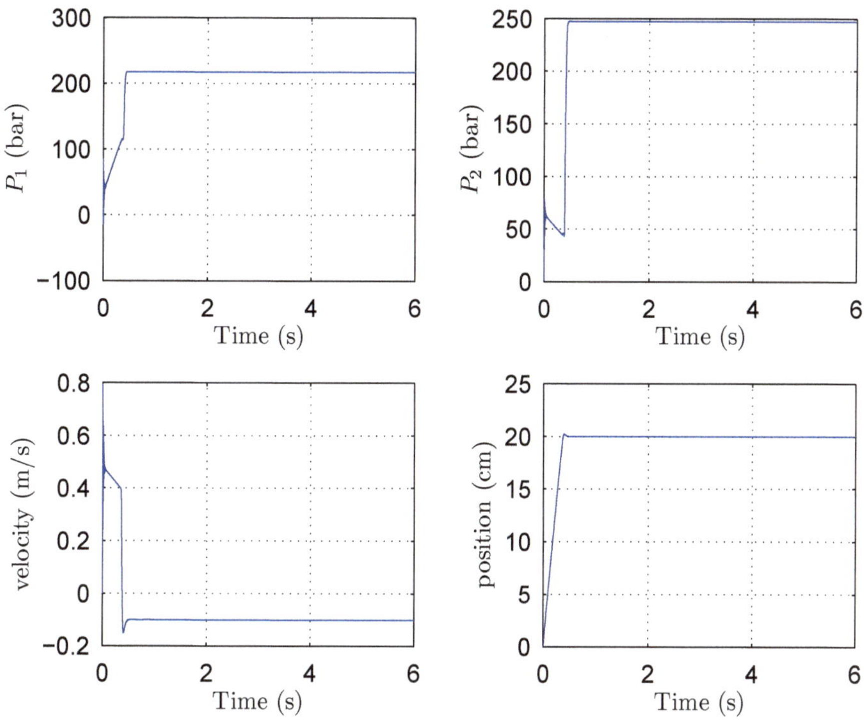

Fig. 16.18 State behavior of the hydraulic servo system under the integral surface sliding mode controller with smooth saturation function

where δk_l represents the 20 % uncertainty of the spring stiffness and f is a nonlinear function which represents the dynamics of E.

For this model we associate the following observer:

$$\dot{\Delta} = L_1 \text{sign}(z_1 - \Delta) \tag{16.60}$$

$$\dot{\hat{v}} = \frac{1}{m + m_0}(\Delta - bz_2 - k_l \hat{y}) + L_2 \text{sign}(z_2 - \hat{v}) \tag{16.61}$$

$$\dot{\hat{y}} = \hat{v} + L_3 \text{sign}(y - \hat{y}) \tag{16.62}$$

where L_1, L_2 and L_3 are the observer gain and z_1 and z_2 are the observer outputs defined by:

$$z_1 = \Delta + (m + m_0)L_2 \text{sign}(z_2 - \hat{v}) \tag{16.63}$$

$$z_2 = \hat{v} + L_3 \text{sign}(y - \hat{y}) \tag{16.64}$$

The observer state Δ is intended to estimate the expression $E = S_1 P_1 - S_2 P_2$. To prove the efficiency of the observer and to show that the controller based on the

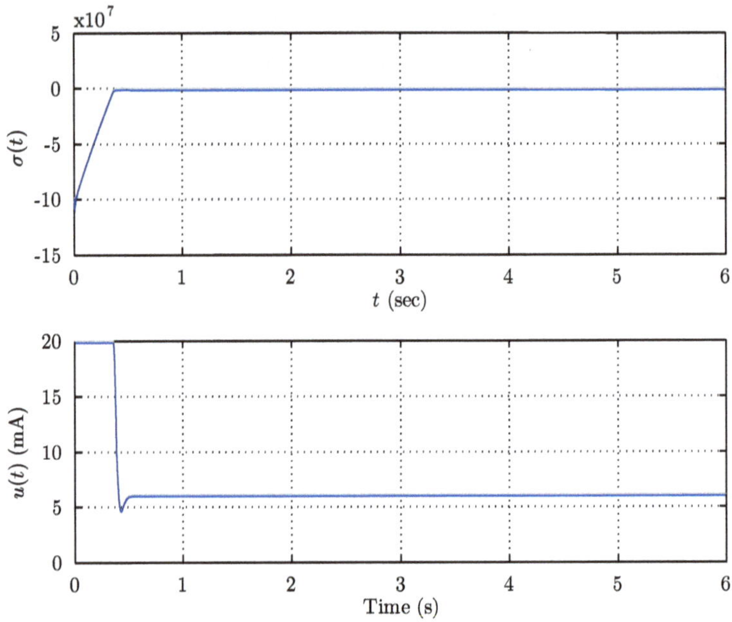

Fig. 16.19 Sliding surface $\sigma(t)$ and the control signal $u(t)$ used to assess the hydraulic servo system with smooth saturation function

estimated states may also achieve accurate positioning in the presence of disturbance and uncertainty, we will conduct a proof step by step.

Step 1: Let $e_y = y - \hat{y}$ and $e_v = v - \hat{v}$, from (16.59) and (16.62) the error dynamics are expressed as follows:

$$\dot{e}_y = e_v + d - L_3\text{sign}(e_y) \tag{16.65}$$

Thus, if we choose L_3 such that:

$$L_3 > \sup_{t>0}\{e_v(t) + d\} \tag{16.66}$$

then the sliding mode is established at the observer sliding surface $e_y = 0$ in a finite time. In addition, we obtain at sliding mode $\dot{e}_y = 0$ and thus from (16.65) we have

$$0 = v - \hat{v} + d - L_3\text{sign}(e_y) \tag{16.67}$$

so

$$v + d = \hat{v} + L_3\text{sign}(e_y) = z_2 \tag{16.68}$$

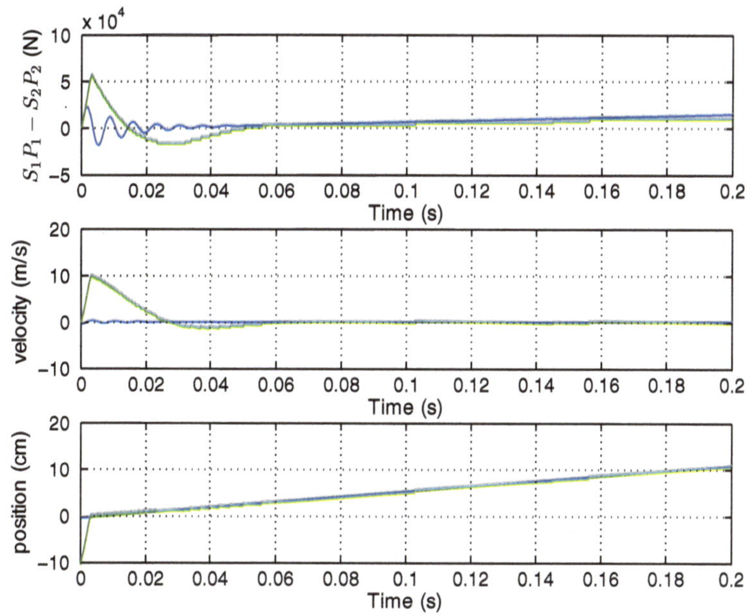

Fig. 16.20 Convergence of the sliding mode observer to the intended values

At this stage, it can be inferred that the observed state $\hat{y}$ will converge in a finite time to the system state and there the observer output z_2 will converge to the perturbed rod velocity $v + d$.

Step 2: Let $e_\Delta = E - \Delta$ and taking into account that $e_y = 0$ and $z_2 = v + d$, then from (16.58) and (16.61) the error dynamics are expressed as follows:

$$\dot{e}_v = \frac{1}{m + m_0}\Big(e_\Delta + bd - \delta k_l y\Big) - L_2 \text{sign}(e_v + d) \tag{16.69}$$

Thus, if we choose L_2 such that:

$$L_2 > \sup_{t>0}\{e_\Delta(t) + bd_{\max} - \delta k_l y(t)\} \tag{16.70}$$

then we can deduce that $e_v + d$ tend to zero and therefore $\hat{v}$ tend to $v + d$. That is the observer state $\hat{v}$ will achieve its output z_2. Moreover, when the error dynamics are on the sliding surfaces $e_y = 0$ and $e_v + d = 0$, then $\dot{e}_v = 0$ and we obtain:

$$E + bd - \delta k_l y = \Delta + (m + m_0) L_2 \text{sign}(z_2 - \hat{v}) = z_1 \tag{16.71}$$

That is the observer will estimate the expression $E = S_1P_1 - S_2P_2$ with a constant difference proportional to the uncertainty δk_l and to the perturbation d. Eventually, when the estimated variables are used to design the sliding surface, then the

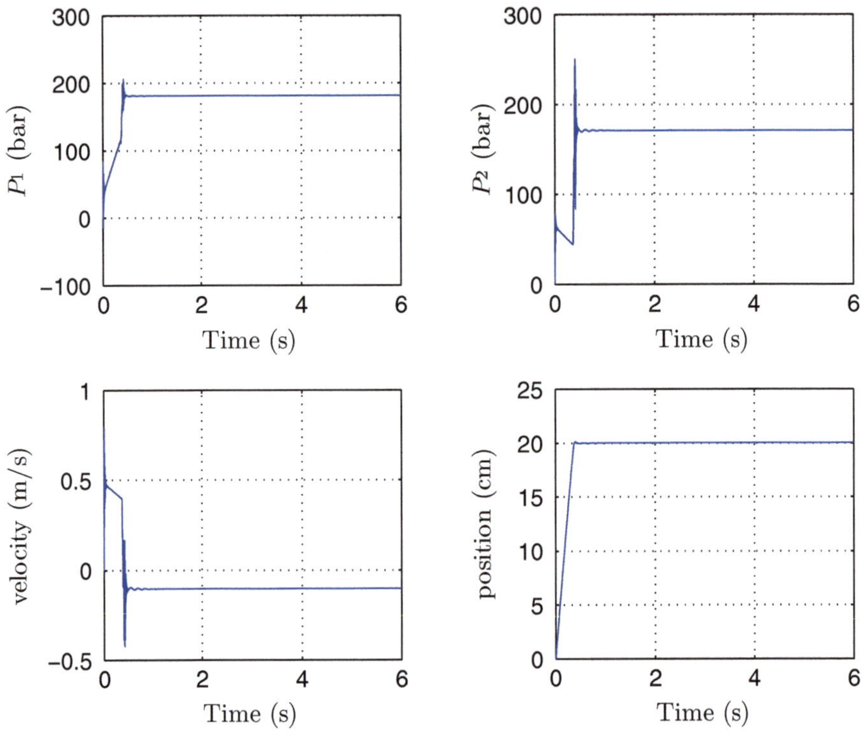

Fig. 16.21 Behavior of the hydraulic servo system controlled with sliding mode controller with estimated states feedback

uncertainty is taken into consideration, therefore when the controller sliding mode is attained the convergence to the reference position is reached despite the existence of the constant perturbation d which will be eliminated by the use of the integral term in the definition of the surface.

Figure 16.20 shows the convergence of the observer state $\hat{y}$ to y despite starting from different initial conditions; $y(0) = 0$ and $\hat{y}(0) = -0.1$ represents respectively the initial conditions of the system and the observer starting at -10 cm. As expected from the previous analysis, $\hat{v}$ tends to $v + d$ and Δ tends to E with a constant difference. We choose the observer gains as $L_1 = 10^7$, $L_2 = 3000$ and $L_3 = 30$. The system behavior is shown on Fig. 16.21 with the sliding surface is calculated using the estimated states and the controller with a smooth saturation function.

16.5 Conclusion

In this chapter, we have detailed the different steps of the design the control law of an electrohydraulic system. In first part, we have presented the effect of the classical

P and PI controller. Then, we have used an anti wind-up approach in order to reduce the settling time. Next, a method has been described to design a complete sliding mode controller with realizable reference to overcome the phenomenon of wind-up due to the saturation of the actuator. The controller achieves zero static error with a significantly shorter response time. This controller is compared to the classical PI controller and sliding mode controller and it has been shown to outperform them in terms of speed. This work has been applied first to a symmetrical system and then secondly to a non-symmetrical system.

Appendix A

See Tables 16.1, 16.2, 16.3, 16.4 and 16.5.

Table 16.1 Numerical values of the parameters of the fluid

Parameter	Value	Unit
B	2.2×10^9	Pa
P_s	300×10^5	Pa
P_r	1×10^5	Pa

Table 16.2 Numerical values of the parameters of the piston

Parameter	Value	Unit
m_0	50	kg
S	1.5×10^{-3}	m^2
S_1	3.1×10^{-3}	m^2
S_2	1.5×10^{-3}	m^2
V_0	0.458×10^{-3}	m^3
V_t	1×10^{-3}	m^3

Table 16.3 Numerical values of the mass parameter

Parameter	Value	Unit
m	20	kg
b	590	kg/s
k_l	125,000	N/m

Table 16.4 Numerical values of the parameters of the servo valve

Parameter	Value	Unit
Symmetric system		
k	5.12×10^{-5}	$m^3s^{-1}A^{-1}Pa^{-1/2}$
γ	8571	s^{-1}
α	4.1816×10^{-12}	$m^3s^{-1}Pa^{-1}$
Non symmetric system		
k	1.46×10^{-5}	$m^3s^{-1}A^{-1}Pa^{-1/2}$
γ	10622	s^{-1}
α	4.605×10^{-13}	$m^3s^{-1}Pa^{-1}$

Table 16.5 Numerical values of controller parameters

Parameter	Value	Unit
Symmetric system		
$u_{\max}$	2	mA
z_{3ref}	0.2	m
k_0	0.05	–
k_l	0.05	–
Non symmetric system		
$u_{\max}$	0.01	mA
y_{ref}	0.2	m
k_0	0.5	–
k_l	0.5	–

References

Ahn, K. K., & Dinh, Q. T. (2009). Self-tuning of quantitative feedback theory for force control of an electro-hydraulic test machine. *Control Engineering Practice*, *17*(11), 1291–1306.

Alleyne, A., Liu, R., & Wright, H. (1998). On the limitations of force tracking control for hydraulic active suspensions. *American Control Conference*, *1*, 43–47.

Aly, A. A. (2011). Pid parameters optimization using genetic algorithm technique for electrohydraulic servo control system. *Journal of Dynamic Systems, Measurement, and Control, Transaction of the ASME*, *2*(2), 69–76.

Chen, H., Renn, J. C., & Su, J. P. (2005). Sliding mode control with varying boundary layers for an electro-hydraulic position servo system. *International Journal of Advanced Manufacturing Technology*, *26*(1–2), 117–123.

Feki, M. (2001). *Synthèse de commandes et d'observateurs pour les systèmes non-linéaires: Application aux systèmes hydrauliques*. Ph.D. thesis, Université de Metz, Metz-France.

Guan, C., & Pan, S. (2008). Nonlinear adaptive robust control of single-rod electro-hydraulic actuator with unknown nonlinear parameters. *IEEE Transaction on Control Systems Technology*, *16*(3), 434–445.

Hanus, R., Kinnaert, M., & Henrotte, J. L. (1987). Conditioning technique, a general antiwindup and bumpless transfer method. *Automatica*, *23*(6), 729–739.

Hwang, C. L., & Lan, C. H. (1994). The position control of electrohydraulic servomechanism via a novel variable structure control. *Mechatronics*, *4*(4), 369–391.

Jian-jun, Y., Duo-tao, D., Gui-lin, L., & Sheng, J. (2012). High precision position control of electro-hydraulic servo system based on feed-forward compensation. *Research Journal of Applied Sciences, Engineering and Technology*, *1*(1), 289–298.

Khebbache, H., & Tadjine, M. (2013). Robust fuzzy backstepping sliding mode controller for a quadrotor unmanned aerial vehicle. *Journal of Control Engineering and Applied Informatics*, *15*(2), 3–11.

Knohl, T., & Unbehauen, H. (2000). Adaptive position control of electrohydraulic servo systems using ANN. *Mechatronics*, *10*(1–2), 127–143.

Kolsi-Gdoura, E., Feki, M., Derbel, N. (2013). Sliding mode-based robust position control of an electrohydraulic system. In *10th International Multi-Conference on Systems, Signals Devices (SSD)*, 1–5.

Merritt, H. E. (1967). *Hydraulic control systems*. USA: Wiley.

Mihajlov, M., Nikolic, V., & Antic, D. (2002). Position control of an electrohydraulic servo system using sliding mode control enhanced by fuzzy pi controller. *Mechanical Engineering*, *1*(9), 1217–1230.

Rossomando, F. G., Soria, C., & Carelli, R. (2014). Sliding mode control for trajectory tracking of a non-holonomic mobile robot using adaptive neural networks. *Journal of Control Engineering and Applied Informatics*, *16*(1), 12–21.

Sastry, S. S., & Isidori, A. (1989). Adaptive control of linearizable systems. *IEEE Transactions on Automatic Control*, *34*(11), 1123–1131.

Sirouspour, M. R., & Salcudean, S. E. (2000). On the nonlinear control of hydraulic servo-systems. *IEEE International Conference on Robotics and Automation, Proceedings ICRA'2000*, *2*, 1276–1282.

Truong, D. Q., & Ahn, K. K. (2009). Force control for hydraulic load simulator using self-tuning grey predictor-fuzzy PID. *Mechatronics*, *19*(2), 233–246.

Yao, B., Bu, F., Reedy, J., & Chiu, G. T. C. (2000). Adaptive robust motion control of single-rod hydraulic actuators: theory and experiments. *IEEE/ASME Transaction on Mechatronics*, *5*(1), 79–91.

MIX
Papier aus verantwortungsvollen Quellen
Paper from responsible sources
FSC® C105338

If you have any concerns about our products,
you can contact us on
ProductSafety@springernature.com

In case Publisher is established outside the EU,
the EU authorized representative is:
Springer Nature Customer Service Center GmbH
Europaplatz 3, 69115 Heidelberg, Germany

Printed by Libri Plureos GmbH
in Hamburg, Germany